Joseph Leidy

Contributions to the extinct vertebrate fauna of the western territories

Joseph Leidy

Contributions to the extinct vertebrate fauna of the western territories

ISBN/EAN: 9783337270421

Printed in Europe, USA, Canada, Australia, Japan

Cover: Foto ©berggeist007 / pixelio.de

More available books at **www.hansebooks.com**

DEPARTMENT OF THE INTERIOR.

REPORT

OF THE

UNITED STATES GEOLOGICAL SURVEY

OF

THE TERRITORIES.

F. V. HAYDEN,
UNITED STATES GEOLOGIST-IN-CHARGE.

IN FIVE VOLUMES.

WASHINGTON:
GOVERNMENT PRINTING OFFICE.
1873.

LETTER TO THE SECRETARY.

Sir: I have the honor to present for your approval and for publication the first part of volume I of the quarto series of reports which are intended to embody the more original and technical results of the survey under my direction. The present memoir on the "Extinct vertebrata of our Western Territories" has been elaborated by Professor Joseph Leidy, the eminent comparative anatomist, and will form one of the most important contributions to the science of extinct organisms ever made in this country. This memoir will be followed by a second part on the same subject by Professor E. D. Cope.

Volume II will embrace the subject of the extinct flora of our western Territories; and it is the purpose to make it as exhaustive as possible. Professor J. S. Newberry is preparing the first part and Professor Leo Lesquereux the second. The well-known reputation of these gentlemen is a sufficient guarantee for the value of their work.

Volume III will include all the materials collected by the survey on the subject of extinct invertebrata, and will be most carefully elaborated by the eminent paleontologist of the survey, Mr. F. B. Meek.

Volume IV will embrace the profiles, sections, maps, and other illustrations, with descriptive text by the geologist in charge.

Volume V will contain separate memoirs on different subjects in recent zoology and botany, prepared by several authors. All the new and imperfectly described species of plants or animals collected by the survey will be studied and fully illustrated. All these volumes are now in an advanced state of preparation. In presenting to the world these important contributions to science, permit me, sir, to extend to you my sincere thanks for your intelligent sympathy and hearty co-operation in the work.

Very respectfully, your obedient servant,

F. V. HAYDEN,
United States Geologist.

Hon. C. Delano, *Secretary of the Interior.*

VOLUME I.

FOSSIL VERTEBRATES.

PART 1.

CONTRIBUTIONS
TO THE
EXTINCT VERTEBRATE FAUNA
OF
THE WESTERN TERRITORIES.

BY

PROF. JOSEPH LEIDY.

TABLE OF CONTENTS.

	Page.
PREFACE	14
EXTINCT VERTEBRATE FAUNA OF THE BRIDGER TERTIARY FORMATION OF WYOMING TERRITORY	15
INTRODUCTION	15
MAMMALIA	27
Order *Perissodactyla*	27
Palæosyops	27
paludosus	28
major	45
junius	57
Limnohyus	58
Hyrachyus	59
agrarius	40
eximius	66
modestus	67
nanus	67
Lophiotherium	69
sylvaticum	69
Trogosus	71
castoridens	71
vetulus	75
Hyopsodus	75
paulus	75
minusculus	81
Microsus	81
cuspidatus	81
Mycrosyops	82
gracilis	83
Notharctus	86
tenebrosus	86
Hopsosyus	90
formosus	90
robustior	93
Order *Proboscidea*	93
Uintatherium	93
robustum	95
Order *Rodentia*	109
Paramys	109
delicatus	110
delicatior	110
delicatissimus	111
Mysops	111
minimus	111
fraternus	112
Sciuravus	113
Order *Carnivora*	113
Patriofelis	114
ulta	114
Sinopa	116
rapax	116
eximia	118

CONTENTS

MAMMALIA—Continued
Order *Carnivora*.
    Uintacyon 118
        edax 118
        vorax 120
Order *Insectivora* 120
    Omomys 120
        Carteri 120
    Palæacodon 121
        verus 121
    Washakius 123
        insignis 123
    Elotherium 124

REPTILIA 125
Order *Crocodilia* 125
    Crocodilus 125
        aptus 125
        Elliotti 126
Order *Chelonia* 132
    Testudo 132
        Coresoni 132
    Emys 140
        wyomingensis 140
    Baptemys 154
        wyomingensis 157
    Baena 160
        arenosa 161
    Chisternon 162
        undatum 162
    Hybemys 174
        arenarius 174
    Anosteira 174
        ornata 174
    Trionyx 176
        guttatus 176
        uintaensis 178
    Remains of Trionyx of undetermined species 180
Order *Lacertilia* 180
    Saniwa 181
        ensidens 181
        major 182
    Glyptosaurus 183
    Chamæleo 184
        pristinus 184

FISHES 184
    Amia (Protamia) uintaensis 185
        media 188
        gracilis 188
    Hypamia 189
        elegans 189
    Lepidosteus 189
        atrox 189
        —— (?) 190
        simplex 191
        notabilis 192
    Pimelodus 193
        antiquus 193
    Phareodus 193
        acutus 193

	Page.
FISHES—Continued.	
REMAINS OF FISHES FROM THE SHALES OF GREEN RIVER, WYOMING	194
Clupea	195
humilis	195
alta	196
DESCRIPTION OF REMAINS OF MAMMALS FROM THE TERTIARY FORMATION OF SWEETWATER RIVER, WYOMING	198
MAMMALIA	199
Order *Ruminantia*	199
Merycochœrus	199
rusticus	199
——— sp. (?)	208
Order *Solidungula*	208
DESCRIPTION OF VERTEBRATE FOSSILS FROM THE TERTIARY FORMATION OF JOHN DAY'S RIVER, OREGON	210
MAMMALIA	211
Order *Ruminantia*	211
Oreodon	211
Culbertsoni	211
superbus	214
Leptomeryx	215
Evansi	215
Agriochœrus	215
antiquus	215
latifrons	216
Order *Artiodactyla*	216
Dicotyles	216
pristinus	216
Elotherium	217
imperator	217
Order *Solidungula*	218
Anchitherium	218
Bairdi	218
Condoni	218
Order *Perissodactyla*	219
Lophiodon	219
Rhinoceros	220
hesperius	220
pacificus	221
Hadrohyus	222
supremus	222
An undetermined carnivore	223
CHELONIA	223
Stylemys	223
nebrascensis	224
niobracensis	225
oregonensis	226
DESCRIPTION OF REMAINS OF VERTEBRATA FROM TERTIARY FORMATIONS OF DIFFERENT STATES AND TERRITORIES WEST OF THE MISSISSIPPI RIVER	227
MAMMALIA	227
Order *Carnivora*	227
Felis	227
augustus	227
imperialis	228
Canis	230
indianensis	230
Order *Proboscidea*	231
Mastodon	231
obscurus	231
mirificus	237
americanus	237

II—G

MAMMALIA—Continued.
Order *Proboscidia*.
Elephas 234
  americanus 234
Megacerops 239
  coloradoensis 239
Order *Solidungula* 242
Equus 242
  occidentalis 242
  major 244
Hipparion 247
Protohippus s. Merychippus 248
Anchitherium 250
    australe 250
    agreste 251
    (?) 252
Order *Ruminantia* 253
Bison 253
  latifrons 253
Auchenia 255
  hesterna 255
Procamelus 258
  virginiensis 259
Megalomeryx 260
  niobrarensis 260
CHELONIA 260
Emys 260
  petrolei 260
FISHES 261
Family *Cyprinidæ* 262
Mylocyprinus 262
  robustus 262
Family *Raiæ* 264
Oncobatis 264
  pentagonus 264
DESCRIPTION OF REMAINS OF REPTILES AND FISHES FROM THE CRETACEOUS FORMATION OF THE INTERIOR OF THE UNITED STATES 265
REPTILES 267
Order *Dinosauria* 267
Poicilopleuron 267
  valens 267
Order *Chelonia* 269
Order *Mosasauria* 270
Tylosaurus 271
  dyspelor 271
  proriger 271
Lestosaurus 276
  coryphæus 276
Mosasaurus 279
Clidastes 281
  intermedius 281
  affinis 283
Order *Lacertilia* 285
Tylosteus 285
  ornatus 285
Order *Sauropterygia* 286
Oligosimus 286
  grandævus 286
Nothosaurus 287
  occiduus 287
FISHES 288
TELEOSTEI 288

	Page
FISHES—Continued.	
Order *Acanthopteri*	288
Sphyrænidæ	288
Cladocyclus	288
occidentalis	288
Enchodus	289
Shumardi	289
Phasganodus	289
dirus	289
Order *Malacopteri*	291
Siluridæ	291
Xiphactinus	291
audax	291
GANOIDEI	292
Pycnodus	292
faba	292
Hadrodus	294
priscus	295
ELASMOBRANCHII	295
Order *Plagiostomi*	295
Ptychodus Mortoni	295
occidentalis	298
Whippleyi	300
Acrodus	300
humilis	300
Galeocerdo	301
falcatus	301
Oxyrhina	302
extenta	302
Lamna a. Oxyrhina	303
Otodus	305
divaricatus	305
Order *Holocephali*	305
Edaphodon	306
mirificus	306
Eumylodus	309
laqueatus	309
NOTICE OF SOME REMAINS OF FISHES FROM THE CARBONIFEROUS FORMATION OF KANSAS	311
Order *Plagiostomi*	311
Cladodus	311
occidentalis	311
Xystracanthus	312
arcuatus	312
Petalodus	312
alleghaniensis	312
Asteracanthus	313
siderius	313
SYNOPSIS OF THE EXTINCT VERTEBRATA DESCRIBED OR NOTICED IN THE PRESENT WORK	315

PHILADELPHIA, *January* 13, 1873.

DEAR SIR: Herewith I transmit to you my report on vertebrate fossils from the Western Territories and States. Many of the specimens were collected during your geological explorations, and were submitted to me for investigation. Others have been collected by different persons living in the West, and sent to me directly, or through the agency of the Smithsonian Institution, for examination. Most of the fossils were obtained in Wyoming, and the others were derived from Oregon, California, New Mexico, Idaho, Colorado, Kansas, and Nebraska.

With respect, I remain, at your service,

JOSEPH LEIDY.

Professor F. V. HAYDEN,
    United States Geologist.

PREFACE.

The present work was commenced in 1870, at which time the amount of materials as subjects of investigation and description was comparatively small. A constant accession of new materials, beyond all anticipation, has greatly extended the work. This will account for the apparent want of system in the arrangement or proper collocation of the subjects of many of the plates.

The interest excited by the numerous discoveries of vertebrate fossils in the Western States and Territories has led to the recent explorations of Professors Marsh and Cope, both of whom have obtained rich collections. The investigations and descriptions, by these gentlemen, of some of the fossils from the same localities, have been so nearly contemporary with my own, that, from want of the opportunity of comparison of specimens, we have no doubt in some cases described the same things under different names, and thus produced some confusion, which can only be corrected in future.

My investigations, in many instances, may appear not so complete as would be desirable, and my excuse for not doing the work more thoroughly is the limited time allowed for the purpose and the little leisure I have had in the intervals of other and necessary professional engagements.

EXTINCT VERTEBRATE FAUNA OF THE BRIDGER TERTIARY FORMATION OF WYOMING TERRITORY.

INTRODUCTION.

The following pages contain a description of fossil remains of vertebrated animals collected in the vicinity of Fort Bridger, a military post situated in the southwest corner of Wyoming Territory.

Many of the specimens were obtained during Professor Hayden's geological explorations of 1869 and 1870, but the greater part of them were collected during the same years and the succeeding one by Dr. James Van A. Carter, residing at Fort Bridger, and by Dr. Joseph K. Corson, United States Army, the surgeon of the post. These gentlemen have diligently explored a wide extent of country in their immediate neighborhood in the search for fossils with the most intelligent interest. The results of their explorations they have liberally placed at the service of naturalists by voluntarily donating all the more characteristic portions of their collections to the Academy of Natural Sciences of Philadelphia.*

After the present work was supposed to be nearly ready for the press, and the accompanying plates from I to XXII were complete, the last summer, the writer received a pressing invitation from his friend Dr. Carter to visit him at Fort Bridger. As the invitation was accompanied with liberal facilities and offers of aid in exploration, the author availed himself of the opportunity of visiting a region of so much interest, and accordingly spent the summer vacation in a trip to the locality.

Fort Bridger occupies a situation in the midst of a wide plain at the base of the Uintah Mountains, and at an altitude of nearly seven thousand feet above the ocean-level. The neighboring country, extending from the Uintah and Wahsatch Mountains on the south and west to the Wind River Range on the northeast, at the close of the Cretaceous epoch, appears to have been occupied

---

* In speaking of this institution hereafter I shall briefly refer to it as the Academy, or the Academy of Philadelphia.

by a vast fresh-water lake. Abundance of evidence is found to prove that the region was then inhabited by animals as numerous and varied as those of any other fauna, recent or extinct, in other parts of the world. Then, too, a rich tropical vegetation covered the country, in strange contrast to its present almost lifeless and desert condition.

The country appears to have undergone slow and gradual elevation; and the great Uintah lake, as we may designate it, was emptied, apparently in successive portions and after long intervals, until finally it was drained to the bottom.

The ancient lake-deposits now form the basis of the country, and appear as extensive plains, which have been subjected to a great amount of erosion, resulting in the production of deep valleys and wide basins, traversed by Green River and its tributaries, which have their sources in the mountain boundaries. From the valley of Green River the flat-topped hills rise in succession as a series of broad table-lands or terraces, extending to the flanks of the surrounding mountains.

The snows of the Uintah, Wahsatch, and other mountain-ranges are a never-failing source to the principal streams; but many of the lesser branches, dependent for their supply on the accumulated snows of winter in ravines of the lower hills and plains, completely dry up as the snows disappear with the approach and advance of summer. The country for the most part is treeless and destitute even of large shrubs, excepting along some of the water-courses. The principal streams are fringed with trees, consisting of cotton-wood (*Populus angustifolia*) and willow, (*Salix longifolia* ;) and the valleys through which they run produce mostly rushes (*Juncus balticus*) and sedges, with some coarse grasses, as *Elymus condensatus* and *Triticum repens.* Hollows of the hills and narrow valleys, favorable to the retention of moisture, support forests of small aspens, (*Populus tremuloides.*) The higher terraces and foot-hills approaching the mountain-ranges are covered with dense forests of aspens, pines, (*Pinus ponderosa* and *P. flexilis,*) and firs, (*Abies Menziesii, A. Engelmanni, A. grandis, &c.*) with a rich undergrowth of herbaceous plants. The great mountains themselves present a broad belt of pines and firs, from which project the rocky summits as bare of vegetation as the wide plains at their base. Many of the lower hill-sides and hollows in certain situations are sparsely covered with cedars, (*Juniperus virginiana,*) most of which are very old in appearance and remarkably distorted, twisted, and broken.

The principal growth of the plains consists of sage-bushes (*Artemesia tridentata*) curiously distorted and split, so as to remind one of the cedars just mentioned. In some places the sage-bushes are mingled with or replaced by the grease-wood, (*Sarcobatus vermiculatus.*) Wide, bare, path-like intervals surround the bushes, or the spaces are occupied by scanty grass, which formerly furnished food to the buffalo, now become extinct in this region and elsewhere west of the Rocky Mountains.*

The fossils which form the subjects of our communication for the most part were derived from the more superficial deposits of the great Uintah basin, which Professor Hayden has distinguished as the Bridger group of beds. These compose the terraces or table-lands in the neighborhood of Fort Bridger, and consist of nearly horizontal strata of variously colored indurated clays and sandstones. As the beds wear away, through atmospheric agencies, on the naked declivities of the flat-topped hills, the fossils become exposed to view and tumble down to the base of the hills among the crumbling *débris* of the beds.

The flat-topped hills or terraces of the Bridger basin, rising from broad valleys and extended plains, form the most conspicuous objects of the landscape. A similar condition of the country, alternating with boundless plains and great mountain heights, forms a characteristic feature of a great part of the region west of the Mississippi.

The flat-topped hills, table-lands, bench-lands, or terraces, as they are variously named, seen from lower levels, are usually called "buttes," especially when they are of limited extent. The name is of French origin, and signifies a bank of earth or rising ground. The name is likewise applied in a more restricted sense to the prominent irregularities of the deeply eroded and naked declivities of the more extended terraces. The buttes therefore vary in extent from a mere mound rising slightly above the level of the plains to

---

* It has already become a question whether the buffalo existed west of the Rocky Mountains at a comparatively recent period. That it did so was amply proved to the writer from his having noticed remains of the animal in a number of places, from ravines skirting the Union Pacific Railroad to the forests high up in the foot-hills of the Uintah Mountains. Judge W. A. Carter, of Fort Bridger, informs us that some of the old trappers and hunters of the district had told him that in their early days they had seen the buffalo in abundance in that country.

hills of varied configuration reaching to the level of the broader buttes or terraces. In the course of ages the wearing away of these has been enormous and still continues under the usual atmospheric agencies, while the detritus is spread out on the plains below.

From the lower plains the neighboring terraces, when of circumscribed extent, appear like vast earth-work fortifications, and when evenly preserved on the declivities for a considerable distance remind one of long railway embankments. Frequently the terraces are so extensively eroded and traversed by narrow ravines that they appear as great groups of naked buttes rising from the midst of the plain, or assembled around the horizon closely facing and flanking the more distant and extended lands as if to protect them. Nothing can be more desolate in appearance than some of these vast assemblages of crumbling buttes, destitute of vegetation and traversed by ravines, in which the water-courses in midsummer are almost all completely dried. To these assemblages of naked buttes, often worn into castellated and fantastic forms, and extending through miles and miles of territory, the early Canadian voyageurs gave the name of "Mauvaises Terres." They occur in many localities of the Tertiary formations west of the Mississippi River.

In wandering through the "Mauvaises Terres," or "Bad Lands," it requires but little stretch of the imagination to think oneself in the streets of some vast ruined and deserted city. No scene ever impressed the writer more strongly than the view of one of these Bad Lands. In company with his friends, Drs. Carter and Corson, he made an expedition in search of fossils to Dry Creek Cañon,* about forty miles to the southeast of Fort Bridger. The cañon, or valley, is bounded by high buttes, and contains a meadow of rushes, traversed by a stream which is liable to be dried up in the latter part of the summer, whence the name of the cañon. On ascending the butte to the east of our camp, I found before me another valley, a treeless barren plain, probably ten miles in width. From the far side of this valley butte after butte arose and grouped themselves along the horizon, and looked together in the distance like the huge fortified city of a giant race. The utter desolation of the scene, the dried-up water-courses, the absence of any moving

---

* The same name is so frequently applied to different places as to lend to considerable confusion. When I speak of Dry Creek Cañon, I refer to a locality forty miles from Fort Bridger; and when Dry Creek is named, it refers to another locality ten miles from Fort Bridger.

object, and the profound silence which prevailed, produced a feeling that was positively oppressive. When I then thought of the buttes beneath my feet, with their entombed remains of multitudes of animals forever extinct, and reflected upon the time when the country teemed with life, I truly felt that I was standing on the wreck of a former world.

The buttes are often specially designated from some supposed resemblance, or other character, as Church Butte, Pilot Butte, Grizzly Butte,* &c.

As before intimated, the more superficial table-lands of the Bridger basin, as they appear in the vicinity of Fort Bridger, are composed of nearly horizontal strata of various colored indurated clays and sandstones. In most localities visited by the writer the clays predominate, and are usually greenish, grayish ash-colored, and brownish. When unexposed they are compact, homogeneous, and of stony hardness. In composition they vary from nearly pure clay to such as are highly arenaceous, and graduate into those in which sand largely predominates, and they usually contain few or no pebbles. They appear to be more or less fissured, and break with an irregular and somewhat conchoidal fracture. Exposed to atmospheric agencies, moisture, and frosts, they readily disintegrate, and the declivities of the buttes, generally entirely destitute of vegetation, are usually invested with crumbling material from a few inches to a foot or more in depth. When this loose material is wet it forms tenacious mud, and along the course of streams in the ravines, the deepest and most treacherous mire. Baked by the sun upon the plains, it fixes the drift-pebbles and other stones as firmly almost as if imbedded in mortar.

In some localities the clays of the buttes abound in fresh-water shells, as Unio, Melania, Planorbis, &c. Less frequently in other places they contain land-shells, as Helix, &c.

The sandstones are more frequently of various shades of green, but are also yellowish and pass into shades of brown. They are compact and hard when unexposed to the weather, and are usually fine-grained, but also occur

---

* This name is applied to an extensive chain of buttes about ten miles to the southeast of Fort Bridger. Judge Carter informed me that the name originated from the circumstance that an old trapper, Jack Robinson, once reported that he had found a petrified grizzly bear on the Butte. From the description of the petrifaction I have no doubt it was that of the animal I have named in the succeeding pages, Palæosyops, the skull of which resembles that of a bear.

of a gravelly constitution. They are fissured in comparatively large masses, which assume a rounded form as they are worn away, so that a ledge of sandstone projecting from the declivity of a butte will appear like a row of cotton bales. As they disintegrate less rapidly than the contiguous clays, masses are often observed resting upon cones and columns of the latter, contributing greatly to the picturesque and sometimes fantastic appearance of the buttes.

Many of the table-lands and lesser buttes in the vicinity of the Uintah Mountains are thickly covered with drift from the latter, consisting of gravel and bowlders of red and gray compact sandstones or quartzites. The drift material is usually firmly imbedded in the surface of the plains so as to appear like a pavement. The bowlders are generally small, but assume larger proportions approaching the Uintahs. In many cases the drift completely covers the terraces or buttes, descending upon the declivities so as entirely to conceal their structure. Usually, however, it accumulates in the ravines of the declivities, leaving bare the intervening ridges of light-colored clays and sandstones. Many of the buttes are nearly or quite free of drift material. Some, again, are strewn with fragments of rock, consisting of the harder materials from the terraces themselves, and these likewise occur mingled with the drift-pebbles and bowlders from the mountain-heights.

The stone-fragments from the buttes consist of harder siliceous and calcareous clays, impure limestones, jaspers, and less frequently agate and chalcedony. In some instances they consist of singularly black incrusted and rounded sandstones, somewhat of the character of septaria. Specimens of these occasionally bear a resemblance to fossil turtles, and when found with the harder crust broken they look like turtle-shells filled with a sandstone matrix.

In the buttes in the vicinity of Carter Station, on the Union Pacific Railroad, I observed many large nodular and cylindroid masses of agate. These have a concentric arrangement of layers resembling that of fossil wood, for which they are taken. Many of the masses contain a nucleus of amber-colored crystals of calcite.

Nodules of chalcedony with dendritic markings occur in some of the buttes. These, together with the condition of many of the fossils of the buttes, indicate the presence of a considerable proportion of soluble silica in the waters of the ancient lake. In some of the sandstones, the fossil shells have had their lime completely replaced by clear chalcedony.*

Occasionally strata of limestone, mostly impure from the admixture of clay and sand, are found in some of the buttes. A frequent constituent also is fibrous arragonite, or satin-spar, in thin seams. Many of the bare mounds of clay among the buttes are thickly strewn with fragments of this arragonite.

The stones imbedded in the surface of the plains and buttes, in some positions favorable for the purpose, are highly polished from the conjoined action of the wind and sand, and when seen in the slanting light of the early morning or evening sun, appear like myriads of scattered mirrors. In many positions, the stones, no matter what may be their composition, are all blackened. The phenomenon I could not explain.

In many places the stone-fragments from the declivities of the terraces, strewn over the lower buttes or distributed over the plains, are splintered or flaked in a remarkable manner. The jaspers especially are often broken in such a way that they appear as spawls from rude implements of art, or even resemble the latter. Some of them are certainly the work of primitive man, but the vast proportion, often scattered over miles of surface, are probably accidental forms. These I suppose to have been produced by stones striking one another in the descent from declivities as they have been carried down, perhaps by glacial movement. The softer rocks of the buttes, those which are too soft for stone works of art, are also observed broken in the same way as the hard ones. In experimenting on some large splintered slabs of jasper from the buttes of Dry Creek Cañon, I found that a quick blow of a hammer would send off, with a ringing sound, a long sharp flake, reminding me of the primitive knives or scrapers of the stone age of man.

Between the well-finished implement and the accidental spawl every gradation of form may be observed among the scattered stones of the plains and

* Perhaps much of this soluble silica may have been supplied by hot springs still so frequent in Wyoming and other Western Territories. Cold springs, slightly alkaline, may have also contributed to the petrifying silica. In Pioneer Hollow, fifteen miles west of Fort Bridger, I observed a dozen springs within the distance of a mile, the water of which reminded me of the congress-water of Saratoga, New York. It is cool and clear, highly carbonated, slightly alkaline, and agreeable to the taste. The springs are circular, from 1 to 15 feet across, and are surrounded with dome-like craters from 1 to 3 feet high. The craters are formed of a siliceous sinter, which has been slowly deposited from the spring-water, and is probably the accumulation of ages. The sinter is brown from the presence of iron, though the water has no perceptibly ferruginous appearance or taste.

buttes. The accompanying figures, from 1 to 12, represent some of the flaked stones, most of which, and perhaps all, are rude works of art.

Many of the accidental forms, as well as those more nearly resembling artificial implements, if they are not actually such, appear greatly to differ in age. Some of the specimens are as sharp and fresh in appearance as if but recently shivered from the parent block, while others are so much worn and so deeply altered from exposure that they look to be of ancient date. In some of these old-looking specimens the jasper, originally brown or black, has become dull white and yellow the depth of one-fourth of an inch from the surface.*

---

* In this relation I may take the opportunity to refer to one of the simplest of stone implements, still in use, and which, if it had alone been found among the flaked materials of the buttes, would certainly have been viewed as an accidental spawl. During my stay at Fort Bridger, the Shoshone Indians made a visit to the post and encamped in its vicinity for a week. Being the first time that I had had an opportunity of seeing a tribe of Indians, I felt much interest in observing them. While wandering through their camp I noticed the women dressing buffalo-skins with a stone implement, the only one of this material I found in use among them. A serrated scraper of iron was also employed, but the stone implement was clearly a common and important one. It was a spawl from a quartzite bowlder made by a single smart blow with another stone. It is circular or oval, plano-convex, and with a sharp edge. The implement is represented in the accompanying figure 13, and according to Dr. Carter, who is quite familiar with the language and habits of the Shoshones, is called by them a "te-sho-a." By a happy accident I learned that it was not a mere recent instrument incidental to the time and place.

While on an excursion after fossils, in company with Dr. Carter, I noticed on the side of a butte a few weathered human bones, to which I directed the attention of my friend. On further examination, we found others, together with some perforated canines of the elk and one of the identical "teshoa" above described. Dr. Carter observed that the Shoshones sometimes buried their dead upon the top of prominent buttes, and these remains had fallen from the grave of a squaw, which in the course of time had become exposed by the wearing away of the edge of the butte. The bones and elk-tusks were much weathered. Their appearance and the probable circumstance that several years had elapsed before the butte could wear away to reach the grave, appear to be sufficient evidence that the "teshoa" was an implement of common use.

To this note I may add a remark relating to the perforated canines of the elk. They are worn as ornamental trophies by the Shoshones and other Indians. In a recent number of the American Journal of Science and Art, for 1872, page 241, in a notice "On fossil man of the cavern of Broussé-roussé, in Italy, by E. Riviere," I notice that, besides a human skeleton associated with the bones of many extinct animals, there were also found several flint knives and a number of *perforated canines of the stag*. In addition to the common form of many of the stone implements, this is a significant fact bearing on the probability of a common origin to the races of man. One of the specimens of perforated tusks of the elk from the Indian grave is represented in Fig. 14, at the end of this introductory chapter.

As the clays and sandstones of the Bridger terraces and buttes crumble away, a variety of remains of terrestrial and fresh-water animals are exposed to view. In some of the buttes they are comparatively abundant; in others, they are rare. The fossils consist of the bones and teeth of vertebrates, and the shells of mollusks. Fragments of silicified wood also occur, though not frequently. Shells of the sandstones are composed of chalcedony; but those imbedded in the indurated clays usually retain their carbonate of lime.

The fossil bones are completely petrified; that is to say, their more perishable constituents have been replaced mainly by siliceous matter. They are frequently as black as ebony; and the teeth are usually black, with the enamel highly lustrous. Often they are brownish, with a greenish aspect, derived from the greenish matrix in which they were imbedded. They are also found of a yellowish clay color and duller aspect.

Many of the bones are more or less crushed and distorted, as a result of the pressure of the superincumbent strata. The fragments are generally but slightly dislocated, showing that the crushing occurred while they were imbedded. The stronger bones are often well preserved, especially the rami of lower jaws and teeth, and the smaller bones of the wrist and ankle. Whole skulls are exceedingly rare, and when discovered are much crushed and distorted. Turtle-shells are among the most frequent fossils, but are usually more or less fractured, crushed, and distorted. In searching over the buttes, little piles of bone-fragments are often seen diverging from a prominent point. These, on examination, generally prove to be the remains of a turtle-shell which, after exposure, has fallen to pieces.

Generally the fossils are sharply preserved; that is to say, they rarely have a rolled or water-worn appearance, indicating that bones and shells were soon enveloped in mud at the bottom of comparatively quiet water. In the gravelly strata rolled fragments of bones are found.

Nearly all the fossils collected from the Bridger beds, and described in the succeeding pages, have been collected as loose specimens picked up on the surface of the buttes. No excavations have been made into the latter in search of fossils, except to exhume a partially exposed bone, or some parts of a skeleton supposed to be contiguous to specimens lying in view on the surface. Usually only a few pieces of a skeleton have been found together, and in no instance has a complete one been discovered which has been brought to my notice. Generally, too, there has been no certainty that bones or fragments found together belonged to the same skeleton, and in most instances they have appeared to belong to several different animals.

The remains of vertebrates thus far discovered in the Bridger Tertiary formation represent all classes except Batrachians, and these no doubt formed members of the ancient fauna; but their delicate bones have, as yet, escaped detection.

The remains of mammals are especially numerous, and they belong to many genera, most of which are extinct, and had not been previously described or found elsewhere. The greater proportion of the mammals were odd-toed pachyderms, whose nearest living allies are the tapirs. Proboscidian and equine forms appear to have been sparsely represented. Even-toed pachyderms were comparatively few; and ruminants, whose remains are so abundant and varied in the later Tertiary formations east of the Rocky Mountains, appear to have been absent. The other remains of mammals belong to rodents, insectivores, and carnivores, nearly all of extinct genera, not previously described nor found in other localities. Primates, bats, marsupials, and edentates are probably represented, but have not been certainly recognized among the fossils which I have had the opportunity of examining. The nature of the formation from which the remains are obtained is such that we do not expect to find evidences of the remaining orders of mammals.

No remains of birds have come under my notice; but Professor Marsh, who has explored the Bridger Tertiary beds with unusual facilities and great diligence, has reported the discovery of specimens which he attributes to half a dozen species of two extinct and previously unknown genera.*

Of reptiles, the remains of turtles are, perhaps, the most abundant fossils met with in the buttes of the Bridger basin. They belong to a number of different genera, several of which are extinct, but others belong to genera still in existence. Most of them are aquatic forms, but one at least was a land-tortoise. The number of species and genera is in striking contrast with the single species, represented by a multitude of individuals in the Tertiary deposits of White River, Dakota, and of Niobrara River, Nebraska.

The turtle remains mostly consist of the shells, often nearly complete, and sometimes including other bones of the skeleton imbedded in the interior matrix.

The remains of crocodiles, which are entirely wanting in the White

---

* Am. Jour. Sc., 1872, p. 256.

River and Niobrara Tertiaries just mentioned, are frequent in the Bridger beds, and represent several species.

Remains of lizards also, allied to the modern iguana and monitor, are found as associates of the Bridger fauna. Professor Marsh has likewise reported the discovery of remains of serpents, which he ascribes to several species and genera.

Multitudes of well-preserved fresh-water fishes are found in the Green River shales. They are chiefly cyprinodonts and herrings, and, for the most part, have been described by Professor Cope.

Black, shining, enameled scales, teeth, and vertebræ of ganoid fishes are frequent among the fossils of the Bridger beds.

The Tertiary strata of Green River and its tributaries, including the latter, as indicated by the character of the vertebrate fossils, are much older than the tertiaries of the Mauvaises Terres of White River, Dakota, and of the Niobrara River, Nebraska. They overlie the cretaceous rocks, with which they are unconformable, and they are probably contemporaneous with the Eocene formations of Europe.

Attention was first directed to the Green River Tertiary formation, which has proved to be so rich in the remains of vertebrates, by the late Dr. John E. Evans, as early as 1856. From Green River he obtained a specimen of shale, with a well-preserved fish, represented in Fig. 1, Plate XVII, of the present work, and briefly described by the writer, under the name of *Clupea humilis*, in the proceedings of the Academy of Natural Sciences of Philadelphia, for October, 1856.

In 1868 Dr. J. Van A. Carter, of Fort Bridger, in correspondence with the author, informed him of the frequent occurrence of the remains of turtles and other animals in the buttes of the neighboring country. The same year Professor Hayden, during his geological explorations, obtained remains of a Trionyx from Church Buttes. Colonel John H. Knight, United States Army, also procured a vertebra of an extinct crocodile from the same formation of Bitter Creek. These remains, together with those of a small insectivorous animal, discovered by Dr. Carter on the Twin Butte, near Fort Bridger, were described by the writer in the Proceedings of the Academy for April, 1869. The little insectivore was named *Omomys Carteri* in honor of its discoverer, and is also described in "The Extinct Mammalian Fauna of Dakota and Nebraska." The specimen upon which it was characterized is represented in

Figs. 13, 14, Plate XXIX, of that work. Subsequently, during 1869 and the following years down to the present time, the Green River basin has been sedulously explored by Professor O. C. Marsh with the most important and fruitful results. In the abundance of fossils and the number of extinct genera and species of vertebrates they represent, his collections are perhaps not exceeded by any obtained from any one locality elsewhere in the world. Professor Marsh has given a succinct account of the geology of the region in the American Journal of Science for 1871, and in the succeeding volumes brief descriptions of the many species and genera of extinct animals discovered by him.

In 1869 Professor Hayden, during his geological exploration of Wyoming, also examined the Green River Tertiary formations, and designated the more superficial ones under the name of the Bridger group. The fossils collected from the latter were submitted to the examination of the writer, and are briefly noticed in the Proceedings of the Academy for 1870, and likewise in Professor Hayden's reports of 1870 and 1871.

During the same and the succeeding years down to the present time, Drs. Carter and Corson explored the buttes in the vicinity of Fort Bridger and discovered many important fossils. Their collections from time to time were transmitted to the author, and by far the greater number of the animals characterized in the following paper are indicated from the specimens of these collections. Most of them have also been briefly noticed in the later volumes of the Proceedings of the Academy, and in Professor Hayden's reports for 1870 and 1871.

I may further remark that during the last summer Professor Cope made an extended exploration of the Green River basin, and obtained large collections of fossils, to a full account of which we look forward with much interest.

FIG. 14. Perforated elk-tusk; one of a number of similar specimens found together with a "teshon" and human bones which had fallen from an old Indian grave, at the edge of a butte, three miles from Fort Bridger.

MAMMALIA.

Order *Perissodactyla*.

Hoofed quadrupeds, with functional toes in the hind feet, and often likewise in the fore feet, in uneven number. Arrangement of the constituent lobes of the crowns of the molar teeth unsymmetrical. Femur with a third trochanter. Astragalus with the fore part divided into two very unequal articular facets.

PALÆOSYOPS.

Among the most abundant and interesting of the mammalian remains from the Bridger Tertiary group, which the writer has had the opportunity of examining, are those of a genus of odd-toed pachyderms to which the above name has been given. The specimens consist of fragments of jaws with teeth, isolated teeth, small portions of other parts of the skull, articular ends of the limb-bones, and some of the smaller bones of the feet.

The anatomical characters of the specimens indicate Palæosyops to have been more nearly related with the tapirs than to any other living animals. The jaws were provided with nearly closed series of teeth in full number, that is to say three incisors, a canine, four premolars, and three molars to each side of both jaws. The canines are as well developed proportionately as in ordinary carnivores, and would lead one to suspect that perhaps Palæosyops used a mixed diet of meat and vegetables.

The genus was originally established in the Proceedings of the Academy of Natural Sciences of Philadelphia for October, 1870, on specimens of teeth discovered at Church Buttes, Wyoming, during Professor Hayden's geological exploration. It was subsequently indicated in Professor Hayden's Preliminary Report of the United States Geological Survey of Wyoming, published in the spring of 1871, and is there arranged among the artiodactyl or even-toed pachyderms. Much additional material, comprising many parts of the skeleton of the same genus, having been received from Drs. Carter and Corson, its characters were more fully ascertained, and its true position as a perissodactyl or odd-toed pachyderm determined. The later account of these is given in Professor Hayden's Preliminary Report of the United States Geological Survey of Montana, &c., published in the spring of 1872.

Since then Professor O. C. Marsh has published a notice in the American Journal of Science of August, 1872, of some remains ascribed to two genera with the names of Palæosyops and Limnohyus. From the notice it would appear he has overlooked the description of Palæosyops in the report last mentioned. He intimates the reference of the genus to the Perissodactyls as if previously unknown, and suggests the reference of specimens to it in which "the last upper molar has two inner cones," though it is distinctly stated in the above report that "the last upper molar of Palæosyops has but a single lobe to the inner part of the crown." Upon this character he founds the proposed genus Limnohyus, which, under the circumstances, appears untenable; but if a pair of lobes to the inner part of the crown of the last molar be considered a distinctive generic character, the name might be transferred to the genus possessing it.

The skull of Palæosyops, and the same may be said of other parts of the skeleton so far as they are known to us, approximates in form and constitution those of its probably contemporaneous ally, the Palæotherium of the Eocene period of Europe. In both genera the skull presents a broad, triangular forehead. In Palæosyops it is more prolonged posteriorly, and is more abruptly curved forward to the root of the muzzle. In both the temporal fossæ are very capacious, indicating masticatory muscles approaching in power those of the great carnivores. In Palæosyops they are separated by a much shorter crest than in Palæotherium. In the former the muzzle is rather abruptly prolonged forward from the base of the forehead; in the latter the convexity of the forehead is continued in the muzzle to the end of the nose. In both genera the muzzle is broad, but in Palæosyops the nasals are longer and project forward as much as the jaws. The lateral nasal notch is nearly alike in both, but is longer in Palæosyops. In both, the orbits are open behind, and are defined from the temporal fossæ by long, angular post-orbital processes. The jaws nearly repeat one another in the two genera. The number of teeth, their kind, relation, and general construction, are likewise the same. In Palæosyops they form more unbroken series in the two jaws, as the hiatus back of the canines, which is comparatively large in Palæotherium, is very trifling in extent in Palæosyops.

PALÆOSYOPS PALUDOSUS.

The species *Palæosyops paludosus* was first indicated under this name in the Proceedings of the Academy of Natural Sciences of Philadelphia in 1870, and was founded on a number of isolated teeth and fragments of others obtained by Professor Hayden at Church Buttes, Wyoming. Of the specimens, a last upper premolar is represented in Fig. 5, Plate V; a fragment of a second upper molar in Fig. 6, Plate XXIII, and two lower premolars and a molar in Figs. 3 to 5, of the same plate. The teeth apparently all belonged to the same individual, which had reached maturity, but had not advanced so far as to have the summits of the tooth-lobes worn through so as to expose the dentine. The enamel is longitudinally wrinkled on the sides of the true molars and in a less degree on the premolars.

The last upper premolar (Fig. 5, Plate V) has a trilolate crown consisting of an outer pair of acute pyramidal lobes, and an inner larger conical lobe embraced by a basal ridge in front and behind.

The fragment of an upper molar (Fig. 6, Plate XXIII) consists of the fore part of the crown, and is composed of an outer crescentoid pyramidal lobe and an inner smaller conical lobe. A strong convex buttress forms the antero-external angle of the crown, and a moderate basal ridge bounds it in front. A conspicuous tubercle, the rudiment of an additional lobe, occupies the angular interval between the principal lobes and the basal ridge.

The lower molar tooth (Fig. 5) has a fore and aft bilobed crown as in Palæotherium and Titanotherium. The lobes are crescentoid pyramidal, and the anterior is the smaller.

The lower premolars have the same essential constitution as the true molar, but are less well developed. In the fourth premolar (Fig. 4) the relative size of the lobes is reversed, the anterior being the larger, and the postero-internal buttress of the crown is obsolete. In the third premolar (Fig. 3) the posterior lobe is still more reduced in size, the anterior lobe is proportionately enlarged, and the inner buttresses of the crown are obsolete.

The measurements of the specimens are as follows:

	Lines.
Fore and aft diameter of second upper molar, estimated	17
Transverse diameter of second upper molar	18
Fore and aft diameter of last upper premolar	9½
Transverse diameter of last upper premolar	12
Fore and aft diameter of second lower molar	16
Transverse diameter of second lower molar	10
Fore and aft diameter of fourth lower premolar	9½
Transverse diameter of fourth lower premolar	6½
Fore and aft diameter of third lower premolar	8½
Transverse diameter of third lower premolar	5

Shortly after the original description of the above specimens, several others were received from Professor Hayden, obtained on Henry's Fork of Green River, Wyoming, which are referred to in the last paragraph of the same article of the Proceedings above mentioned as the former ones. The additional specimens consist of several small jaw-fragments, with teeth, belonging to an individual past maturity, as indicated by the worn condition of the latter.

One of the specimens, a much-worn last upper premolar, is represented in Fig. 4, Plate V. It agrees with the corresponding tooth above described both in form and proportions. The summits of the three lobes of the crown are worn down so as to expose large tracts of dentine.

A second specimen consists of an upper-jaw fragment retaining a portion of the first molar and the complete second one. The former was so much worn as to have a great part of the enameled triturating surface removed. The second molar, represented in Figs. 8, 9, Plate V, has a low trapezoidal crown composed of four lobes, of which the anterior two agree in constitution and proportions with the fragment of the corresponding tooth above described. The outer pair of lobes are cresecntoid pyramidal and bounded externally by strong convex buttresses. The inner lobes, of which the anterior is much the larger, form broad cones. A strong basal ridge bounds the crown in front. The enamel is worn smooth and is abraded from the summits of the outer lobes so as to expose broad dentinal tracts. The fore and aft diameter of the crown of the second upper molar is $16\frac{1}{2}$ lines; its transverse diameter is 18 lines. The remaining specimen consists of an upper-jaw fragment containing the last molar, represented in Figs. 6, 7, Plate V. The tooth is fractured and its parts somewhat dislocated, so as to extend its breadth. It has the same constitution as the former tooth, except that it has but a single internal lobe, which in great part is broken away in the specimen.

Many more complete specimens referable to *Palæosyops paludosus* have been received from Drs. Carter and Carson. One of the most important of these consists of the facial portion of a skull containing nearly all the molars and the canines of both sides. The specimen submitted to my examination by Dr. Carter, represented in Fig. 51, Plate XVIII, was discovered in a greenish friable sandstone of the Grizzly Buttes. The face is entirely broken away at its upper part and fore extremity. The molar teeth, of which a full series is represented in Figs. 3, 4, Plate IV, are for the most part preserved entire, but the canines, of which one is represented in Figs. 2, 3, are broken off at the crown. The specimen pertained to an individual past maturity, as indicated by the worn condition of the teeth. The enamel is abraded from the summits of the outer lobes of the last premolar and the molars and the summits of the inner lobes of the first molar, so as to expose tracts of dentine. Elsewhere the enamel is worn smooth, but remains of its original rugose condition are yet visible in the last molar. In anatomical character and proportions the teeth agree in all respects with those corresponding among the specimens above described.

The molar series consists of seven teeth which successively increase in size from first to last. The molars or true molars approach in character those of Titanotherium, and in a less degree those of Anoplotherium and Chalicotherium. The crown is broad and low and rather rhombic in outline. It is composed of four principal lobes expanding in a common base. The outer lobes are the larger and have the shape so common in many allied animals as Palæotherium, Anchitherium, Anoplotherium, Oreodon, Cervus, &c. They are three-sided pyramids with crescentoid summits, the anterior extensions of which form stout external buttresses to the crown. The inner lobes are broad cones less prominent than the outer lobes. The anterior is the larger, and is situated opposite the angular recess of the outer lobes; the posterior occupies a position opposite the postero-internal face of the contiguous outer lobe at the inner back corner of the crown.

A strong basal ridge occupies the fore part of the crown, and elements of the same are found at the bottom of the outer faces of the external lobes. A tubercle exists in the angular interval of the anterior lobes and the basal ridge in front, which looks as if it were the rudiment of the large antero-internal lobe in Anoplotherium and its homologue in ordinary ruminants.

In the last molar the postero-internal lobe, as existing in the molars in advance, is absent or is substituted by a small tubercle extending outwardly as a posterior basal ridge to the crown.

In the unworn condition of the upper molars of Palæosyops the external lobes of the crown have acute crescentoid summits which conjoin on the summit of the median outer buttress. As the teeth were worn away in mastication, a W-shaped tract of dentine appeared on the outer lobes, and this gradually widened with the progress of abrasion. As the summits of the inner lobes were worn away, circular islets of dentine made their appearance, which likewise gradually expanded as a result of mastication. A continuance of the process would unite the inner and outer tracts, and in an advanced condition of abrasion the distinction of the four lobes with the intervening valleys would be obliterated, leaving a broad concave dentinal surface bordered by the enamel at the sides of the crown.

The upper molars of Palæosyops, while presenting considerable resemblance to those of Palæotherium, also exhibit well-marked differences. They differ especially in the greater prominence and more robust character of the external buttresses of the outer lobes, in the form and more complete isolation of the inner lobes, and in the absence of the deep pit at the termination of the oblique valley of the crown.

In comparison with the upper molars of Anoplotherium, those of Palæosyops especially differ in having proportionately stouter buttresses to the crown externally; in possessing but a rudiment of the antero-internal lobe as existing in the former, and in the different shape and relationship of position of the postero-internal lobe, which in Anoplotherium has the form nearly of the contiguous outer lobe and embraces it as in the deer.

In comparison with the corresponding teeth of Chalicotherium, several important differences are observable. Of the outer buttresses of the crown in this genus, the posterior is the larger, but in Palæosyops the anterior is the larger. The antero-internal lobe is proportionately less prominent, and the postero-internal lobe has a different shape, being nearly like that in front of it, and it is completely isolated. In Chalicotherium it is more like that in Anoplotherium, and it joins the fore part of the postero-external lobe. In the last molar of Palæosyops the postero-internal lobe is obsolete, but in Chalicotherium is proportionately as well developed as in the other molars.

As previously intimated, it is to the upper molars of Titanotherium that those of Palæosyops have most resemblance. The abrupt and deep pit near the centre of the crown is absent. The rudimental lobe at the fore part of the crown between the anterior principal lobes is proportionately less developed, and yet is more isolated from the basal ridge. In the last molar the postero-internal lobe is nearly suppressed, while in Titanotherium it is still a conspicuous element of the crown, though less well developed than in the other molars.

The premolars of Palæosyops undergo a successive reduction forward and assume a more and more elemental condition.

The fourth premolar has an oblong square crown with the transverse diameter exceeding that fore and aft, and with the inner part nearly semi-circular. The crown is composed of three lobes, corresponding with the outer pair of the molars, and apparently the large inner one situated opposite the recess of the former. The outer lobes are like those of the molars, with the back one proportionately less well developed, with the outer median buttress of the crown suppressed, and with the outer median fold of the antero-external lobe more prominent. The inner lobe is a single, broad, undivided cone less prominent but rather larger than the outer ones. It appears to be homologous alone with the anterior of the inner cones of the molars, and at least does not appear to be a connate pair as in the corresponding tooth of Titanotherium. The conspicuous pit in the center of the crown in this genus is absent in Palæosyops. A thin basal ridge starting in front and back of the internal lobe festoons the crown outwardly and at the bottom externally of the outer lobes.

The third premolar is a diminished representative of the one behind, but has its antero-external lobe proportionately a little larger, and the postero-external lobe proportionately reduced. The teeth of the two sides are not symmetrical in the specimen. That opposite to the one represented in the figure has its fore part broken away, but the postero-external lobe is considerably longer than in the entire tooth.

The second premolar has a trihedral crown, in which but two lobes are conspicuous. In comparison with the premolars behind, the internal lobe is greatly reduced in size, and the antero-external lobe is much enlarged so as to become the main portion of the crown, while the postero-external lobe is obsolete.

The first premolar is a small tooth separated from the others by a slight interval. It has a simple short conical crown with the base slightly extended backward, and is inserted by a pair of fangs. The other premolars and the molars are inserted with three fangs, of which the inner one in the latter teeth consists of a connate pair.

The canine teeth of Palæosyops were powerful and efficient weapons, and resembled those of ordinary carnivores more than they do those of nearly allied living animals. Though imperfect in the specimen under consideration, the remaining portions, as represented in Figs. 2, 3, Plate IV, indicate teeth of the form and proportions of those of living bears. They also appear to have nearly the same relative position with the other teeth and the same direction as in the latter. The fang is robust and gibbous, and comes from the alveolus in a direction downward and forward with a greater degree of divergence than usual among carnivora. The face of Palæosyops, judging from the imperfect specimen, a side-view of which is given in Fig. 51, Plate XVIII, in its complete condition, would appear to resemble in shape that of the Elotherium of White River, Dakota, except that the muzzle was proportionately shorter. Among living animals, it appears to have resembled that of the bears more than in those nearly related to it. The zygomatic arches are of robust proportions and widely divergent at their anterior attachment to the face. The malar portion of the zygoma is divided by an acute ridge curving from the anterior orbital margin outward, downward, and backward. The surface above this ridge curves outwardly and downward from the floor of the orbit continuously. The surface beneath is a broad trilateral plane looking forward, downward, and outward, and is roughened for the attachment of a powerful masseter muscle. The space behind the anterior abutment of the zygoma indicates a temporal fossa of large capacity.

The orbit appears low, and is directed obliquely forward and outward. In advance of the prominent antorbital margin the side of the face is nearly vertical. The infra-orbital foramen is rather large, and is situated over the position of the last premolar. In front of the foramen begins the swell of the large canine alveolus, and below its position is the alveolar border, marked by the vertical ridges of the molar fangs. The hard palate is well arched, and nearly parallel at the sides. Its surface in the specimen is obscured by the attachment of rocky matrix. The breadth of the face at the zygomata appears to have about equaled the length.

The measurements of the specimen are as follows:

	Inches.	Lines.
Breadth of face at zygomata on line with middle of last molars	7	8
Breadth of face outside of last molars	4	6
Breadth of face outside of canine alveoli	3	6
Breadth of face at infra-orbital foramina	3	2
Breadth of hard palate between last molars	1	7
Breadth of hard palate between last premolars	1	5
Distance from back of last molar to fore part of canine	6	8

For comparison, the measurements of the teeth will be given after the description of the following series.

Some additional specimens, which I suppose to belong to *Palæosyops paludosus*, notwithstanding certain differences hereafter to be mentioned, consist of most of the upper teeth, with small attached jaw-fragments, obtained by Dr. Carter on Henry's Fork of Green River. Of these specimens a complete series of nearly perfect molar teeth is represented in Figs. 5, 6, Plate IV. The teeth in their abraded condition indicate an older animal than that to which the facial specimen above described belonged. The summits of the constituent lobes of the teeth are nearly all worn to such a degree as to exhibit tracts of dentine, and the enamel is everywhere smooth, except on the external faces of the outer lobes near the basal ridge.

The molars are almost identical in character with those above described in the facial specimen. Trifling differences consist in the less production of the median fold on the outer face of the external lobes, and perhaps the less degree of prominence of the tubercle in the interval anteriorly of the anterior pair of lobes. The last premolar is likewise identical with those above described, except that its crown is rather more square, or is not quite so wide. The anterior three premolars depart considerably from their character in the facial specimen, and their differences may probably indicate a different species. The third premolar is a diminished representative of the one behind it, the three lobes of the crown holding nearly the same proportionate development; whereas in the corresponding tooth of the facial specimen the postero-external lobe is considerably reduced in its proportions. In the second premolar the crown still retains a postero-external lobe reduced in proportion to the others, but in the corresponding tooth above described it is obsolete. The retention of this lobe gives the crown a greater fore and aft breadth than that contained in the facial specimen. The first premolar has the same form as that of the latter, but it is much larger.

The mutilated canine, accompanying the molars first described, is represented in Fig. 1, Plate IV, and is but little more than half the size of those contained in the facial specimen.

An isolated incisor, represented in Fig. 8, accompanying the molars and canine just described, is regarded as an upper one. The crown is mutilated, but when complete appears to have had a short, conical crown, bounded behind by a strong basal ridge. The fang is laterally compressed, and is about an inch in length.

Comparative measurements of the series of teeth of the two individuals of Palæosyops, indicated by the facial specimen, with teeth, from Grizzly Buttes, and the specimens of teeth from Henry's Fork, just described, are as follows:

	Lines.	Lines.
Space occupied by the entire molar series	69	71
Space occupied by the true molar series	41	41
Space occupied by the premolar series	28	32

	Antero-posterior.	Transverse.	Antero-posterior.	Transverse.
	Lines.	Lines.	Lines.	Lines.
Diameter of first premolar	5	3	7	4
Diameter of second premolar	6	7	8	7
Diameter of third premolar	7	8	8	9
Diameter of fourth premolar	10	10½	8	10
Diameter of first molar	12	13	12	12¼
Diameter of second molar	16	17	15	16
Diameter of last molar	17	18½	17	16½

	Lines.	Lines.
Length of fang of upper canines	28	18
Antero-posterior diameter of canines	12½	7½
Transverse diameter of canines	10½	7

The question arises whether the differences which have been indicated in the premolars and canines of the two different series of teeth above described indicate more than one species. The differences are clearly in degree of development and size, and these may probably be of a sexual character. The individual with the more powerful canines I suppose to have been a male, in which, with a greater proportionate degree of development of these organs than in the female, there appears to have been a reduction in the degree of development of the anterior premolars.

Another specimen submitted to my examination by Dr. Carter, and represented in Figs. 6, 7, Plate XXIV, belonged to an older animal than the former, as indicated by the more worn condition of the teeth. The latter consist of the anterior three premolars and a portion of the fang of the canine, and they have the same form and proportions as the corresponding teeth above described. The first premolar is close to the others, or is not separated by a conspicuous interval as in the former specimen. The lobes of the second and third premolars are worn nearly to a level with their base. The outer surface of the maxillary, as seen in Fig. 6, is defined by an oblique ridge at the nasal border, within which the suture of the premaxillary pursues its course over the position of the fang of the canine. Just outside of the nasal border the surface of the maxillary is depressed.

The measurements of the specimen are as follows:

	Lines.
Space occupied by the anterior three premolars	21
Antero-posterior diameter of first premolar	$6\frac{3}{4}$
Transverse diameter of first premolar	$5\frac{1}{4}$
Antero-posterior diameter of second premolar	7
Transverse diameter of second premolar	$6\frac{3}{4}$
Antero-posterior diameter of third premolar	8
Transverse diameter of third premolar	9
Diameter of fang of canine	8

Fragments of half a dozen lower jaws referable to Palæosyops, collected in various localities in the vicinity of Fort Bridger by Drs. Carter and Corson, have been submitted to my examination.

A well-preserved specimen, consisting of the greater part of the jaw, was discovered by Dr. Carter imbedded in a greenish gravel thirteen miles southeast of Fort Bridger. The right ramus is represented in Fig. 11, Plate V, and it contains the molars and the back two premolars, which are also represented with a view of the triturating surfaces in Fig. 10 of the same plate. The teeth, corresponding with those in part upon which the species *Palæosyops paludosus* was originally indicated, are identical in anatomical character and so nearly in size that the jaw may be regarded as pertaining to the same species.

In advance of the teeth retained in the jaw there are indications of two additional premolars verging close upon the remains of the canine alveolus, and thus the specimen shows that the number of the lower molar series of Palæosyops is seven.

The lower molars of Palæosyops resemble those of Palæotherium and Anchitherium, but even more closely those of Titanotherium. The crowns are proportionately wider and lower, or appear more robust than in the former genera.

The crown of the anterior two molars is quadrately oblong oval, with the fore and aft diameter largest and the depth less than the width. It is composed of two divisions or lobes, one in advance of the other. The last molar has the same form and construction, with the addition of a third but smaller lobe.

In the unworn molars the principal constituent lobes present acute crescentoid summits embracing a concavity which opens to its bottom by an angular notch on the inner side of the crown. The contiguous arms of the crescentoid summits conjoin in a strong conical eminence situated just in advance of the middle of the crown internally. The point of this eminence is simple or undivided; in Anchitherium it is deeply indented and appears to be composed of a connate pair of eminences.

The fore part of the summit of the anterior lobe in Palæosyops curves downward and inward, and ends in a slight prominence at the anterior inner corner of the crown. The hind part of the summit of the posterior lobe ends in a prominence like that in advance, but smaller, and situated at the posterointernal corner of the crown.

Externally the lobes of the crown are angularly convex, and include deep angular recesses sloping outwardly and downward, and bounded by festooned elements of a basal ridge. The inner surface of the crown is nearly vertical, smooth, and without a basal ridge. The latter is especially well developed at the fore and back part of the crown, except in the last molar, in which the additional lobe takes its place. This lobe is a much reduced likeness of those in advance, with the arms of its crescentoid summit contracted and conjoined with the posterior conical eminence of the crown internally.

The molars undergo a rapid reduction forward, and they are inserted by two fangs. The crown of the last premolar is a reduced representative of that of the succeeding molar, with the posterior lobe proportionately, in comparison with the anterior lobe, less well developed. In the crown of the third premolar there has been a further proportionate reduction in the back lobe, but the anterior remains nearly the same, except that it appears more robust from its connation with the homologue of the anterior of the inner conical eminences of the teeth behind.

In Palæotherium and Anchitherium the corresponding premolars with those described repeat the form of those of the molars, and in this respect greatly differ from Palæosyops. The inferior premolars of Titanotherium in a perfect condition are not sufficiently well known to institute a comparison with those of Palæosyops.

The lower molars of Palæosyops in wearing would assume the same appearance as those of Palæotherium and Anchitherium at the same stages of attrition.

The space occupied by the entire molar series is estimated at about $6\frac{1}{2}$ inches, of which the true molars occupy rather less than 4 inches.

The measurements of the molar teeth contained in the lower jaw are as follows:

	Antero-posterior.	Transverse.
	Lines.	Lines.
Diameter of third premolar	$8\frac{1}{4}$	$5\frac{1}{2}$
Diameter of fourth premolar	9	$6\frac{1}{4}$
Diameter of first molar	$12\frac{1}{2}$	8
Diameter of second molar	15	$9\frac{1}{2}$
Diameter of last molar	19	10

The premolars are inserted by a pair of fangs, except the first, which has but a single fang.

The lower jaw of Palæosyops, as seen in Fig. 11, Plate V, approximates in form that of the tapir and hog, though presenting important differences. The dentary portion of the ramus is proportionately shorter and deeper than in either of those animals, and the alveolar border is more ascending posteriorly. The base is more convex fore and aft than in the hog but less than in the tapir, and is more obtuse than in either. The outer surface is vertical, with a slight outward slope at the fore part.

The back part of the jaw is of more uniform breadth than in the tapir, and is more like that in the hog. Toward the angle the outer surface is a vertical plane, with the lower border or base more directed downwardly than in the hog. The upper or ascending portion presents a masseteric fossa about as deep as in the tapir but of considerably greater width.

The condyle is large and thick, and much like that in the tapir, but is less inclined inwardly. It has about the same proportionate elevation above the level of the base of the jaw, but less above the level of the teeth.

The border of the jaw below the condyle behind is at first slightly concave and then convex, as in the hog, but to a less degree. The coronoid process is about as long as that of the tapir, but the fore part curves upward and backward without any inclination forward. The notch behind hardly descends below the level of the condyle.

The mental foramen is smaller than in the tapir, and is situated below the interval of the second and third premolars.

The length of the lower jaw, from its back border to the fore part of the second premolar, is 9¾ inches, and in the complete condition it measured about 2 inches more.

Portions of several lower jaws, apparently all referable to Palæosyops, were obtained by Dr. Corson at Grizzly Buttes. The specimens exhibit some variation in character, and may, perhaps, belong to more than one species of the genus. One of the specimens consists of a dentary fragment containing the true molars and the fangs of the two premolars in advance. The retained teeth are like those previously described, but are in a trifling degree smaller. The series measures 3¾ inches. The jaw-fragment nearly agrees with the corresponding portion of the specimen above described, but is of more uniform depth.

Another specimen consists of a right ramus, without the chin and back part, and broken into three pieces. It contains the fang of the canine and most of those of the molars. The jaw is of more uniform depth below the position of the teeth than in the more complete specimen first described, and more robust than in either of the former specimens. The retained portion of the fang of the canine indicates a larger tooth than existed in the first-described specimen—one, also, that would accord in its robust character with those of the facial specimen referred to *Palæosyops paludosus*. The presence of the fang of the canine produces a strong bulge at the side of the chin, which appears to have been comparatively feeble in the first-described specimen. Two mental foramina are situated below the position of the second and third premolars. The first premolar appears to have had a single fang consisting of a connate pair. It was separated from the canine and second premolar by conspicuous intervals, the posterior of which is the larger. A portion of the chin being retained in the specimen, it would appear in the entire condition to form a broad slope defined at the sides by the convexities of the canine alveoli. The rami were completely co-ossified at the symphysis without the suture of union being apparent.

The remaining specimen consists of a portion of the jaw containing the fangs of the last two molars, and the portion immediately behind extending toward the angle. The dentary portion of the bone is considerably deeper than in the corresponding portion of the preceding specimens. The base below the position of the last molar tooth is rather more conspicuously tuberous and roughened for muscular attachment, and the concavity back of this is more posterior and deeper than in the first-described specimen.

Comparative measurements of the lower-jaw specimens, including the one first described, are as follows:

	No. 1.	No. 2.	No. 3.	No. 4.
	Lines.	*Lines.*	*Lines.*	*Lines.*
Space occupied by the entire molar series................	.....	.....	.....	78
Space occupied by the molars and last two premolars.....	64	65	60	62
Space occupied by the true molars	46	48	45	45
Distance from last molar to back of jaw...............	49	48	.....	.....
Width of ramus back of condyle	44	41	.....	.....
Depth of ramus at middle of last molar	31	32	33	33
Depth of ramus below last premolar	23	23	26	29
Thickness of base below second molar	13	11	12	13
Antero-posterior diameter of last molar...............	19	20	19	19

A small fragment of the chin of a lower jaw, referable to Palæosyops, retains part of the alveolus of a large canine, and portions of the fangs of three incisors of the same side, thus indicating the number of the latter teeth in the animal. The canine alveolus has been about an inch in diameter. In the ramus of the jaw above described, retaining the fang of a canine, this tooth has been nearly in proportionate size to that of the alveolus just mentioned.

Small fragments from three different skulls, attributable to Palæosyops, consisting of portions of two sagittal crests and the supra-occipital, indicate capacious temporal fossæ, separated by a short, thick crest and a broad occiput.

The fragments of sagittal crests are from the fore part, retaining the suture and notch for the summit of the frontal. The upper surface of the crest is a flat triangle, slightly depressed at the middle, with the notch for the frontal in its base. In the latter position it is $1\frac{1}{2}$ inches wide; and a couple of inches back of this position the crest is $\frac{3}{4}$ of an inch thick.

The occipital fragment on each side in front presents a wide, sloping surface, which contributes to the temporal fossa. The posterior surface in general appearance resembles that in the rhinoceros. The upper portion forms a broad, even concavity, undivided by any trace of a vertical ridge, and only roughened at the summit for the attachment of the nuchal ligament. The lateral processes are angular and divergent, and the space between them is 4½ inches in width. The lower portion of the occipital surface approaching the occipital foramen is convex. The height of the occiput from the latter is about 4¾ inches.

A lumbar vertebra was found by Dr. Corson at Grizzly Buttes. It presents the ordinary ungulate form. The body is 2 inches long, but somewhat shortened below. It is concave fore and aft, at the sides, and beneath, where it is also slightly carinate. The anterior extremity is slightly convex, 1½ inches transversely, and a little less in depth to the prominence beneath. The posterior surface is flat, or feebly depressed. The transverse process springs from the upper level of the body. A well-developed metapophysis projects from the position of the anterior zygapophysis. The diameter of the spinal canal is about an inch.

Besides the skull-fragments and vertebra of Palæosyops, a number of isolated carpal and tarsal bones, and many fragments of the long bones and other portions of the skeleton have been collected by Dr. Carter, Dr. Corson, and Professor Hayden's party. Many of the bones had been fractured, or more or less crushed, while they lay in their bed, and many have been further injured after exposure through the influence of the weather and other causes. The bones nearly resemble in size and construction the corresponding ones of the American tapir.

The distal extremity of a humerus, represented in Fig. 3, Plate XIX, was found by Dr. Corson in the vicinity of Fort Bridger. The breadth of the specimen between the supra-condyloid eminences is 3⅓ inches. A deep supra-condyloid fossa occupies the front of the humerus, opposed to the deeper and more capacious olecranon fossa. The articular trochlea is 2½ inches wide in front, and narrows an inch less behind.

A mutilated femur, without the head and trochanters, represented in Fig. 1, Plate XIX, was obtained on Henry's Fork, of Green River, by Dr. Carter. In its complete condition it has approximated 15 inches in length. The shaft is three-sided, and at the middle is 16 lines in diameter from before backward, and 19 lines transversely. The median trochanter projects from the outer border of the prismoid shaft, and is higher up in position than in the tapir. The distal extremity resembles the corresponding part in the latter, but the trochlea for the patella is of less breadth.

Fig. 2, Plate XIX, represents a much better preserved distal extremity of the femur than that of the former. It was obtained by Professor Hayden's party at Grizzly Buttes. At the supra-condyloid eminences it is $3\frac{1}{4}$ inches in diameter. The width at the condyles is $2\frac{3}{4}$ inches. The trochlea for the patella, where widest, measures 16 lines.

The detached head of a femur, in perfect condition, found by Dr. Carter near Fort Bridger, measures about 2 inches in diameter. A deep cup-like pit for the round ligament approaches the center of the head much more closely than in the tapir.

A nearly entire femur of Palæosyops, received from Dr. Carter since the above was written, is represented in Fig. 5, Plate XXIX. It nearly repeats the form of that of the tapir, but rather resembles that of the Indian tapir, or Baird's tapir, of Guatemala, than that of the American tapir. In comparison with that of the *Tapirus Bairdi*, it is rather larger, and the upper extremity is proportionately somewhat wider. The inner trochanter is longer or more prominent, but the third trochanter is neither so long nor so hook-like.

The measurements of the specimen are as follows:

	Inches.
Length externally from summit of great trochanter	$15\frac{1}{2}$
Width between head and great trochanter	$4\frac{1}{2}$
Width at third trochanter	$2\frac{3}{4}$
Diameter of head	$2\frac{1}{2}$
Diameter fore and aft of shaft at middle	$1\frac{1}{2}$
Width at condyles	$3\frac{1}{2}$

Fig. 1, Plate XX, represents a nearly entire tibia, obtained by Professor Hayden's party at Grizzly Buttes. The upper condyles are in some degree pressed toward each other, and the extremity of the internal malleolus is broken off. The bone is not quite so long as that of a tapir with which it was compared, but is somewhat stouter. The tuberosity for the ligament of the patella is of more robust proportions, and extends lower on the shaft than in the tapir. The ridge descending from it is thicker than in the latter—straighter, and is obtusely rounded. The length of the tibia is 9 inches; the breadth of its distal end over 2 inches.

Fig. 2, Plate XX, represents a calcaneum, obtained by Dr. Corson near the stage-route at the crossing of Smith's Fork of Green River. It is nearly like that of the tapir, but is stouter in proportion to its length. The tuberosity of the calcaneum is less compressed and is more obtuse in front. The sustentaculum is of much greater extent vertically, and sustains a long elliptical facet for the astragalus. The anterior articular facet for the latter is of much less extent than in the tapir, and is more distinctly separated from it by the interosseous sinus. The articulation for the cuboid is of greater depth but less width than in the tapir.

The extreme length of the calcaneum is 4¼ inches. The length of the tuberosity is nearly 3 inches. The breadth of the anterior extremity of the bone is 2¼ inches.

Of two additional calcanea obtained by Dr. Carter, one was found on Henry's Fork of Green River; the other near Millersville.

Fig. 3, Plate XX, represents an astragalus found by Dr. Carter at the bluffs, three miles from Millersville. The trochlea for the tibia is of less extent fore and aft than in the tapir; and the anterior extremity of the bone is of less width but greater depth. The length of the astragalus is 2 inches; the breadth of the trochlea twenty lines; the breadth of the anterior extremity is the same, and its thickness is an inch.

Another astragalus, slightly larger, was obtained by Professor Hayden's party at Church Buttes.

Fig. 4, Plate XX, represents three tarsal bones obtained by Professor Hayden's party at Church Buttes. They pertained to the same individual, and consist of the cuboid, scaphoid, and the outer cuneiform.

The cuboid is more cubical and stouter than in the tapir. The upper surface is more regularly square and nearly a third wider than in that animal. The articular facet for the calcaneum has about the same depth, but is nearly twice the width. The facet for the first metatarsal bone is also of equal depth, but a third greater in width.

The scaphoid is of rather less breadth than in the tapir, nearly of equal depth, but not quite so thick. The articular facet for the astragalus is of about the same extent, less breadth, but proportionately more uniform depth, and it is less concave. The articular facet for the outer cuneiform is of about the same depth, but of much less breadth than in the tapir. The facets for the inner two cuneiforms have about the same extent as in the latter.

The external cuneiform is about the same depth as in the tapir, but of considerable less breadth and of greater thickness.

The metatarsal articular facets of the cuboid and external cuneiform appear to indicate that the outer toe of Palæosyops was as large as the middle toe, and that this was much smaller than in the tapir. This appears to be confirmed by the specimen represented in Fig. 5, Plate XX, which I suppose to be a middle metatarsal of Palæosyops. It was found by Dr. Corson in the vicinity of Fort Bridger. It resembles the corresponding one of the tapir, but is shorter and of more slender proportions. It has about the size of the lateral metatarsals of the tapir.

Figs. 6 and 7, Plate XX, represent a first and second phalanx, probably of Palæosyops. The specimen of the first was obtained by Dr. Carter on Henry's Fork of Green River; the specimen of the second was found near Fort Bridger.

A specimen of a metacarpal, which I suppose to belong to Palæosyops, was obtained by Dr. Corson at Grizzly Buttes. It has about the same length as the middle metatarsal attributed to Palæosyops, but is somewhat wider. If it corresponds with the second of the series of four toes of the fore foot of the tapir, it exhibits a corresponding reduction in relation with the contiguous toes that the middle metatarsal does to the others of the hind foot.

PALÆOSYOPS MAJOR.

A larger species of Palæosyops is apparently indicated by some fragments of large bones obtained by Dr. Carter at Grizzly Buttes and other localities in the vicinity of Fort Bridger. Several of the specimens consist of portions of limb-bones, but too much mutilated either for description or representation. Even the best specimen, consisting of a fragment of the lower jaw, represented in Fig. 8, Plate XX, is barely more than sufficient to render it probable that it pertained to Palæosyops. The jaw-specimen is furthermore in some degree abnormal in form, due to inflammation or some other affection connected with the second molar tooth. The bone outside the position of the latter is much swollen, and the alveolar border is hollowed out and irregular. The alveolus is also filled with the clay matrix, so that the tooth was perhaps lost before the death of the animal. In its proportions, the jaw, in a normal condition, would appear to be of more robust character than in *Palæosyops paludosus*. In its present state, the base is more convex fore and aft than in the latter, and the alveolar border more ascending posteriorly.

The remains of the molar fangs at the entrance of the alveoli appear to indicate teeth of the same form and construction as in *Palæosyops paludosus*, for which reason the fragment was referred to the same genus. The true molars appear to have occupied a space of 4¾ inches, though this is probably somewhat exaggerated, as the interval occupied by the last intermediate molar appears proportionately somewhat too large. The crown of the last molar, which was clearly trilobate as in *Palæosyops paludosus*, had an antero-posterior diameter of 2 inches.

The former existence of a larger species than *Palæosyops paludosus*, and probably the same as that indicated under the name at the head of the present chapter, is apparently confirmed by more characteristic materials placed at my disposal by Dr. Carter in my recent visit to Fort Bridger. One of the best preserved specimens consists of the greater part of the left ramus of the lower jaw, containing six molar teeth, as seen in Fig. 1, Plate XXIII. The jaw is considerably more robust than in those referred to *Palæosyops paludosus*, though not to the degree I supposed from a view of the diseased specimen above described. At the side it is more rounded toward the base, and is more convex in a curving line from the root of the coronoid process beneath the true molars, and is more bent inward and convex from the position just indicated toward the technical angle of the bone. Rugosities of the surface in several positions indicate stronger attachment of the soft parts, in accordance with the greater bulk of the animal, than in *Palæosyops paludosus*.

The true molars have the same form and proportions as in the species just named. Trifling differences appear to be dependent only on a difference in the robust character of the species. The external basal ridge is slightly better developed, as is also the case with the median ridge, descending on the inner slope of the external lobes of the crown. The back lobe of the last molar is also rather better developed, and incloses a shallower fossa on its inner side.

The first premolar, situated immediately behind the canine, is inserted by a single fang, and is separated from the second premolar by a hiatus about a third of an inch in extent. Below the hiatus, the jaw externally presents a small concavity.

The last premolar has the same form as that in the jaw referred to *Palæosyops paludosus*, though, from its worn condition, it looks different. Independent of this, it exhibits no difference except that the base in advance of the anterior lobe is produced externally in a strong ridge.

The third premolar also is like that of *Palæosyops paludosus*, excepting that it exhibits a tendency to the production of a basal ridge not evident in the former.

The second premolar, not present in the jaw-specimen of *Palæosyops paludosus*, is a reduced form of the tooth behind it. A portion of the canine alveolus retained in the specimen indicates a tooth of moderate size in comparison with the size of the jaw itself.

Another series of lower molar teeth, attached to small jaw-fragments, and represented in Fig. 2, Plate XXIII, also appear to me to be referable to *Palæosyops major* notwithstanding certain differences presented by the premolars. The teeth are considerably more worn than in the preceding specimen; most of the summits of the constituent lobes of the crown of the molars and last premolar being so worn as to exhibit islets of exposed dentine. The second molar is most worn, and presents on the summits of the outer or principal lobes broad, depressed, shield-shaped tracts of dentine.

The molars have the same constitution as in the preceding specimen. The last one is smaller, but the others are nearly of the same size, except that the first one is thicker, especially at its fore part, and is therefore of more uniform diameter. The basal ridge of the anterior two molars is better developed externally than in the former specimen. In the first molar the anterior lobe, being proportionately rather better developed than in the corresponding tooth of the previous specimen, its anterior ridge curving inwardly, is stronger, and it embraces a more conspicuous fossa.

The last premolar differs somewhat in proportions from that of the former specimen, but is otherwise nearly the same, except so far as it is altered in appearance from being more worn. It is of less breadth fore and aft, and is thicker, and it does not present the ridge at the fore part of the base, externally, of the anterior lobe, being in this respect more like the corresponding tooth in the jaw-specimen of *Palæosyops paludosus*.

The third premolar differs from that of the former specimen very much in the same manner as the succeeding tooth. The crown is of less breadth fore and aft, and is thicker. It has exactly the same constitution, but looks different on account of its more worn condition, its difference of proportion, and from the absence of a basal ridge occupying the fore part of the crown, externally, in the former specimen.

Comparative measurements of the teeth and jaws of the specimens just referred to *Palæosyops major*, and the jaw-specimen, with teeth, of *P. paludosus*, are as follows:

	Palæosyops major.		Palæosyops paludosus.
	Lines.	*Lines.*	*Lines.*
Depth of jaw at middle of last molar	37	......	32
Depth of jaw at middle of last premolar	26	......	23
Thickness of jaw below interval of last two molars	16	......	13
Thickness of jaw below third premolar	11	......	$10\frac{1}{2}$
Distance from canine alveolus to back of last molar	92	......	?77
Length of the complete molar series	90	......	....
Length of the molar series, excluding the first premolar	82	......	....
Length of the molar series, excluding the first two premolars.	72	68	61
Length of the premolar series	38	......	....
Length of the true molar series	53	51	$46\frac{1}{2}$
Breadth of second premolar	9	......	....
Thickness of second premolar	5	......	....
Breadth of third premolar	$8\frac{3}{4}$	8	$8\frac{1}{2}$
Thickness of third premolar	$5\frac{3}{4}$	6	$5\frac{1}{2}$
Breadth of fourth premolar	$9\frac{1}{2}$	$8\frac{3}{4}$	9
Thickness of fourth premolar	7	$7\frac{1}{4}$	$6\frac{1}{4}$
Breadth of first molar	$13\frac{1}{2}$	$13\frac{1}{2}$	$12\frac{1}{2}$
Thickness of first molar	9	$9\frac{1}{4}$	8
Breadth of second molar	$16\frac{1}{2}$	$16\frac{1}{2}$	15
Thickness of second molar	$10\frac{3}{4}$	$10\frac{1}{2}$	$9\frac{1}{2}$
Breadth of third molar	24	22	19
Thickness of third molar	$13\frac{1}{2}$	12	10

Some additional specimens, found by Dr. Corson in the buttes of Dry Creek Cañon, appear to belong to the larger Palæosyops. These consist of some upper teeth, comprising a canine, a second and last premolar, and the second and third molars.

The latter are represented in Figs 10, 11, Plate XXIII, and they agree in character with the corresponding smaller teeth described under the head of *Palæosyops paludosus*. They are but slightly worn at the summits of the lobes of the crown, and the enamel is conspicuously wrinkled.

The last premolar, represented in Fig. 9, of the same plate, likewise agrees with the corresponding tooth described under the head of *Palæosyops paludosus*, except that it is of larger size. The tooth is but slightly worn, and exhibits a much less wrinkled condition of the enamel than the true molars

The second premolar, represented in Fig. 8, resembles in form that of the second series of specimens of upper molar teeth, described under the head of

Palæosyops paludosus. It is larger, less worn, and has, comparatively with the true molars, smooth enamel.

An upper canine tooth, represented in Fig. 7, is of less size than that in the facial specimen of *Palæosyops paludosus*, the reverse of the condition in this respect of the molar teeth. The canine tooth resembles, in its form and proportions, the corresponding weapon of the bear. The crown is of moderate length, and curved conical. It is provided with a subacute ridge in front and behind, defining the smaller inner face from the outer one, and has the base slightly thickened internally. The enamel is nearly smooth, and is somewhat worn on the anterior face. The fang is considerably longer than the crown, less curved, and is in some degree gibbous.

A lateral incisor, represented in Fig. 5, Plate XXIV. is a strong tooth, somewhat resembling that of the tapir. The crown is conical, with the inner and outer faces defined by ridges, with the base thickened in front, and a strong basal ridge internally. The fang is about twice the length of the crown, and is somewhat sigmoid.

The measurements of the upper teeth of *Palæosyops major*, in comparison with those of *P. paludosus*, are as follows:

	Palæosyops major.	Palæosyops paludosus.
	Lines.	Lines.
Antero-posterior diameter of last upper molar	18	18
Transverse diameter of last upper molar	19	16¾
Antero-posterior diameter of second upper molar	19	16
Transverse diameter of second upper molar	18	16
Antero-posterior diameter of last premolar	10½	9
Transverse diameter of last premolar	12½	11
Antero-posterior diameter of second premolar	9	8
Transverse diameter of second premolar	7½	7
Length of crown of canine	18	
Antero-posterior diameter of base of canine	10	
Transverse diameter of base of canine	9¼	
Diameter of fang of canine	11	12½
Length of crown of lateral incisor	10	
Diameter of base of crown of incisor	7	
Diameter of fang of incisor	8	

A small collection of teeth belonging to the larger Palæosyops was obtained by Dr. Carter in a butte ten miles distant from Dry Creek Cañon. Among the specimens there is a series of upper premolars, from the second to the last, inclusive, represented in Fig. 12, Plate XXIII. The crowns of the teeth are worn, and also somewhat eroded, but not to such an extent as to obscure their characters.

The last upper premolar agrees with that previously described and referred to *Palæosyops major*, except that it is of more uniform width.

The third premolar is a reduced representative of that behind it, but is also proportionately of less width transversely.

The second premolar is like the corresponding tooth above described and referred to *Palæosyops major*, but is considerably narrower fore and aft.

A much worn upper true molar, partially broken away externally, is rather smaller than the specimen of a second upper molar above described and referred to *Palæosyops major*. It sufficiently accords with it to be the first of the series of true molars.

Another specimen consists of a mutilated canine, intermediate in size to the more perfect one above described, and the larger one, contained in the facial specimen described under the head of *Palæosyops paludosus*. The fang toward the extremity is more curved than in either of the other specimens.

An upper lateral incisor was about the size of the one previously described, but has a stouter fang. Its crown, broken toward the point, is deeply worn away internally.

Another incisor, a lower one, is represented in Fig. 15, Plate XXIII. It has a short, conical crown, with a strong basal ridge posteriorly.

The measurements of the specimens are as follows:

	Lines.
Antero-posterior diameter of upper true molar, estimated	16
Transverse diameter of upper true molar	18
Antero-posterior diameter of upper last premolar	8¾
Transverse diameter of upper last premolar	12¼
Antero-posterior diameter of upper third premolar	8
Transverse diameter of upper third premolar	10¼
Antero-posterior diameter of upper second premolar	7
Transverse diameter of upper second premolar	8¾
Length of fang of canine	25
Diameter of fang of canine	12
Length of fang of upper incisor	22
Diameter of fang of upper incisor	7½
Length of crown of lower incisor	6
Antero-posterior diameter of lower incisor	6
Transverse diameter of lower incisor	5

An important and instructive specimen pertaining to *Palæosyops major* is represented in Fig. 16, Plate XXIII, and Fig. 1, Plate XXIV. It consists of a cranium, discovered by Dr. Carter in the buttes of Dry Creek Cañon. The specimen was broken into many pieces, but these have been united so as to give us a good idea of the shape and construction of the cranium. This is of remarkable form, and exhibits more resemblance to that of a bear than to that of its nearer relative the tapir.

The forehead, as seen in the upper view of the cranium, Fig 16, Plate XXIII, forms a long triangle, with the apex prolonged backward and expanded at the summit of the occiput. Its fore part more abruptly widens as it extends outwardly upon the conspicuous postorbital processes. Its surface from the apex forward is strongly convex, but approaching the muzzle between the position of the postorbital processes it becomes in the same direction concave. Transversely it is nearly straight between the boundaries of the temporal fossæ, but is convex between the postorbital processes. The latter are strong and unusually prominent, trihedral, hook-like projections. Their upper acute border forms the anterior extension of the temporal boundary from the forehead. Their supra-orbital margin curves from the face backward and outward to the point. Their anterior or facial surface is depressed or concave.

The postorbital process preserved in the specimen is broken at the end, but is there so narrow as to make it appear that it did not meet an ascending process from the malar bone as to form a postorbital arch. The strongly arched supra-orbital border is directed outward with a moderate backward inclination, indicating a more forward direction for the orbit than in the tapir and rhinoceros.

The short postorbital eminence of the malar bone in the facial specimen referred to *Palæosyops paludosus*, and represented in Fig. 51, Plate XVIII, would also indicate that the orbits were open behind in Palæosyops, notwithstanding the great length of the postorbital process of the frontal in the specimen under consideration.

The base of the muzzle, or the face, between the position of the orbits is broad and convex.

The specimen exhibits no evident traces of the sutural conjunctions of the parietals, frontal, the maxillaries, and the nasals.

The cranial crest separating the temporal fossæ is exceedingly short compared with that of the tapir. It is formed by the approach of the temporal boundaries, which appear in this position as two obtuse ridges squeezed together, and leaving between them a narrow groove extending from the forehead to a transverse concavity at the summit of the occiput.

The temporal fossæ are of huge proportions, and appear even to exceed those of the greatest living carnivores, as the lion and the Bengal tiger. The zygomata are as prominent as in these, but are proportionately of greater strength, being both deeper and thicker. Excepting in their greater extension outwardly from the posterior root, as in the latter animals, in their sigmoid direction downward and forward they are more like those in the tapir. Their outer surface is convex, and is directed obliquely upward.

The temporal surface at the side of the cranium, and extending on the zygomatic root, forms a deep excavation or concavity slightly overhung by the upper part of the temporal ridge. It exhibits a comparatively feeble swelling about the position of the squamous suture, but much less conspicuous than that in the tapir. The great hollow of the temporal surface is in striking contrast with the swelling of the corresponding surface in the great living carnivores, and while it is expressive of an equal if not greater extent of powerful muscles, it is further expressive of a proportionate decrease in the capacity of the cranium and therefore of a much smaller brain.

The cranium is constricted at the sides at the lower part of the temporal fossæ, just in advance of their middle, and the fore part, independent of the extension of the zygomatic roots, appears nearly as wide as the back part.

The squamosals are large, and reach half way up the temporal surface. A conspicuous group of neuro-vascular foramina occupy their upper back part, including the contiguous part of the parietals. The occiput is wider than high, is strongly concave above, but at the lower part slopes backward to the margin of the occipital foramen. Its sides below are bent forward, as in the tapir, and the lateral borders above, as in the latter animal, are produced in wing-like expansions. The basal angles of the occipital triangle are formed by comparatively short, wide processes, composed of the conjoined paramastoid and post-tympanic processes. These extend from within the position of the occipital condyles and reach outwardly a considerable width beyond them, but do not project much below the root of attachment of the condyles. The occipital condyles are of greater proportionate width but less depth than in the tapir or the bear; and they project from the occipital surface backward more than in either of those animals. The occipital foramen is transversely oval.

The general plane of the occiput is intermediate in its degree of inclination to that of the tapir and our large carnivores, and is indeed nearly vertical. The occipital condyles project posterior to the general surface, and thus form the most prominent portion of the occiput, whereas in the tapir, bear, and cats the summit of the occiput is most prominent backward.

The articular surfaces of the condyles extend forward on the basi-occipital, and approach quite near each other, as in the bear.

A large venous foramen occupies the course of the occipito-temporal suture, about the center of the lateral plane of the occiput.

The auditory archway is high and narrow compared with that of the tapir. It is widest above and has its sides converging inwardly.

The post-glenoid tubercle, compared with that in the tapir and bear, is very thick and strong. It is broad and mammillary, and is directed obliquely outward and projects downward below the post-tympanic process. The base of the cranium is very broad compared with that of the tapir, and in this respect is more like that of the great carnivores.

The basi-occipital is broad and thick. It narrows forward from the position of the paramastoid processes. Its sides are concave from before backward, slope strongly from the upper edge toward each other, and are separated by a median carina which expands behind and ends in front in a prominence. The basi-sphenoid, completely co-ossified with the basi-occipital, appears as a narrowed extension of this, and is transversely convex.

Large vacant spaces, occupied with the matrix of the fossil, are situated below the position of the petrosals. The tympanics are lost.

The glenoid articular surface is broad and nearly horizontal above, and extends obliquely downward, outward, and backward on the robust post-glenoid tubercle.

The anterior condyloid foramen is situated about three-fourths of an inch in advance of the occipital condyle.

The root of the pterygoid process is pierced with an ali-sphenoid canal, and the oval foramen occupies a position just above it.

Measurements of the cranium are as follows:

Length of cranium from the concavity at the summit of the occiput to a line between the post-orbital processes, following the curvature of the forehead .... 9 inches.
Breadth across the face, following the convexity between the ends of the post-orbital processes .... 8½ inches.
Distance between the orbits across the face above .... 6 inches.

Thickness of the short cranial crest separating the temporal fossæ posteriorly	⅞ inch
Breadth of temporal fossæ from the occipital border to the end of the post-orbital process	9 inches.
Vertical extent in advance of zygomatic root	5 inches.
Breadth of cranium outside of zygomata	11 inches.
Height of occiput	5⅞ inches.
Breadth of occiput at post-tympanic processes	6½ inches.
Breadth of cranium at ends of post-glenoid processes	8 inches.
Transverse diameter of occipital foramen	23 lines.
Vertical diameter of occipital foramen, estimated	16 lines.
Breadth at occipital condyles together	47 lines.
Depth of occipital condyles	18 lines.
Breadth of occipital condyles	19 lines.
Width of basi-occipital at anterior condyloid foramina	18 lines.
Width of basi-occipital at conjunction with basi-sphenoid	15 lines.
Distance between glenoid articular surfaces	54 lines.

An upper-jaw fragment, from the same individual as the cranium just described, contains the last two molars, of which the penultimate one is represented in Fig. 3, Plate XXIV. This tooth closely resembles the corresponding one of the same species represented in Fig. 10, Plate XXIII, and also that of *Palæosyops paludosus* as represented in Figs. 3 to 5, Plate IV, and Fig. 9, Plate V. The last molar, as far as it is preserved, likewise resembles the corresponding tooth represented in the same plates. The inner part of the crown presents a single conical lobe.

The infra-orbital border forms a thick, obtusely rounded ledge projecting obliquely forward on the face. In *Palæosyops paludosus* the corresponding ridge presents an acute anterior edge defining it from the facial surface beneath. The outer part of the thick infra-orbital ridge rises in a short, blunt, conical eminence or postorbital process. The orbital floor is concavely depressed within the prominent margin, and forms a long, triangular platform terminating behind in the thick posterior boundary of the maxilla.

The measurements of the specimen are as follows:

	Lines.
Space occupied by the last two molars	32
Breadth of second molar	20
Width of second molar	20
Height of anterior orbital margin from the molars	26

Fragments of both sides of the lower jaw with all the teeth broken away, except portions of the last molars, also accompany the preceding specimens. The best preserved fragment partially restored from the corresponding portion of the opposite side is represented in Fig. 4, Plate XXIV. It agrees in form and proportions with the same portion of the jaw in *Palæosyops paludosus*, excepting that the masseteric fossa is much deeper. The preservation of the angle of the jaw, not retained in any of the previous specimens of Palæosyops, permits the determination of its character. It presents a nearly semi-circular border projecting moderately below the base of the bone, and in a less degree posteriorly. Toward the base it is somewhat bent inward.

The last molar, in a restored condition, of the natural size, is represented in Fig. 14, Plate XXIII, but, unfortunately, the artist has made its thickness in front proportionately too great.

The measurements of the specimen are as follows:

	Palæosyops major.	Palæosyops paludosus.
	Lines.	Lines.
Distance from last molar to back border of jaw	64	49
Depth from condyle to bottom of angle	72	....
Depth of jaw below fore part of last molar	35	30
Thickness of jaw below fore part of last molar	16	14
Breadth of last molar tooth	22	19
Width at fore part of last molar tooth	13	10¼
Width at middle part of last molar tooth	11½	9¼

Fragments of another jaw similar to the above, and presenting the same comparatively deep masseteric fossa, were found by Dr. Corson at Grizzly Buttes.

Fig. 2, Plate XXIV, represents a mutilated facial portion of a skull apparently referable to *Palæosyops major*. The specimen was found on one of the buttes of the Bridger formation by a Shoshone Indian, and brought to Dr. Carter, by whom it was presented to the writer. Though much distorted in form, it gives us a fair idea of the shape and construction of a portion of the skull of Palæosyops that we had not previously had the opportunity of examining. It is crushed in such a manner that the upper part of the face is pressed downward and toward the right side, and the orbit has its roof brought near to the floor, so that it looks as if it were closed behind by the presence of a postorbital bridge.

The specimen shows that the form and construction of the face of Palæosyops are very similar to what they are in Palæotherium. The upper part of the face appears to have been directed in a moderately sigmoid course, nearly horizontally from the bottom of the convex forehead to the end of the muzzle. The nasals are large, thick, and strong. Proportionately they exceed in length those of the Palæothere, the rhinoceros, and the tapir. If I mistake not a fracture for a suture, their posterior extremity reaches as far as the position of the fore part of the orbits, and their free extremity projects quite as much as the jaws. They are strongly arched transversely and are more abruptly rounded and thick at the lateral borders. They gradually narrow forward and terminate in a blunt extremity, which is nearly straight but rounded at the outer angles. Posteriorly, they include a deep and wide angular notch, which receives a corresponding angular prolongation of the frontals.

The lateral nasal notch resembles that of the rhinoceros, but is proportionately of greater depth, and in this respect also resembles that of the Palæothere. Its exact extent cannot be determined on account of the mutilated condition of the specimen.

The upper jaw in its form and proportions is nearly like that of the Palæothere. It is of greater proportionate depth below the orbit, and exhibits a greater swell at the border of the nasal notch, due to the greater size of the canine teeth. The infra-orbital foramen is large, and is situated over the position of the last premolar. The hard palate is flat along the middle anteriorly. Its posterior part is destroyed in the specimen, so as to prevent the determination of its extent.

The incisive foramina appear to be comparatively small and widely separated. They appear also to be circular, and continuous with grooves descending forward to the incisive alveoli.

The teeth form a series as unbroken nearly as in Anoplotherium. They are all mutilated in the specimen, but the crown of the last premolar, and the molars are sufficiently well preserved to exhibit the characteristics of Palæosyops as already described.

Measurements of the specimen, for the most part approximative on account of its distorted condition, are as follows:

Length of jaw from back of last molar to front of incisive alveoli	9 inches.
Length of face from the anterior orbital margin to the end of the nose...	7 inches.
Length of the nasals in the median line.	5½ inches.
Breadth of the nasals together at their middle........................	3¾ inches.
Length of space occupied by the molar series...............	6¼ inches.
Length of space occupied by the true molars.........................	3¾ inches.
Breadth of last premolar.................................	9 lines.
Width of last premolar.	12 lines.

Breadth of first molar.... 14¼ lines.
Width of first molar.... 14½ lines.
Breadth of second molar.... 16 lines.
Width of second molar.... 16 lines.
Breadth of last molar.... 19½ lines.
Width of last molar.... 18 lines.
Width of palate between canines.... 28 lines.
Width of canine alveoli.... 9 lines.

PALÆOSYOPS JUNIUS.

Dr. Carter recently sent the writer several small fragments of the right side of a lower jaw, together with a sketch of a larger fragment of the left side, containing the last premolar and the succeeding molars. The specimens were obtained from the Bridger beds, and appear to indicate a small species of Palæosyops, though it is not improbable that they pertain to a small variety of *P. paludosus*.

The parts agree closely with the corresponding parts of the lower jaw and teeth of the latter, except in size. They have been viewed as representatives of a species with the name of *Palæosyops junius*.

The measurements of the teeth in comparison with those of *P. paludosus* are as follows:

	Palæosyops junius.	Palæosyops paludosus.
	Lines.	Lines.
Space occupied by the last premolar and molars	48	55
Space occupied by the molars	30½	46
Breadth of last premolar	8	9
Thickness of last premolar	5½	6¼
Breadth of first molar	10	12½
Breadth of second molar	12	15
Breadth of third molar	17	19
Thickness of third molar at middle	7	9½

LIMNOHYUS.

This genus was originally named by Professor Marsh, in a communication published in the American Journal of Science for August, 1872, and was applied to Palæosyops under the misapprehension that this genus had not been distinguished by the possession of one or two cones to the inner part of the crown of the last upper molar tooth. As it was as clearly demonstrated as the nature of the specimens would admit, that the last upper molars of

Palæosyops possessed but a single cone to the inner part of the crown, the name subsequently proposed by Professor Marsh on account of this characteristic was untenable. Under these circumstances, though I previously viewed the difference as simply specific, I would adopt the generic name of Limnohyus for those forms of Palæosyops, as recognized by the general constitution of the teeth, in which the last upper molars have two cones to the inner part of the crown.

Fig. 13, Plate XXIII, represents an upper molar tooth, apparently the first of the series of true molars, resembling in form the corresponding teeth of *Palæosyops paludosus*. The enamel of the tooth is, however, comparatively smooth, a condition which is clearly independent of its age, as the tooth is but moderately worn. As a considerable degree of variation is observed in the extent of wrinkling of the enamel of the teeth of *Palæosyops paludosus*, independent of wearing, it is not improbable the specimen may pertain to an individual variety of the same, though it probably may indicate another species.

The specimen was found by Dr. Corson in association with the large tusks originally referred to *Uintamastix atrox*, described in a later chapter, and represented in Figs. 1 to 3, Plate XXV.

Since writing the description of the smooth, enameled molar tooth, Professor Marsh, who has inspected the specimen, informs me that it pertains to the same animal he has described under the name of *Palæosyops laticeps*, (Am. Jour. Sc., Aug., 1872.) As this is stated to have four lobes to the crown of the last upper molar, for reasons already given, it would belong to the genus Limnohyus.

Fig. 8, Plate XXIV, represents the crown of an upper molar tooth, which was found, together with some small fragments of other molars, both upper and lower, by Dr. Corson on the buttes of Dry Creek Cañon. The specimen I supposed to belong to a small species of Palæosyops, and so referred it, under the name of *P. humilis*, in a letter to the Academy, published in its Proceedings for July 30, 1872. Under the impression that it was perhaps the last tooth of the series, in view of the distinction suggested by Professor Marsh between Palæosyops and Limnohyus, I subsequently ascribed it to the latter. Professor Marsh informs me that he has a number of specimens which lead him to regard the tooth as pertaining to the temporary series of Palæosyops.

HYRACHYUS.

An extinct genus of odd-toed pachyderms, under the above name, was originally inferred from specimens of fossils obtained during Professor Hayden's exploration in Wyoming, in 1870. One of the specimens, represented in Fig. 11, Plate II, consists of the greater portion of a ramus of the lower jaw, without teeth, found on Smith's Fork of Green River. The other specimen, represented in Fig 12, consists of a lower-jaw fragment, with several teeth, of a young animal, from Black's Fork of Green River.

Hyrachyus is closely related with the extinct tapiroid genus Lophiodon, the remains of which belong to the early Tertiary formation of Europe. In a less degree, also, it is related with the rhinoceros-like Hyracodon of the Mauvaises Terres of White River, Dakota. Among living animals, it is most nearly allied to the tapir, and more remotely with the rhinoceros.

The dental series of the true Lophiodon, if the *L. isselense* of Issel, France, be viewed as the type of the genus, or of Tapirotherium, as it had been previously named by De Blainville, consists of three incisors, a canine, three premolars, and three molars. The living tapir at maturity has one premolar more to the upper series.

In one species of Hyrachyus at maturity there are four premolars to the series above and below, as in Hyracodon. Apparently, in a second species there are four premolars in the upper series, and three in the lower, as in the tapir.

The last lower molar of Lophiodon has a trilobate crown. In Hyrachyus, as in the tapir, it has a bilobed crown.

The crowns of the lower molars are intermediate in character with those of Lophiodon and Hyracodon.

The upper molars of Hyrachyus closely resemble those of Lophiodon. In both genera the upper back two premolars have a single lobe to the inner part of the crown representing the inner pair of lobes of the crowns of the succeeding molars in a connate condition. In Lophiodon a ridge proceeds from the inner lobe of the crown of the premolars mentioned to the antero-external lobe. In Hyrachyus, in the corresponding teeth, a pair of ridges proceed from the inner lobe of the crown to both the outer lobes.

The lower jaw of Hyrachyus has nearly the form and construction of that of the tapir.

HYRACHYUS AGRARIUS.

This species, originally indicated and named from the specimen represented in Fig. 11, Plate II, consisting of a ramus of the lower jaw without teeth, we have now the opportunity of illustrating by many well-preserved and more characteristic specimens. Most of these were collected by Dr. J. Van A. Carter, during the last summer, on Henry's Fork of Green River, near Lodge-Pole Trail, at Bridger Butte, and other localities in the vicinity of Fort Bridger, Wyoming. A few others were obtained by Dr. Joseph K. Corson, from Grizzly Buttes, Wyoming.

The specimen represented in Fig. 12, Plate II, being one of those upon which the genus Hyrachyus was originally proposed, was referred to another species from the former one, with the name of *Hyrachyus agrestis*. This I now regard as of the same species. The specimen, a lower-jaw fragment, belonged to a young animal, which still retained its temporary teeth. Of these, the fossil contains the first premolar, the fangs of the succeeding two, and the molar tooth. Behind this the first molar of the permanent series is inclosed within the jaw.

Professor Marsh has described remains apparently of the same animal under the name of *Lophiodon Bairdianus*. The specimens, which he observes are among the most common of the mammalian fossils of the Wyoming Tertiary, were found at various localities near Fort Bridger, and also on the White River, in Eastern Utah.

The dental series of *Hyrachyus agrarius*, in the mature condition, consists of three incisors, a canine, four premolars, and three molars, in both jaws.

A well-preserved series of upper molar teeth, considerably worn, is represented in Figs. 9 and 10, Plate IV, from a specimen discovered by Dr. Carter near Lodge-Pole Trail, about eleven miles from Fort Bridger. Fig. 11, of the same plate, represents an upper second molar, which was obtained by Dr. Carter on Henry's Fork of Green River.

Of the upper molars, or true molars, the middle one is the largest, and the others are nearly equal in size. Four principal lobes enter into the constitution of their crown, which is inclosed by a basal ridge, except externally, and at the most prominent portion of the inner lobes internally. Of the outer lobes, which are conjoined, the posterior is the wider and is pyramidal; the anterior is the more prominent externally and is conical. This is also strengthened in front by a large conical buttress continuous with the comparatively wide anterior basal ridge of the crown. In the last molar the posterior of the outer lobes is proportionately less well developed than in the molars in advance. The inner lobes of the crown are conical internally, and are extended obliquely outward so as to form ridges continuous with the fore part of the outer lobes. The oblique valley separating the inner lobes is closed externally by the conjunction of the outer lobes. A wide, angular recess occupies the interval of the posterior lobes of the crown and the posterior basal ridge.

In the unworn or moderately worn condition of the molars, as seen in Fig. 11, Plate IV, a narrow but conspicuous ridge or fold is observed projecting from the antero-external lobe into the median valley of the crown. In the worn condition of the molars, as seen in Fig. 10, c to g, they exhibit a tract of exposed dentine extending along the summits of the outer lobes including the abutment in front, and prolonged inwardly in two pouch-like extensions upon the summits of the inner lobes.

The upper premolars not only exhibit from behind forward a successive diminution in size, but also a reduction to greater simplicity. The latter condition is induced through connation and disappearance of constituent elements as they are observed to exist in the back teeth. Thus if we compare the back two premolars, Fig. 10, c, d, with the molars behind, it will appear that the most striking difference is due to the connation internally of the inner lobes. From this arrangement the premolars appear to have a single lobe to the inner part of the crown, from which a pair of ridges proceed to join the outer lobes. A central pit represents the median valley opening internally in the crown of the molars. The basal ridge extends around the inner part of the crown.

The abutment so conspicuous at the antero-external angle of the crown of the molars is successively reduced forward in the premolars and disappears in the anterior two.

In the crown of the second premolar, Fig. 10, b, the outer lobes are more connate than in those behind, and the inner lobe appears more isolated from the absence of the intervening ridges.

The crown of the first premolar, Fig. 10, a, about half the size of that of the tooth behind, is conoidal with an oval base. For the most part it is homologous with the outer lobes of the other premolars in a completely connate condition. A small offset internally is a rudiment of the inner lobe of the succeeding premolar.

A basal ridge exists at the outer back part of the crown of the second premolar; and, less produced, exists in the same position in the third. No ridge occupies the inner prominence of the inner lobe of the second premolar.

A specimen of an upper left last premolar, found at Grizzly Buttes by Dr. Corson, is represented in Fig. 12, Plate IV. It is larger than in the entire series of Fig. 10 and is less worn. It exhibits a basal ridge externally interrupted at the middle; and internally the ridge is also interrupted or nearly obsolete at the middle. The posterior ridge or fold between the inner and postero-external lobes, though smaller, is more defined from the lobes than the anterior ridge. The latter appears rather as a prolongation of the inner lobe to the fore part of the base of the antero-external lobe. The posterior ridge has the appearance of an introduced piece defined from the lobes by constrictions or grooves. The arrangement is badly represented by the artist; nor is it obvious if it existed in the corresponding more worn tooth of the series of Fig. 10.

In a much mutilated specimen, obtained by Dr. Corson at Grizzly Buttes, containing the remains of the last two premolars and succeeding two premolars, the basal ridge is better developed at the inner part of the crown than in any of the preceding. The last premolar exhibits the same condition of the posterior ridge intervening to the internal and postero-external lobes of the crown as that described in the isolated tooth. The same tooth, barely worn, exhibits the summit of the inner lobe of the crown slightly divided into two points, so that it presents a less degree of connation than in the preceding specimens.

The upper molars and premolars, except the first one, are inserted by three fangs, of which the inner one is a connate pair; the connation being most complete in the premolars. The first premolar has two fangs. The space occupied by the upper molar series is about $3\frac{3}{4}$ inches.

Fig. 13, Plate IV, represents a specimen, found by Dr. Carter, in company with the upper molar teeth of Figs. 9 and 10, and evidently pertaining to the same individual. The specimen consists of the anterior extremity of the lower jaw, retaining the incisive alveoli, the canines, and on one side the four premolars. A view of the triturating surfaces of the latter is given in Fig. 14. Figs. 15 and 16 represent a second molar, and Fig. 18 an incisor from the same individual.

Fig. 25, Plate XX, was drawn from a specimen consisting of the greater part of a lower jaw, including both rami, obtained by Dr. Carter, at Bridger Butte, seven miles west of Fort Bridger. The left ramus contains the posterior three premolars and the succeeding two molars, of which a view of the triturating surfaces is given in Fig. 26.

The lower molars, including the last one, in *Hyrachyus agrarius* have all bi-lobed crowns. These are oblong square, and bounded by a basal ridge in front, behind, and in a more or less interrupted condition externally. The constituent lobes have somewhat curved rectangular summits as in Hyracodon and rhinoceros. The summit of the anterior lobe curves forward and inward, and becomes continuous with the basal ridge of the fore part of the crown. As the acute summits are worn, tracts of dentine become exposed crossing the teeth. In the progress of attrition the expanding dentinal tracts extend in an irregular L-like manner, and finally the contiguous tracts of each tooth become continuous, as in rhinoceros at the same stage of wear.

The crowns of the premolars present the same essential constitution as those of the molars, with the constituent lobes, successively, from behind forward, becoming more reduced or rudimental. The posterior lobe becomes proportionately more reduced than the anterior, and in the first premolar has disappeared.

The crown of the last premolar resembles those of the molars, with the posterior lobe proportionately more reduced than the anterior one.

The crown of the third premolar, in the specimen represented in Fig. 25, Plate XX, has the same form as in the last premolar, and is simply reduced in size. In the specimen represented in Fig. 13, Plate IV, the tooth looks different, from the oblique ridge or summit of the anterior lobe of the crown as existing in the former, being contracted in this into a conical and somewhat more elevated point. This gives such a remarkable difference to these teeth in the two specimens, that, had they been found isolated, without a knowledge of their collocation, they would have been attributed to different genera of animals.

The anterior two premolars have an oval crown elevated into a median conical point and presenting offsets behind and in front, in which may be detected the rudiments of the posterior lobe and anterior extension of the anterior lobe of the better developed crowns of the teeth behind.

All the lower premolars, as well as the molars, are inserted by a pair of fangs. The space occupied by the lower molar series in several specimens ranges from 3 inches and 5 lines to $3\frac{3}{4}$ inches.

The inferior canine teeth are quite like those of the tapir in appearance. They curve upward and forward, with a slight inclination outward. The crown is laterally compressed conical, subacute in front and behind, but worn in both these positions in the specimen under examination.

The upper canines are unknown, unless the specimen represented in Fig. 17, Plate IV, is one. This was found at Grizzly Buttes by Dr. Corson, in association with some upper premolars of Hyrachyus; all of which look as if they had belonged to the same individual. The crown of this specimen of an upper canine is short, and worn off to a considerable extent at its fore part. It is compressed conical, and has the inner and outer surfaces defined by an acute ridge posteriorly. The fang is double the length of the crown, and is laterally compressed.

The incisor tooth, represented in Fig. 18, appears to be the second of the series of the lower jaw. It resembles the corresponding tooth of the tapir. Its chisel-like crown is worn off at the cutting edge.

No characteristic portions of the upper jaw of Hyrachyus have come under our notice. In one specimen the infra-orbital foramen is observed to occupy a position above the third premolar.

The lower jaw resembles, in its form and proportions, that of the Hyracodon and the tapir. The anterior extremity, in the construction of the chin, the contraction between the position of the canines and molar series, and other features, repeats the condition observed in the tapir. A similar wide hiatus separates the canines from the molar teeth. The free border of the hiatus, upward of an inch in length, is concave fore and aft, and acute.

The body of the lower jaw is less robust or thick, in relation with its depth, than in the tapir. It is also less convex externally, and at the base fore and aft. The outer surface, in comparison, appears quite vertical.

The ascending portion of the ramus rises vertically at its fore border, and is deeply impressed on the outer surface just back of the latter.

The condyle projects less externally and more posteriorly than in the tapir. Its articular surface is more flat, and in a less degree inclined inwardly.

In the specimen represented in Fig. 13, Plate IV, five small mental foramina are observed, in a row extending from the position of the third premolar to that of the canine tooth. In the specimen represented in Fig. 25, Plate XX, a large mental foramen is situated below the interval of the third and fourth premolars; and in advance of this several small ones exist. In the specimen represented in Fig. 11, Plate II, the mental foramen is situated below the first premolar.

Measurements from several specimens of lower jaws of *Hyrachyus agrarius* are as follows:

	Lines.	Lines.	Lines.
Length of space occupied by the molar teeth		42	44
Distance from incisive alveoli to first premolar	25	24	....
Length of hiatus between canine and first premolar	13	13	....
Depth of jaw below last molar		18	20
Depth of jaw below last premolar		15	16
Breadth of jaw at canine alveoli		14	13	....
Breadth of jaw below hiatus	12	10½	....
Length of symphysis	28	22	....

Measurements of upper molar teeth are as follows:

	Inches.	Lines.
Length of the entire upper molar series	3	9
Length of series of premolars	1	8½
Length of series of molars	2	1½

	Fore and aft.	Transverse.
	Lines.	Lines.
Diameter of first premolar	4	3
Diameter of second premolar	5	6
Diameter of third premolar	5½	7
Diameter of fourth premolar	6½	8½
Diameter of first molar	8½	9½
Diameter of second molar	9¼	10
Diameter of third molar	7½	9¼

Comparative measurements of two upper last premolars are as follows:

	Lines.
Fore and aft diameter in both	7
Transverse diameter in one	8½
In the other	9

Measurements of lower molar teeth are as follows:

	Lines.	Lines.	Lines.
Length of the entire lower molar series	....	42	44
Length of series of premolars	20	19	20
Length of series of molars	....	24	24½

	Fore and aft.		Transverse.	
	Lines.	Lines.	Lines.	Lines.
Diameter of first premolar	3½	....	2	....
Diameter of second premolar	4½	4½	2½	2½
Diameter of third premolar	6	5½	3½	3½
Diameter of fourth premolar	6½	5½	4½	4
Diameter of first molar	....	7¼	....	5
Diameter of second molar	....	8½	....	5½
Diameter of third molar	....	8½	....	5

Hyrachyus agrarius was about the size of the common collared peccary, *Dicotyles torquatus*.

HYRACHYUS EXIMIUS.

A supposed larger species of Hyrachyus than *H. agrarius* is inferred from several specimens, consisting of small fragments of a lower jaw with a tooth and portions of others. These were obtained by Dr. Carter on Henry's Fork of Green River.

The best and most characteristic specimen is represented in Figs. 19, 20, Plate IV, consisting of a lower-jaw fragment containing the last premolar and portion of the first molar much worn. Both the teeth and the jaw agree in form with the corresponding parts in *Hyrachyus agrarius* and differ only in size. The specimen also agrees with the corresponding part of Lophiodon sufficiently to belong to the same genus, so that until more ample material is discovered it must remain uncertain whether it really pertains to Hyrachyus.

Comparative measurements are as follows:

	H. eximius.	H. agrarius.
	Lines.	Lines.
Depth of jaw at last premolar	18	14½
Thickness of jaw at last premolar	9	7
Diameter fore and aft of last premolar	7	5¾
Diameter transversely of last premolar	5½	4
Diameter transversely of first molar	6½	5
Diameter transversely of second molar	7	5¼

Figs. 9, 10, Plate XXVI, represent a tooth recently obtained by Dr. Carter on the buttes of Dry Creek. It would appear, from its proportions, to be the left lower penultimate molar of *Hyrachyus eximius*. It is a nearly unworn and perfect specimen, and agrees in its anatomical characters with the corresponding tooth of *H. agrarius*. The crown measures an inch antero-posteriorly and 7½ lines transversely.

The specimens above described indicate an animal about the size of the common American tapir.

HYRACHYUS MODESTUS.

Under the impression that teeth of like form with those of *Hyrachyus agrarius*, from the Bridger Tertiary formation, pertain to the same genus, I now view the tooth represented in Fig. 13, Plate 11, which I previously referred to *Lophiodon modestus*, as belonging to Hyrachyus. The specimen was obtained during Professor Hayden's exploration of 1870, on Smith's Fork of Green River, near Fort Bridger.

The tooth is a first or second upper molar, and differs in size and proportion from the corresponding teeth of *Hyrachyus agrarius* sufficiently to indicate a smaller species. The only other difference observable, one, however, which may prove not to be constant in additional specimens, is in the internal surface of the antero-internal lobe of the crown, being strongly wrinkled instead of being elevated in a single conspicuous fold as in *H. agrarius*.

The comparative measurements of the specimen are as follows:

	H. modestus.	H. agrarius.
	Lines.	Lines.
Diameter fore and aft of second upper molar	7	9
Diameter transversely of second upper molar	6¾	10

Hyrachyus modestus was about a third less in size than *H. agrarius*.

HYRACHYUS NANUS.

Portions of two lower jaws I have referred to a small species of Hyrachyus with the above name. One of the specimens was obtained at Lodge-Pole Trail, by Dr. Carter; the other, represented in Fig. 14, Plate II, and Fig. 42, Plate VI, was found at Grizzly Buttes, by Dr. Corson.

In both specimens, which belonged to animals at maturity but not advanced in life, the number of teeth in the molar series is six, or one less than in Hyrachyus, and the same number as in Lophiodon and the tapir. The last molar, however, has a bilobed crown as in the latter, but the premolars, in their less degree of development in comparison with the molars, are more like those of Lophiodon.

The suppression of an anterior premolar may perhaps be regarded as a less important generic character than the others which have been indicated as separating Hyrachyus from Lophiodon and Tapirus. Under the circumstances, notwithstanding the reduction in the number of premolars, I view the two jaw-specimens above indicated as pertaining to Hyrachyus.

Professor Marsh has described several specimens, from Grizzly Buttes, under the name of *Lophiodon nanus*, which I suspect to belong to the same animal as the lower-jaw fragments above indicated. He observes that the most characteristic of the specimens is a right upper jaw containing a series of four premolars and three molars. If, then, this really belongs to the same animal, it would give with the lower-jaw specimens, as the formula of the molar series, seven teeth above and six below, as in the tapir. The upper premolars, however, present a greater amount of difference from the molars than in the latter, the difference being mainly due to a less degree of development of the premolars and in the connation of the inner lobes of their crowns.

The molar teeth and the portion of the jaw containing them are almost repetitions of form of the corresponding parts in *Hyrachyus agrarius*. The mental foramen is situated below the first premolar. *Hyrachyus nanus* was about half the size of *H. agrarius*.

Measurements from two lower-jaw specimens are as follows:

Space occupied by the complete series of molar teeth	2 inches.
Space occupied by the premolars	9¾ lines.
Space occupied by the molars	14 lines.
Depth of jaw below last premolar	10 lines.

	Fore and aft.	Transverse.
	Lines.	Lines.
Diameter of first premolar	2½	
Diameter of second premolar	3½	2¼
Diameter of last premolar	3¾	2¾
Diameter of first molar	4	3
Diameter of second molar	4½	3¼
Diameter of last molar	5½	3¼

Fig. 11, Plate XXVI, represents the greater part of the right ramus of the lower jaw of *Hyrachyus nanus*, which I found, together with a fragment of the opposite side and several other bones of the skeleton, near the Lodge-Pole Trail, crossing Dry Creek Valley. The specimen was found in part exposed and partially imbedded in the indurated clay of a butte, in company with quite a profusion of well-preserved shells of *Helix wyomingensis*.

The jaw resembles in its form that of *Hyrachyus agrarius*, and also that of the recent tapir. It contained a series of six molars, of which it retains the back four. The molars are separated by a wide hiatus from a continuous arch of alveoli, for the accommodation of the incisors and canines, which correspond in number with those of the tapir.

The depth of the jaw is rather less than in the fragments previously described, while the dimensions of the molar series is nearly the same. The measurements of the specimen are as follows:

	Lines.
Length of space from incisive alveoli to back of last molar	42
Length of space occupied by the molar series	24
Length of space occupied by the true molars	14½
Antero-posterior diameter of last molar	4½
Length of symphysis	16
Length of hiatus in advance of molars	12
Depth of jaw below molars	9½

An upper-jaw fragment, recently sent to me by Dr. Carter, I suppose to pertain to *Hyrachyus nanus*. It contains the fangs of the anterior three premolars, and the entire last one, which is represented in Figs. 21, 22, Plate XXVII, magnified two diameters. This premolar resembles the corresponding tooth of *H. agrarius*, but the ridge in the latter, which represents the postero-internal lobe in the true molars, is reduced to the smallest rudiment.

The space occupied by the four premolars measures 11½ lines. The breadth of the last premolar is 3.2 lines; the width transversely is 4 lines.

LOPHIOTHERIUM.

LOPHIOTHERIUM SYLVATICUM.

The genus *Lophiotherium* was proposed by Gervais, from some fragments of several lower-jaws with molar teeth, which were found in association with remains of true Palæotheria, in a formation of France which he regards as belonging to the upper Eocene Tertiary. The genus is viewed as a tapiroid pachyderm closely allied to Lophiodon, though the molar teeth appear very unlike those of the latter.

During Professor Hayden's exploration of 1870, a specimen was found on

Henry's Fork of Green River, which appears to pertain to a species of Lophiotherium. The specimen, represented in Fig. 33, Plate VI, consists of a lower-jaw fragment containing the last premolar and the first and last true molars—the crown of the intervening true molar having been lost. The teeth appear closely to resemble in form and constitution those of *Lophiotherium cervulum*, as represented in Plate II of Gervais's *Zoologie et Paléontologie françaises*. The only apparent difference, which, nevertheless, is an important generic one, if it really exists, is the division of the summit of the antero-internal lobe of the crown of the teeth into two points in the American fossil.

The anterior teeth, Fig. 34, of the latter have oblong quadrate crowns, slightly narrower at the fore part and otherwise alike in form. They are quadrilobate, the lobes being tri-laterally pyramidal and connate at base.

The last molar, Fig. 35, is prolonged behind in the manner so common in allied animals of the same order. This prolongation is mainly due to the addition of a fifth lobe to the crown, which is narrowed posteriorly in the reverse direction to the teeth in advance.

A strong basal ridge incloses the crowns of the teeth, excepting internally. In the last molar it is less well developed and does not exist posteriorly. The constituent lobes of the crowns are nearly of uniform size. The antero-internal lobe, as before intimated, has its summit divided into two points. The division extends so short a distance that it would be early obliterated from the wearing of the teeth in the trituration of the food. It is hardly perceptible, even in the unworn condition in the last molar, and in the specimen is most distinct in the first true molar. As a character, it may be inferred to be most obvious in the anterior two true molars, and less so in the premolars of like form and in the last true molar.

The postero-internal lobe of the crowns has a simple pointed summit. The inner lobes have the crescentoid summit declining from a central point inwardly, so common in the corresponding teeth of allied animals. The fore arm of the summit of the antero-external lobe is a thick ridge curving to the base of the antero-internal lobe in front. The back arm is a short ridge directed inwardly to the anterior division of the summit of the antero-internal lobe. The fore arm of the summit of the postero-external lobe reaches the middle of the antero-internal lobe. The back arm joins the posterior basal ridge, producing an elevated point at its middle. From the inner side of the same lobe a third but less conspicuous ridge extends directly to the lobe within.

In the last molar the fifth lobe has a crescentoid summit declining from a median point. The outer arm of the summit joins the contiguous arm of the lobe in advance, and the inner arm joins the base of the postero-internal lobe.

The minutely detailed description of these teeth, and the same may be said of those of other fossils, is essential to the distinction of generic characters.

From the back molars of Lophiodon, those of Lophiotherium especially differ in the distinction of four instead of two lobes to the crown: though the two lobes in the teeth of Lophiodon and Tapirus represent the four of Lophiotherium in a connate condition.

The jaw-fragment of the fossil referred to *Lophiotherium sylvaticum* presents nothing peculiar. The outer vertical surface is slightly convex, and the base fore and aft is also moderately convex.

The measurements of the fossil are as follows:

	Lines.
Depth of lower jaw below middle of last premolar	5¾
Depth of lower jaw below middle of last true molar	6½
Antero-posterior diameter of last premolar	3½
Antero-posterior diameter of first true molar	3¾
Antero-posterior diameter of second true molar	3¾
Antero-posterior diameter of last true molar	5¼
Transverse diameter of last premolar	2¼
Transverse diameter of true molars	2½

Should the duplication of the summit of the antero-internal lobe of the crown of the lower back molars not be a character present in the *Lophiotherium cervulum* of France, it would probably be a concomitant of other characters in the upper teeth, now unknown to us, which would distinguish the American animal as generically distinct from Lophiotherium.

In the American Journal of Science for 1871, Professor Marsh notices some remains, from Grizzly Buttes, which he attributes to a species about two-thirds the size of the former, and names it *Lophiotherium Ballardi*.

TROGOSUS.

TROGOSUS CASTORIDENS.

One of the most curious of the extinct mammals of the Bridger Tertiary fauna is an odd-toed pachyderm about the size of the larger living peccary, which, with the usual complement of molar teeth, was apparently devoid of canines, and was provided with a large pair of incisors like those of rodents.

The singular character of the animal was first recognized in a fossil specimen, consisting of a mutilated lower jaw, discovered by Dr. Carter in the vicinity of Fort Bridger, and sent to the writer in the spring of 1871. The specimen, represented in Figs. 1 to 3, Plate V, besides the two large incisors, contains the remains of most of the molar teeth, but none in an entire condition. The best preserved is the second molar of the left side, and this is so much worn as to have the distinctive features of its triturating surface, as seen in Fig. 2, completely obliterated.

The specimen was originally described in the proceedings of the Academy of Natural Sciences in May, 1871, and from its peculiarities the animal to which it belonged was named *Trogosus castoridens*, or the beaver-toothed gnawing hog. Professor Marsh had previously described an isolated tooth, of the same animal, from Grizzly Buttes, which he referred to a species of Palæosyops with the name of *P. minor*. From the description, I supposed it not to differ from *P. paludosus*. An examination of the specimen has satisfied me that it belonged to the same animal as the jaw referred to Trogosus.

The isolated tooth belonged to a younger animal, and is not so worn as to have the characteristic arrangement of its masticating surface destroyed. On seeing it I was struck with its resemblance to another isolated molar tooth which I had formerly described under the name *Anchippodus riparius*. This tooth was discovered by Dr. Knieskern in a Tertiary formation, supposed to be of Eocene age, in Monmouth County, New Jersey. It was given to Mr. T. Conrad, by whom it was presented to the Academy of Natural Sciences of Philadelphia. The same formation has yielded the remains of a peccary, an Elotherium, and a rhinoceros.

A comparison of the tooth of the New Jersey Anchippodus with the corresponding one in the jaw-specimen and with the isolated molar, would appear to indicate that the Wyoming fossils belong to the same genus, and indeed the teeth are sufficiently alike in form and size to pertain to the same species. Should further discovery prove this to be the case, it would, perhaps, indicate the contemporaneous character of the Bridger Tertiary formation of Wyoming and that of Monmouth County, New Jersey. The New Jersey fossil, in its general appearance of color and condition, so closely resembles the Wyoming fossils that it would readily pass for one of them.

It is by no means positive that Trogosus and Anchippodus are the same, for we have examples enough of different genera having the lower molars alike, while the upper ones and the premolars are unlike.

The jaw of Trogosus retains evidences of the existence of six molar teeth, and there may have been another small premolar, but this cannot be ascertained from the mutilated condition of the specimen. The first of the series of six molar teeth approached so close upon the large incisor as to leave but a small interval for the introduction of other teeth.

The best preserved tooth of the molar series, the second molar, presents a bilobed crown, in which the anterior lobe is the longer or least worn. The triturating surface, represented in Fig. 2, Plate V, exhibits a wide tract of exposed dentine with a yoke-like outline of enamel. Its fore and aft measurement is $9\frac{1}{2}$ lines. The thickness of the anterior lobe at base is 8 lines.

In the less worn specimen of the corresponding tooth described by Professor Marsh, he gives the antero-posterior diameter as 10 lines; the transverse diameter at the summit of the lobes of the crown as 5 and $5\frac{1}{2}$ lines wide.

The constitution of the lower molars of Anchippodus is apparently the same as in Trogosus, as observed in the New Jersey tooth represented in Figs. 45, 46, Plate XXX, of "The Extinct Mammalian Fauna of Dakota," &c. From this it will be seen the crown is composed very nearly on the same plan as that of the corresponding teeth in Anchitherium, Palæotherium, &c. It is composed of a fore and aft pair of lobes with crescentoid summits, convex externally and with a recess internally. The size of the tooth is the same as that retained in the jaw-specimen above described. The fangs of the last molar in the lower jaw indicate a trilobed crown, as in Anchitherium, Palæotherium, &c. The premolars, so far as can be ascertained by their remains in the jaw, are inserted by two fangs. Canines most probably do not exist in Trogosus, their absence being fully compensated by the large incisors.

The incisor teeth on both sides together are four in number; but while the lateral ones are developed to the proportions of those of rodents, the intermediate pair were quite small. The latter are lost from the specimen, leaving the alveoli occupied by matrix. The space they occupied was about the fourth of an inch from side to side.

The large lateral incisors are wonderfully like the incisors of rodents, not only in form, position, and structure, but they were also alike in their perpetual mode of growth. They do not extend so far back within the jaw as in most rodents, and in this respect are more like those in the rabbits, or, as in their nearer relatives, the peccary and hog. They extend beneath the premolars, but the bottom of the alveolus does not reach the position of the first molar.

The incisors are convex in front, and not flat, as usual in rodents. The anterior convexity is invested with thick enamel longitudinally striated, the striæ being wrinkled. Externally the edge of the enamel appears proportionately more prominent than in rodents; that is to say, it projects more above the level of the contiguous exposed dentine. In transverse section the incisors are ovoid, with the narrow extremity behind. The fore and aft diameter of the section is 10 lines; the transverse diameter at the edges of the enamel layer is $6\frac{1}{2}$ lines. The anterior convexity covered with enamel is 4 lines; the posterior convexity is $\frac{1}{2}$ an inch.

The cutting edges of the incisors are broken, but the extremities of the teeth are sufficiently well preserved to exhibit the manner of wearing. They were not only worn in a sloping manner backward, as in rodents, but also externally, so that it appears the upper incisors were more divergent than the lower ones, and held a position related to them more like the condition observed in the peccary.

The rami of the jaw, as usual in pachyderms, are completely co-ossified. The symphysis is remarkably strong and deep, and in the median line is nearly 3 inches in length. The rami just back of the symphysis are nearly an inch thick. The chin forms a long, broad slope, defined at the sides by the prominences of the large incisor alveoli, curving from the base of the jaw parallel with each other upward and forward. The chin resembles that of the peccary or rhinoceros, but is more convergent, as in the beaver. Approaching the exit of the large incisors from their alveoli the intermediate space is deeply grooved, as represented in Fig. 3.

The body of the ramus is short, deep, and thick. Its outer surface is vertically convex. The base is thick and convex fore and aft as well as transversely.

The masseteric fossa is deep, and extends downward to about the middle line of the body of the ramus. Two mental foramina on one side, and three on the other, occupy a position in advance of that of the last premolar.

Measurements taken from the lower-jaw specimen are as follows:

	Inches.	Lines.
Distance from incisive alveoli to back of last molar	4	10
Space occupied by the molar teeth, estimated	4	0
Space occupied by the true molars	2	7
Fore and aft diameter of last molar	1	1
Depth of jaw at second molar	1	6
Thickness of jaw below second molar	0	10
Estimated length of lateral incisors	3	3
Depth of symphysis following its slope	2	10

TROGOSUS VETULUS.

An apparent smaller species of *Trogosus* is indicated by the fragment of an incisor tooth, represented in Fig. 43, Plate VI. The specimen was discovered by Dr. Carter in the vicinity of Fort Bridger, and sent to the writer last summer. It consists of the exserted portion of the tooth, and agrees in form and proportions with the corresponding portion of the incisors in the jaw-specimen above described. The enamel is smoother, but invests the tooth to the same relative extent. The antero-posterior diameter of the tooth has been about 8 lines; the transverse diameter 4 lines.

HYOPSODUS.

HYOPSODUS PAULUS.

One of the smallest of pachyderms, referred to a genus and species above named, is established on many specimens, chiefly consisting of portions of lower jaws with teeth, (Figs. 1 to 9. Plate VI.) It was originally indicated, from a lower-jaw fragment with teeth (Figs. 1, 2) of an old animal, discovered by Professor Hayden, in 1870, in the vicinity of Fort Bridger, Wyoming. Since then the writer has received a number of more characteristic specimens, obtained by Dr. J. Van A. Carter and Dr. Joseph K. Corson, at Grizzly Buttes, Henry's Fork of Green River, Lodge-Pole Trail, and other localities in the neighborhood of Fort Bridger, Wyoming.

The animal was rather less in size than the *Aphelotherium Ducernoyi* of Gervais, the remains of which were found in the gypsum quarries of Paris, France. It also appears to have been allied to this, as indicated by the number, relation, and constitution of the teeth. Both Aphelotherium and Hyopsodus possessed unbroken arches of teeth to the jaws, as in the Anoplotherium, whose remains are found in association with those of the first-named genus.

The number of teeth in Hyopsodus appears to be three incisors, a canine, and seven molars to the series on each side of both jaws.

Neither incisors nor canines are preserved in any of the specimens we have the opportunity of examining. Two lower-jaw specimens retaining portions of the incisive, canine, and premolar alveoli, and the true molars, apparently prove the number of teeth to be as above indicated.

The canine tooth of Hyopsodus is comparatively of small size, though larger than the incisors or the first premolar. It appears to have about the same size in relation with the other teeth as in Aphelotherium and Anoplotherium.

The premolars successively increase in size from the first to the fourth. The first possesses a single fang; the others two fangs. The anterior two premolars are lost from all the specimens under examination.

The inferior true molars (Figs. 1 to 9, Plate VI) of Hyopsodus have oblong quadrately oval crowns, with the fore and aft diameter exceeding the transverse, which is about equal to the depth. They are inserted in the usual manner in pachyderms by a pair of fangs, the posterior of which in the last tooth is widened backwardly, as is commonly the case in congeneric animals.

The crowns are composed of four principal lobes, connate at base; but the crown of the last tooth has an additional or fifth lobe at its back part as well developed as some of the lobes in advance. A rudiment of this fifth lobe is recognized in the other true molars as a small tubercle, occupying a corresponding position.

The four principal lobes of the crown of the true molars are arranged in pairs not quite transverse, but slightly oblique, so as to appear somewhat alternating. The fifth lobe of the last molar is opposite the interval of the pair of lobes in advance.

Of the four lobes, the outer are demi-conoidal, and the posterior one is slightly the larger. The inner lobes are simply conical, and the anterior is the larger. The outer lobes in the unworn condition have acute crescentoid summits, or form V-like ridges, with the arms declining from the pointed angle. The inner lobes in the same condition have pointed summits.

The contiguous horns of the crescentoid summits of the outer lobes join the antero-internal lobe. The anterior horn of the crescentoid summit of the antero-external lobe curves inwardly to the base of the antero-internal lobe.

The posterior horn of the crescentoid summit of the postero-external lobe ends in the tubercle at the back of the crown, and in the last molar, in the homologous fifth lobe. The latter is joined by an acute ridge, descending to the base of the postero-internal lobe.

A thin basal ridge exists at the fore and back parts of the crown of the first and second molars, and the fore part in the last molar. An element also exists at the interval, externally, of the outer principal lobes, and in some specimens is more or less produced around the bottom of the antero-external lobe.

As the crowns of the true molars are worn away, circular islets of dentine appear at the summits of the inner lobes, and crescentic islets at the summits of the outer lobes. In the progress of attrition the dentinal surfaces expand, and the horns of the crescentic islets become united with the circular islets. In an advanced stage of wear the triturating surface of the molars presents two elliptical surfaces crossing the crown, with a slight obliquity, and united by a median isthmus, the whole bordered by a band of enamel. Such a condition is seen in the specimen represented in Figs 1, 2, Plate VI, which is that upon which the genus was originally proposed. By comparing this with the others in different stages of wear, represented in Figs. 3 to 9, of the same plate, its correspondence with these, which preserve more characteristic generic marks, can be readily recognized.

The last lower premolar of Hyopsodus (Figs. 5, 8, Plate VI) is smaller than the true molars, and like them is inserted by a pair of fangs. Its crown is proportionately of greater depth than in the true molars, and, as in these, is widest fore and aft. The outer fore part of the crown is composed of a demiconoidal lobe, which is the principal one, and it corresponds with the antero-external lobe of the true molars. It has the same form as the latter lobe, but is better developed. The anterior horn of its crescentoid summit forms a curved ridge, defining the fore part of the triturating surface of the crown. The posterior horn of the crescentoid summit terminates in a small conical lobe occupying the middle of the crown internally. The back of the crown is formed by a broad heel, skirted by a basal ridge externally, and divided by another ridge, which descends from the summit of the principal lobe of the crown, and borders the heel posteriorly and internally. A thin basal ridge occupies the fore part of the crown. In the wearing of the crown of the last premolar, the exposed dentine assumes the form of the Greek letter ε, lying on its right side.

The penultimate lower premolar (Fig. 8) is a reduced form of the one behind, with the internal median conical lobe obsolete.

The lower jaw has its two rami co-ossified at the symphysis. It is thick and rounded at the base, which is convex fore and aft beneath the molar series. The chin is rounded transversely. The masseteric fossa is well marked and defined anteriorly by a prominent ridge descending from the front border of the coronoid process to the lower third of the side of the jaw.

The ramus of the jaw would appear to have increased in depth and assumed a more robust condition in the advance of age, for in those specimens in which the teeth are least abraded, the jaw is shallowest, and in that in which they are most worn it is deepest. Specimens exhibiting the teeth in an intermediate state of wear have the jaw of intermediate depth and strength to the others.

All the specimens represented in Figs 1, 3, 4, 6, and 7 I attribute to the same species, notwithstanding the difference in the proportion of length to depth in the different ones.

Two or three mental foramina occupy a slightly variable position beneath the premolars.

During the summer of 1871, Dr. Carter discovered, at Grizzly Buttes and Lodge-Pole Trail, several specimens, consisting of fragments of upper jaws with well-preserved teeth, which are of a size and form that would adapt them to the lower-jaw specimens of Hyopsodus. The specimens from Grizzly Buttes were accompanied by one of the lower-jaw specimens upon which the latter was founded, and this looks sufficiently like several of them in general appearance to have belonged to the same individual.

One of the specimens represented in Fig. 18, Plate VI, contains a series of three premolars and the succeeding molars. In advance of the series there remains a portion of an alveolus which apparently belonged to another premolar. If such is the case, the number of premolars would be the same as in the lower jaw of Hyopsodus.

The teeth (Figs. 18 to 22) increase in size from the first to the sixth, the seventh being again reduced to the size of the fifth. The second premolar is inserted by a pair of fangs of which the posterior is wider than the other. The succeeding premolars and molars are inserted with three fangs, of which the inner one of the molars is a connate pair.

The crowns of the molars (Figs. 19 to 21) are quadrate, wider transversely, and about half the depth of the breadth. They are composed each of six lobes expanding and continuous at base. The outer and inner pair of lobes are nearly equal; the intermediate pair is smaller.

The outer lobes of the molars are conical, and united where contiguous, but they do not form an external buttress by their union, the intervening surface externally being concave. In the last molar the posterior of the outer lobes is proportionately less well developed than in the others.

The inner lobes of the crown are likewise conical, and united where contiguous. The posterior of these lobes is the smaller, and in the last molar is entirely suppressed, or appears only as a slight elevation of the basal ridge occupying the back of the crown. The summit of the antero-internal lobe is prolonged obliquely to join the antero-median lobe. The summit of the postero-internal lobe is prolonged outwardly back of the postero-external lobe, so as to appear as a basal ridge to this part of the crown.

The median lobes hold a slightly more advanced position than the including lobes. The back one is isolated or free to its base; the front one, by prolongation, is associated with the antero-internal lobe and the fore part of the base of the antero-external lobe.

A strong basal ridge occupies the fore part of the crown, and also, less well developed, festoons the outer part.

In the wearing of the upper molars (Figs. 19 to 21) islets of dentine first appeared at the summits of the six lobes of the crown. Those of the two outer lobes soon became continuous; followed by those of the antero-median and anteror internal lobes With the widening of these two tracts, the islet of the postero-internal lobe next became continuous with that in advance. At this stage there would appear three dentinal tracts: one for the outer pair of lobes, a second for the internal and antero-median lobes, and a third as a circular islet on the postero-median lobe.

The posterior two premolars (Figs. 19, 21) have bilobed crowns, reminding one of the premolars of ruminants. The sudden reduction from the six lobes of the crown of the molars to the single pair of the crown of the premolars is a remarkable anatomical character. The lobes are pyramidal, and so far spread apart as to give the crown a greater width transversely. The summit of the inner lobe forms a crescentoid ridge, embracing the bottom of the outer lobe. A strong basal ridge bounds the fore and back part of the crown, and festoons it externally.

In the third premolar the inner lobe is less well developed at its fore part than in the fourth.

The crown of the second premolar (Figs. 19, 22) is formed of a single conical lobe, corresponding with the external lobe of the succeeding tooth. Postero-internally, it presents a feeble rudiment of an inner lobe. The crowns of the premolars were worn away mostly in a slanting manner posteriorly. The exposed dentine in the two lobed crowns became continuous at the back of the crown.

The molars above described resemble, in construction and in the relative position of the six lobes of the crown, those of Hyracotherium and Pliolophus, two extinct genera of pachyderms, described by Professor Owen from remains found in the London clay, an Eocene formation of the estuary of the Thames. In both the genera named the last molar is proportionately better developed; and in all the molars the postero-internal lobe and the basal ridge are likewise proportionately better developed. The upper premolars are quite different. In Hyracotherium the back two premolars have five lobed crowns, and in Pliolophus the last premolar has a similar constitution.

Too small a portion of the jaw (Fig. 18) containing the teeth above described has been preserved to ascertain anything of importance as to the shape of the face. The infra-orbital foramen is situated immediately above the interval of the back two premolars.

Measurements from some of the upper-jaw specimens and teeth above described are as follows:

Specimens in Plate VI.	Figs. 18,19.	Fig. 20.	
	Lines.	Lines.	Lines.
Space occupied by series of six molar teeth	9		
Space occupied by true molars	$5\frac{1}{4}$	5	$5\frac{1}{2}$
Space occupied by three premolars	$3\frac{3}{4}$		
Breadth of second premolar	1		
Width of second premolar	1		
Breadth of third premolar	$1\frac{1}{4}$		
Width of third premolar	$1\frac{1}{2}$		
Breadth of fourth premolar	$1\frac{1}{4}$		$1\frac{1}{4}$
Width of fourth premolar	2		$2\frac{1}{4}$
Breadth of first molar	$1\frac{3}{4}$	$1\frac{1}{2}$	$1\frac{3}{4}$
Width of first molar	$2\frac{1}{4}$	$2\frac{1}{4}$	$2\frac{1}{4}$
Breadth of second molar	2	2	2
Width of second molar	$2\frac{1}{4}$	$2\frac{3}{4}$	3
Breadth of last molar	$1\frac{3}{4}$	$1\frac{1}{2}$	$1\frac{3}{4}$
Width of last molar	2	$2\frac{1}{4}$	$2\frac{1}{4}$

HYOPSODUS MINUSCULUS.

A smaller species of Hyopsodus than the one described in the preceding pages appears to be indicated by a specimen discovered by Dr. Carter in the buttes of Dry Creek. The specimen consists of an upper-jaw fragment containing the true molars and part of the last premolar, which are represented in Fig. 5, Plate XXVII. The teeth differ in no essential character and only in size. Their comparative measurement with those of *H. paulus* are as follows:

	Hyopsodus minusculus.	Hyopsodus paulus.
	Lines.	Lines.
Length of space occupied by the last premolar and molars..	5.4	6.45
Length of space occupied by the molars...................	4.4	5.2
Breadth of last premolar..................................	1.05	1.4
Width of last premolar....................................	1.8	2.2
Breadth of first molar....................................	1.6	1.8
Width of first molar......................................	2.0	2.4
Breadth of second molar..................................	1.6	2.0
Width of second molar....................................	2.1	2.8
Breadth of third molar...................................	1.25	1.6
Width of third molar.....................................	1.7	2.35

MICROSUS.

MICROSUS CUSPIDATUS.

The genus Microsus is obscurely determined and is uncertain in its distinction from the previous genus. It was originally inferred from the lower-jaw fragment with the back two molars, represented in Fig. 10, Plate VI, at the same time that *Hyopsodus paulus* was characterized from the only specimen then in our possession, represented in Fig. 1 of the same plate. The well-marked difference in the form and proportion of the corresponding portion of the jaw led me to view it as pertaining to a different genus from Hyopsodus. Subsequently I have had the opportunity of examining many and more characteristic specimens referable to the latter, but none which with any certainty could be ascribed to Microsus.

The jaw-specimen referred to the latter was obtained by Professor Hayden on Black's Fork of Green River. The jaw would appear to be narrower and weaker than in Hyopsodus. The fragment as seen in Fig. 10, in comparison with corresponding portions of the latter, as seen in Figs. 1, 3, 4, 7, is proportionately of less depth, and at the base beneath the molars curves much more upwardly in a backward direction.

The teeth, represented in Figs. 10 and 11, are unworn, and they have the same size and constitution as those of Hyopsodus. In the specimen the constituent lobes of the crown appear more prominent and their intervals more deeply angular than in those of Hyopsodus, but this difference is probably due to difference in age. In all the specimens ascribed to Hyopsodus the teeth are more or less worn, and only in that attributed to Microsus are they unworn. Observing, also, that the proportionate depth of the lower jaw in the different specimens of Hyopsodus holds some relationship with the age of the animal, as indicated by the extent of wearing of the teeth, I have supposed that the variations observed in the jaw-fragment of Microsus might be due to the same cause, and that it therefore really pertains to *Hyopsodus paulus*. Thus by comparison of Fig. 10, representing Microsus, with that of Fig. 6, representing a specimen of Hyopsodus, in which the teeth are least worn, the resemblance is observed to be greater than with the other and older specimens of Hyopsodus.

Comparative measurements of the specimen referred to Microsus with specimens of Hyopsodus represented in Figs. 1, 3, and 6 are as follows:

	Microsus.	Hyopsodus.		
	Fig. 10.	Fig. 1.	Fig. 3.	Fig. 6.
	Lines.	*Lines.*	*Lines.*	*Lines.*
Depth of lower jaw at fore part of last molar......	2½	3½	3½	3¾
Space occupied by last two molars	4⅘	4¼	4½	4
Antero-posterior diameter of last molar............	2¾	2¾	2½	2½
Antero-posterior diameter of second molar.........	2½	2	2	2

MICROSYOPS.

An extinct genus of small mammals, to which the above name was given, was originally founded on several lower-jaw fragments containing molar teeth. Though classed among the pachyderms, it is not positive that such is the true position of the genus. The more complete of the specimens upon which it was characterized indicates a series of six molar teeth following closely after a well-developed canine. The number of incisors is unknown.

In the genus Limnotherium, as established by Professor Marsh, from the typical species *L. tyrannus*, the dental formula consists of two incisors, a canine and seven molars.

MYCROSYOPS GRACILIS.

The more characteristic specimen upon which this species was named consists of a portion of the left ramus of the lower jaw, represented in Fig. 14, Plate VI, of the natural size. The specimen was discovered by Dr. Carter on Grizzly Butte. Besides the fang of the canine and those of the premolars, it contains the true molars entire. These are moderately worn at the summits of the constituent lobes of their crowns, and their triturating surfaces are represented in Fig. 15 of the same plate, magnified four diameters. The jaw-fragment retains part of the rough sutural surface of the symphysis, showing that its union was ligamentous with the other ramus. The basal portion beneath the molars is broken away. Below the premolars it is of moderate depth and thickness, and soon curves upward with the fang of the canine. The mental foramen is situated below the second premolar. The fang of the canine indicates a proportionately larger tooth than in Hyopsodus. It is laterally compressed and curves upward, forward, and outward. The transverse section is oval with the long diameter directed obliquely forward and outward, and measuring 1.8 line, while the short diameter is 1 line.

The premolars successively increase in size. The first is separated by a slight hiatus from the canine, and was inserted by a single fang. The others have two fangs.

The molars have oblong quadrately-oval crowns of nearly uniform size. They are inserted with two fangs, with the back one of the last molar widened, as usual in most ungulates.

The crown of the molars is composed of two divisions, in addition to which the last one has a large posterior tubercle. The anterior division of the crown is smaller than the posterior, and appears of the same form in a more contracted condition. Each division consists of an external crescentoid conical lobe and an internal rudimental conical lobe or tubercle, which is placed opposite the back part of the former lobe.

The front arm of the anterior crescentoid lobe ends in a thickening in advance of the antero-internal lobe. The corresponding arm of the better developed posterior crescentoid lobe terminates at the base of the lobe in front of it. The back arm of the same lobe forms a slight thickening contiguous to the postero-internal lobe. In the last molar this thickening appears to be developed into the large tubercle back of the second division of the crown. Feeble traces of a basal ridge occupy the interval of the outer lobes and the back of the crown.

Measurements of the lower-jaw specimen of *Microsyops gracilis* are as follows:

	Lines.
Depth of lower jaw below last molar	4
Thickness of lower jaw below last premolar	2¼
Distance from canine alveolus to back of last molar	10
Space occupied by the entire molar series	9¾
Space occupied by the premolars	4
Space occupied by the molars	5¾
Breadth of first molar	1¾
Breadth of second molar	2
Breadth of third molar	2

The specific name of *M. gracilis* was originally given under the impression that the remains referred by Professor Marsh to *Hyopsodus gracilis* pertained to the same animal. A specimen exhibited to the writer by Professor Marsh would indicate that *M. gracilis* is the same as the animal named by him *Limnotherium elegans*. As Microsyops is generically distinct from Limnotherium, as characterized from the typical species *L. tyrannus*, the specific name of the former would be *Microsyops elegans*.

Another specimen, originally referred to *Microsyops gracilis*, is represented in Fig. 16, Plate VI, and was found by Dr. Carter near Lodge-Pole Trail, about ten miles from Fort Bridger. It consists of a portion of the left ramus of the lower jaw, containing the penultimate molar and part of the last one. The only remaining entire molar, a view of the triturating surface of which is given in Fig. 17, closely resembles the corresponding tooth in the specimen first described, except that it is a little larger. (The artist has made it appear different by exaggerating the proportions of the tubercle between the posterior lobes, and leaving it out altogether in the corresponding view of Fig. 15.) The remaining portion of the last molar also agrees with the corresponding portion in the first-described specimen. The lower jaw is comparatively deep, and is nearly straight along the base. The fore part with the symphysis is lost, but it would appear not to have been so shallow and thick as in the former specimen, which leads me to suspect that it perhaps belongs to a different animal. The mental foramen holds the same relative position as in the other specimen. The ridge bordering the fore part of the coronoid process terminates in a tubercle at the fore part of the moderately deep masseteric fossa.

Another foramen, perhaps not constant to the species, is situated below the position of the fore part of the first molar.

The measurements of the specimen are as follows:

	Lines.
Depth of jaw below last molar	4¾
Depth of jaw below last premolar	4¾
Space occupied by last premolar and molars	7¼
Space occupied by the molars	6
Breadth of penultimate molar	2¼
Breadth of last molar	2¼

The only specimen of an upper tooth which may, with any probability, be supposed to belong to Microsyops, is contained in a small fragment of the jaw, found by Dr. Carter on Dry Creek. The tooth, apparently a first or second molar, is represented in Figs. 19, 20, Plate XXVII. The crown is not so square and is proportionately of less breadth fore and aft than in the corresponding tooth of Hyopsodus. It narrows inwardly more than in the latter, the reduction taking place posteriorly, where the crown is concave. The constitution of the tooth is nearly as in Hyopsodus, and the principal difference is found in the condition of the postero-internal lobe. In Hyopsodus, this is a reduced form of the lobe in advance, being crescentoid. In the supposed tooth of Microsyops, the postero-internal lobe appears as a conical tubercle springing from the base postero-internally of the larger crescentoid lobe in front. In Hyopsodus, the postero-median lobe is a simple cone, but in the tooth in question it is pyramidal.

The antero-posterior diameter of the crown externally is 1.6 lines; internally, 1.2 lines; the transverse diameter anteriorly is 2.2 lines.

UNDETERMINED.

Fig. 12, Plate VI, represents a specimen found by Dr. Carter on Henry's Fork of Green River. It consists of a lower-jaw fragment with the last premolar and the fangs of the molars of a mature animal of undetermined character, but, from the form of the remaining tooth, evidently allied with Hyopsodus. The premolar, Fig. 13, is unlike the corresponding one of the latter genus, as seen by comparing it with Figs. 5 and 8, but resembles the true molars. Suspecting that it might be a last temporary molar, notwithstanding its slightly worn condition and its association with the full series of molars in functional position behind, I examined the jaw beneath, but found no trace of a successor.

The portion of jaw is of more uniform depth, and the base less convex than in Hyopsodus. It is also more impressed and concave below the position of the back molars. The space occupied by the molars is about equal to that in the smaller specimens of Hyopsodus.

Perhaps the specimen may pertain to Microsus, or probably may belong to a genus different from either of those just named. Its measurements are as follows:

	Lines.
Space occupied by the last premolar and molars	7½
Space occupied by the molars	5½
Depth of jaw at fore part of last molar	3½
Depth of jaw at last premolar	3½
Antero-posterior diameter of last molar	2¼
Antero-posterior diameter of last premolar	1½

NOTHARCTUS.

NOTHARCTUS TENEBROSUS.

A small extinct pachyderm, referred to a genus with the above name, judging from the anatomical characters of the specimen upon which it was founded, was probably as carnivorous in habit as the raccoon and bear. The specimen to which I allude, represented in Fig. 36, Plate VI, consists of the right ramus of a lower jaw with most of the teeth. It was discovered during Professor Hayden's exploration of 1870, on Black's Fork of Green River. I at first viewed it as pertaining to a carnivorous animal, and thus referred it; but the anatomical relations of the specimen with those of remains of other animals which have been found in association with it have led me to view the jaw as having belonged to a pachyderm. The ramus of the jaw contained a series of seven molar teeth, all of which are preserved except the first premolar. A well-developed canine occupies a position immediately in advance of the molar series, and the incisors filled the interval between the canines of the two sides. Thus the teeth of the lower jaw of Notharctus form an unbroken arch. The incisors are lost from the specimen, and the condition of the alveoli is such that the number of them cannot be ascertained.

The canine tooth of Notharctus in its relative position, form, and proportions resembles that of ordinary carnivores. It curves from the opening of the alveolus slightly backward with an inclination outward. The crown is considerably elevated from an increased protrusion of the fang, such as is observable in carnivorous animals past maturity. The fang is gibbous and feebly curved.

The molar teeth, represented in Fig. 37, magnified two diameters, are considerably worn in the specimen, all of them exhibiting exposed tracts of dentine, due to the wear of mastication.

The four premolars successively increase in size, and are inserted by a pair of fangs, except the first, in which they appear to have been connate. The crowns of the premolars from behind forward exhibit a successive reduction to a simpler form from that of the molars.

The crowns of the second and third premolars, and no doubt also that of the first one, which is lost from the specimen, have the conical form of the corresponding teeth in carnivores, though they appear less prominent, due to their worn condition. They are slightly thicker behind than in front, and a basal ridge internally forms a slight offset or heel posteriorly, and a still feebler one in the third premolar anteriorly.

The crown of the fourth premolar is intermediate in character with those in advance and those of the molars behind. Its fore part consists of a conical lobe like the crown of the anterior premolars; its back part is a broad heel corresponding with the back lobe of the molars. The summit of the principal lobe is extended obliquely inward and backward and is continuous with the inner basal ridge of the crown. Externally, the latter is embraced by a basal ridge.

The crown of the second premolar is worn away along its posterior slope; the crown of the third to a greater degree in a corresponding position, and also to a less degree along its anterior slope.

The molars are nearly alike in form and constitution, and are inserted with two fangs. The crown of the molars bears a certain degree of resemblance in construction to those of the raccoon, and in a less degree to those of the opossum, but certainly enough resemblance to both to indicate a relation which is not merely accidental.

In the unworn condition of the lower molars of Notharctus, the crown would appear to be composed of two divisions. The anterior division presents three prominent points continuous in an acute crescentoid ridge. The principal point is central and external, the second is nearly as well developed and internal, and the third point, feebly developed, occupies the fore part of the crown. The posterior division presents two elevated points, conjoined in a crescentoid ridge. The anterior extremity of this ridge joins the front division of the crown. Its more elevated point is external and posterior, and its less developed one occupies the postero-internal corner of the crown. A basal ridge incloses the crown, except internally.

Each division of the crown of the molars incloses in the arms of its crescentoid ridge a depression, which is largest in the posterior division. The crown of the last molar is more prolonged backward than the others, arising from the greater degree of development in this direction of its posterior division.

In the worn condition of the molars, as seen in Fig. 37, the crescentoid ridges of the divisions of the crown have been so much abraded as to expose broad crescentoid tracts of dentine continuous on the two divisions of the crown.

The rami of the lower jaw of Northarctus appear to be co-ossified at the symphysis, and the specimen under consideration was broken off just to the left of the latter. The chin is narrow and convex transversely, and it forms a nearly straight or slightly convex slope of about 45°. The body of the bone is nearly of uniform depth; the relation of depth to length being much greater than in the raccoon and more in proportion with the measurements in the hog and peccary. The outer face of the body is nearly vertical. The base is thick and slightly convex fore and aft. Near the middle, directed inwardly, it exhibits a strong impression for muscular attachment.

The angle of the jaw, the back border of the bone, and the coronoid process are lost. The outer face of the ascending portion of the ramus is depressed into a masseteric fossa extending nearly or quite to the base, but shallow compared with that of ordinary carnivores. The condyle is remarkably short, and resembles that of some of the monkeys more than that of ordinary pachyderms. It is transversely oval, with the breadth less than twice the fore and aft diameter, which is directed obliquely from without inward and backward. The articular surface is transversely convex and inclines more outwardly than inward.

The form of the condyle clearly indicates more varied movements in the jaw than exists in the carnivora, and would rather be favorable to the proper reduction of the food of an omnivorous animal.

A mental foramen occupies a position about midway between the third premolar and the base of the jaw. Two small foramina are situated below the position of the first premolar.

In the American Journal of Science for 1871, Professor Marsh has described the lower jaw, with teeth, of a small pachyderm, under the name of *Limnotherium tyrannus*, which would appear at least to belong to the same genus as the former. The specimen was found near Dry Creek, Wyoming. According to the description and measurements, the jaw nearly accords with that of *Notharctus tenebrosus*. The teeth in an interrupted series consist of two incisors, a canine, four premolars, and three molars. If we suppose Notharctus to have two incisors, the number, character, and relative position of the teeth agree in both. In Limnotherium the first and second premolars are observed to have a single fang. This character alone would be insufficient to distinguish a genus, and, perhaps, would hardly be regarded as a specific character. The description of the molars of Limnotherium would apply to those of Notharctus.

The measurements of the lower-jaw specimen of *Notharctus tenebrosus* are as follows:

	Lines.
Length of jaw from condyle to incisive alveoli	30
Depth of jaw below last molar	5¾
Depth of jaw below last premolar	6
Length of symphysis	8
Breadth of condyle	3
Antero-posterior diameter of condyle	2
Length of dental series	19½
Length of space occupied by the molar series	16
Length of space occupied by the true molars	8¾
Diameter of canine at base of crown	1¾
Fore and aft diameter of second premolar	1½
Transverse diameter of second premolar	¾
Fore and aft diameter of third premolar	1¾
Transverse diameter of third premolar	1¼
Fore and aft diameter of fourth premolar	2
Transverse diameter of fourth premolar	1½
Fore and aft diameter of first molar	2¼
Transverse diameter of first molar	2
Fore and aft diameter of second molar	2¾
Transverse diameter of second molar	2
Fore and aft diameter of third molar	3¼
Transverse diameter of third molar	2

In many respects the lower jaw of Notharctus resembles that of some of the existing American monkeys quite as much as it does that of any of the living pachyderms. Notharctus agrees with most of the American monkeys in the union of the rami of the jaw at the symphysis, in the small size of the condyle, in the crowded condition of the teeth, and in the number of incisors, canines, and true molars, which are also nearly alike in constitution. Notharctus possesses one more premolar and the others have a pair of fangs. The resemblance is so close that but little change would be necessary to evolve from the jaw and teeth of Notharctus that of a modern monkey. The same condition which would lead to the suppression of a first premolar, in continuance would reduce the fangs of the other premolars to a single one. This change, with a concomitant shortening and increase of depth of the jaw, would give the characters of a living Cebus. A further reduction of a single premolar would give rise to the condition of the jaw in the Old World apes and man.

HIPPOSYUS.

Hipposyus formosus.

Several small fragments of jaws with teeth, discovered by Dr. Carter in the vicinity of Fort Bridger, are suspected to belong to a different genus of pachyderms from any of those indicated in the preceding pages. One of the specimens consists of an upper-jaw fragment with the molars in a mutilated condition. The first and second molars are the best preserved, and are represented in Fig. 41, Plate VI, magnified three diameters. The first one is nearly entire, but in the figure is represented in a restored condition by the addition of the antero-external angle marked by the zigzag black line.

The upper molars bear a general resemblance in the construction of their crowns to those of Anchitherium. The outer lobes are like those in the latter, but have their outer buttress-like ridges proportionately thicker. The antero-internal lobe is larger than that behind and conjoins it. In Anchitherium the inner lobes are nearly equal and isolated from each other. The antero-median lobe, as existing in Anchitherium, in the present fossil is completely connate as part of the antero-internal lobe, and the postero-median lobe of the former is nearly obsolete, or appears as a mere rudiment in Hipposyus. A strong basal ridge incloses the crown in front, behind, and internally, but is absent in Anchitherium in the latter position.

The measurements of the specimen are as follows:

	Lines.
Space occupied by the three molars	7½
Breadth of first molar	2½
Breadth of second molar	2¾
Breadth of third molar	2½
Width of first molar transversely	3¼

The other specimens accompanying the former, and suspected to belong to the same species, consist of fragments of three lower jaws, containing each one or two molar teeth. One of the specimens contains the first and second molars and the remains of the last one. The portion of jaw is like the corresponding portion in Notharctus, but is thicker and the teeth are stouter. The teeth are considerably worn, as represented in the view of the triturating surface of the second molar in Fig. 38, Plate VI, magnified two diameters. The crown is oblong square, and consists of two divisions, each of which, in the unworn condition, presented an acute crescentoid summit. In the abraded condition the divisions present two broad, semi-lunar tracts of dentine continuous with each other. The tracts embrace, internally, shallow enameled recesses, of which the posterior is much the larger. The contiguous horns of the tracts are continuous upon a tubercle at the inner part of the crown just in advance of the middle. The posterior horn of the posterior crescent is likewise continuous, with a tubercle at the postero-internal angle of the crown. Externally the latter is bounded by a basal ridge, and an element of the same occupies the postero-internal angle of the crown.

Measurements of the specimen are as follows:

	Lines.
Depth of jaw below the middle molar	5¼
Space occupied by the three molars	9
Breadth of crown of first molar	2¼
Width of crown of first molar	2
Breadth of crown of second molar	3
Width of crown of second molar	2¼

A second specimen consists of a nearly corresponding portion of another jaw containing the first and second molars. The jaw-fragment is of greater depth than in the former, but otherwise is about as robust, and the teeth are nearly of the same size. The measurements are as follows:

	Lines.
Depth of jaw below middle molar	5⅞
Breadth of crown of first molar	2¼
Width of crown of first molar	2¼
Breadth of crown of second molar	2¾
Width of crown of second molar	2¼

Two additional specimens consist of portions of both rami of the lower jaw of a younger animal than the preceding, but only one contains a single tooth, the first molar, which is represented in Fig. 39, Plate VI, magnified two diameters. The rami of the jaw are of more slender proportions than indicated by the fragments above described, but are nearly as thick; and the retained tooth is of the same size and form as its fellow in the fragments of older jaws.

One of the rami contains the fangs of the complete molar series, together with part of the canine alveolus, which is close to the former. The number of premolars I cannot determine with certainty. If three, the first of the series is larger than the second, and has its fangs more widely separated. If the number is four, the anterior two have each a single fang.

Perhaps the latter is the true condition, which accords with that attributed to Limnotherium by Professor Marsh.

Three vasculo-neural foramina are situated at the outer part of the ramus: one just back of the position of the canine alveolus; a second below the interval of the back two premolars, and the third beneath the first molar. In the opposite ramus the latter is below the last premolar, and it occupies the same position in the former two specimens.

The first molar tooth retained in one of the rami agrees with the description of those in the older jaw-fragments. Fig. 38, Plate VI, represents the right second molar much worn; and Fig. 39 represents the first left molar in a much less abraded condition.

Measurements from the two rami of the lower jaw just described are as follows:

	Lines.
Space occupied by the premolar and molar series	16½
Space occupied by the molar series	9
Depth of jaw below the premolars	5
Depth of jaw below the middle molar	4¾
Breadth of the first molar	3
Width of the first molar	2¼

Figs. 1, 2, Plate XXVII, represent a specimen of a tooth recently discovered by Dr. Carter on Grizzly Buttes. It appears to be a first upper true molar of *Hipposyus formosus*, and is scarcely worn. It was found isolated and unaccompanied by any other pieces which could be reasonably attributed to the same animal. From the comparative perfection of its crown, its constitution is more evident. It resembles in miniature the corresponding teeth of Anchitherium, and differs especially in the less proportionate development of the median lobes of the crown, in the greater degree of production of the basal ridge, in the more intimate union of the inner lobes and their more sloping character externally, in the more isolated condition of the postero-median lobe from the contiguous inner one, and in the more wrinkled condition of all the lobes approaching the base of the crown. The transverse diameter of the latter is ⅛ of an inch; its fore and aft diameter externally ¼ of an inch.

Hipposyus robustior.

A lower-jaw fragment containing a single tooth, obtained by Professor Hayden on Henry's Fork of Green River, apparently indicates a more robust species of the same genus as the former. I at first attributed the specimen to a species of Notharctus, with the name of *N. robustior*, but a comparison of the tooth, represented in Fig. 40, Plate VI, with those of *Hipposyus formosus*, Figs. 38, 39, will at once suggest the probability of its pertaining to a larger species of the latter genus. Perhaps the specimen may belong to a more robust individual of the same species.

The jaw-fragment is too imperfect to ascertain anything in regard to its anatomical characters other than its thickness. Below the second molar it is ¼ of an inch thick; in the specimens attributed to *H. formosus* it ranges in the same position from 3¼ to 3½ lines in thickness. The second molar tooth is 3½ lines broad and 2¼ lines wide.

Order *Proboscidea?*

Large quadrupeds with five toes to the feet; molar teeth with transverse ridges; femur without a third trochanter; nose prolonged into a cylindrical trunk or proboscis.

UINTATHERIUM.

While encamped in Dry Creek Cañon, forty miles to the east of Fort Bridger, Drs. Carter and Corson spent a day in traversing a most desolate region to some buttes about ten miles farther to the east. They returned to camp after sundown laden with fossils, among which were the remains of the largest animal which had yet been brought to our notice from the Bridger Tertiary beds. These remains consist of the cranial portion of a skull with fragments of both jaws attached to the same matrix, a nearly complete arm-bone, and fragments of other limb-bones. A notice of these remains, attributed to a pachyderm with the name of *Uintatherium robustum*, was communicated in a letter to the Academy of Natural Sciences of Philadelphia, and was published August 1, 1872.

On the previous day to the discovery of the remains of Uintatherium, while engaged in the search for fossils along the buttes, about a mile to the east of our camp, Dr. Corson called my attention to a large tusk which he had found mingled with some drift-pebbles that had fallen from the top of the butte. In the tusk I thought I recognized the canine of a large carnivore related to the extinct saber-toothed tiger of Brazil. On further search, we found a portion of the opposite tusk, an isolated molar supposed to belong to Uintatherium, another of Palæosyops, and the scale of a ganoid fish.

In the same letter above mentioned, the large tusks were described and attributed to a carnivore with the name of *Uintamastix atrox*.

On our return to Fort Bridger, while examining and discussing the fossils collected in our expedition, the question arose whether the large tusks did not pertain to the same animal I have named Uintatherium. Our specimen of the skull of the latter did not assist the determination of the question, as the facial portion was wanting, excepting small fragments of the back of the jaws containing the last molar teeth. While admitting the probability of the tusks pertaining to Uintatherium, from their being so unlike those of any known pachyderm, and from their near resemblance, both in form and size, to those of the great extinct Machairodus of Brazil, I thought the weight of evidence was in favor of their reference to a carnivore. The finding of a molar tooth of Uintatherium in association with the tusks appeared to me not to outweigh this evidence any more than the association with them of a molar of Palæosyops.

Professor Marsh has published several notices in the American Journal of Science of the remains of large mammals from the Bridger Tertiary formation, which appear to be related with Uintatherium.

In June, 1871, he reported the discovery of bones of a large animal which he referred with doubt to Titanotherium, with the name of *T. anceps*. From some additional remains, in a foot-note of July 22, 1872, he refers them to a proboscidean under the name of *Mastodon anceps*. This is corrected in an erratum of August 19, referring the animal to a new genus with the name of *Tinoceras anceps*. September 21, he published a notice of a new species with the name of *Tinoceras grandis*, founded on portions of a skull and teeth, &c. Of this he observes, "The skull is proportionately very small, and indicates one of the most remarkable animals yet discovered. It supports a pair of short horns, and has also two powerful tusks, which, in size, shape, and direction, resemble the canines of the walrus."

More recently, September 27, Professor Marsh has published a "notice of some remarkable fossil mammals," which are referred to two species of a new genus with the names of *Dinoceras mirabilis* and *D. lacustris*. Of the skull of this genus, he observes that it presents a most remarkable combination of characters. "It is wedge-shaped, elongated, and quite narrow, especially in front, and was armed with horns and huge decurved canine tusks. The top of the skull, moreover, is deeply concave, and has around its lateral and posterior margins an enormous crest. On the frontal bones, above the orbits, and in advance of the lateral crest, there is a pair of very large horn-cones, just behind and above the canines. These are directed upward and outward, and their summits are obtuse and nearly round. They are solid, except at the base, which is perforated by the upper extremity of the canine. Near the anterior margin of the nasals there is still another pair of horn-cones, which are near together, and have obliquely compressed summits. The nasal opening is small. The premaxillaries are slender and without teeth. The upper canines are greatly elongated, slightly curved, and compressed longitudinally. The lower portion is thin and trenchant. Behind the canine is a long diastema, followed by a series of six small teeth. The molars have their crowns composed of two transverse ridges, separated externally, and meeting at the inner extremities. The skull measures about $28\frac{1}{2}$ inches long and $8\frac{1}{2}$ inches in width over the orbits. The canine is $9\frac{1}{4}$ inches in length below the jaw, 64 millimeters in longitudinal diameter at base, and 25 millimeters in transverse diameter. The last upper molar has an antero-posterior diameter of 36 millimeters."

It appears to me that the brief description of the skull and molar teeth of Dinoceras applies so closely to the corresponding parts of Uintatherium as to render it probable they are of the same genus. The description of the tusks of the former also equally well apply to those of Uintamastix, so as to lead me to suspect that this may likewise be the same as Uintatherium. It is probable, too, that should the latter not be the same as Dinoceras it may prove to be the same as Tinoceras, or perhaps the Eobasileus s. Loxolophodon of Professor Cope.

The characters of Uintatherium, as expressed in the material at our command, are so peculiar and unlike those of any other known animal as to render its ordinal affinities obscure. From the form and constitution of the molar teeth alone, I should have viewed the genus as pertaining to the odd-toed pachyderms. If the remains noticed by Professor Marsh under the name of Dinoceras belong to the same animal, the presence of horns in pairs to the head would render such a reference improbable. Professor Marsh observes of Dinoceras and the related Tinoceras, that they have the vertebral and limb bones very similar to those of recent proboscideans, but refers them to a new order with the name of Dinocerea.

The form of the thigh-bone and the short tarsal bones of Uintatherium would appear to indicate limbs and feet most nearly constructed like those of the elephant. I have provisionally placed the animal in the order of Proboscidea, leaving to Professor Marsh the determination of its true position from the more abundant materials at his command.

Uintatherium robustum.

The remains which are specially to be regarded as characteristic of the animal above named, and from which it was originally indicated, consist of a mutilated cranium, to the matrix of which there adhered portions of both jaws, containing all the last molars and an isolated molar. A nearly complete humerus, together with some less well preserved specimens found in association with the former, are supposed to have pertained to the same individual.

A small fragment of the upper jaw, containing the last molar tooth, is represented in Fig. 8, Plate XXV. The tooth, also represented in Figs. 6, 7, of the same plate, and Fig. 30, Plate XXVII, has the crown composed of a pair of wide pyramidal lobes projecting from a broad expanded base. The lobes extend across the crown, conjoining internally and diverging externally in a V-like manner. They project at their outer extremities in prominent points, and also form together a prominent point at their conjunction internally. The outer extremity of the anterior lobe is the most prominent of the three points of the crown. The outer extremity of the posterior lobe is the least prominent of the three points, while that at the conjunction of the lobes is scarcely more so. The acute summits of the lobes between the points are transversely concave, and are worn off on their anterior slope so as to present narrow tracts of exposed dentine. The posterior slope of the lobes is slightly concave; and the valley between them is triangular, and opens outwardly.

From the posterior slope of the inner part of the back lobe of the crown there projects a rounded tubercle about half-way between the basal ridge and the pointed conjunction of the lobes. A second rounded tubercle occupies the entrance of the triangular valley between the lobes.

A stout basal ridge embraces the crown in front and behind, and in a reduced condition continues interruptedly on the inner and outer parts. In outline the base of the crown is ovoidal, with the narrower extremity corresponding with the outer part of the anterior lobe. The tooth is inserted by a pair of fangs widely compressed, conical, and convergent internally. The transverse diameter of the crown of the last upper molar is 20 lines; its fore and aft diameter is nearly 18 lines. The description of the upper molars of *Dinoceras mirabilis*, and the size of the last one, as given by Professor Marsh, so well apply to the tooth above described as to lead me to suspect that the animal so named is the same as *Uintatherium robustum*.

The fragments of both sides of the lower jaw of the latter, represented in Fig. 11, Plate XXV, and Figs. 32, 33, Plate XXVII, contain the last molar tooth, also represented in Figs. 9, 10, of the former plate, and Fig. 31 of the latter. The tooth has an oblong square crown, rounded at the corners and moderately constricted at the middle laterally. It is inserted in the jaw by a pair of wide, compressed conical fangs.

The crown is composed of three lobes, with oblique intervening valleys, which receive the pair of lobes of the corresponding upper tooth when closed upon the lower one.

The anterior lobe forms nearly half the crown, and rises internally in a point, which is the most prominent part of the tooth. The front and back surfaces are sloping, and the former is transversely concave, and bounded by a short, oblique basal ridge. The inner and outer surfaces of the extremities are convex, and extend to the bottom of the crown. The acute summit curves downward and outward from the inner point. It is worn off on the posterior slope with a more forward direction externally, and exhibits a narrow tract of exposed dentine. The prominent point of the inner extremity is notched just below the summit postero-internally.

The posterior and middle lobes of the crown are nearly of the same size and prominence. The posterior lobe is separated from the anterior lobe internally by a deep, angular notch, and diverges from it externally. It forms the posterior convex surface of the crown, and has an anterior sloping surface defined from it by a ridge curving from the inner side backward and outward, and then becoming continuous, with a basal ridge sweeping downward to the bottom of the middle lobe of the crown externally. The middle lobe appears like an ovoidal wedge introduced from the outer side, and separating the anterior and posterior lobes. Its summit is worn off with a slight posterior slope, and exhibits an exposed tract of dentine.

A thin, inconspicuous basal ridge occupies the inner half of the back part of the crown; a thicker festoon extends from the summit of the posterior lobe externally to the bottom of the middle lobe; and a short, prominent ledge occupies the middle of the front of the crown.

The fore and aft diameter of the crown of the last lower molar is $1\frac{1}{2}$ inches; the transverse diameter in front is 14 lines; behind, $12\frac{1}{2}$ lines.

Associated with the other specimens referred to Uintatherium, there was found the isolated tooth represented in Fig. 12, which I suppose to be a first upper molar. It has the same constitution as the last upper molar above described, but is smaller. In the present condition of the crown, the posterior lobe is more prominent than the anterior, and it exhibits a broad horseshoe-shaped exposed tract of dentine extending upon the summits of both lobes. The dentinal surface is concave from side to side, and inclines forward. The outer extremity of the anterior lobe, broken in the specimen, is considerably thicker than that of the posterior lobe. Back of the inner conjunction of the lobes, just below the summit, the rounded tubercle is visible, such as exists in a corresponding position in the last molar. It is worn so as to exhibit a small circular islet of dentine.

The basal ridge, as in the last upper molar, is thick in front and behind, but feeble upon the inner and outer sides.

The first molar was inserted by a pair of fangs. The antero-posterior diameter of the crown is 16 lines; the transverse diameter at the hinder lobe is $15\frac{1}{2}$ lines.

The upper molars of Uintatherium above described bear considerable resemblance to the last upper molar of *Lophiodon parisiense*, as represented in Fig. 3, Plate XVII of Gervais's Paléontologie. They differ especially in the absence of the offset from the middle of the anterior part of the front lobe of the crown.

The upper molar teeth, attributed by Gervais to *L. parisiense*, represented in his Figs. 3, 4, so nearly resemble the corresponding teeth of Uintatherium and so decidedly differ from those of Lophiodon, as characterized from the typical species *L. isselense*, that it may be questioned whether it belongs to the same genus. The characters presented by the teeth referred to *L. parisiense*, are sufficiently distinct and well marked to consider them as indicating a genus differing from Lophiodon and Uintatherium, and probably more nearly related with the latter than the former.

The cranium of Uintatherium, represented in Fig. 1, Plate XXVI, is of remarkable form and unlike that of any other known animal. The specimen, though much mutilated, is yet sufficiently well preserved to give us some notion of the peculiarities of the skull.

The top of the cranium presents a deep basin-like concavity separated on each side from the temporal fossæ by a wide projecting crest. The entire extent of this cannot be determined from the broken condition of its edge in the specimen, but on one side it projects obliquely outward and upward for three inches beyond the inner surface of the temporal fossa. Posteriorly, the crest is continuous with a thick broken border extending across the top of the occiput so as to make it appear as if the lateral projections of the cranium were continuous behind. The depth of the supra-cranial hollow in the specimen is upward of several inches, and was, no doubt, greater in the complete skull.

The temporal fossa is a long deep concavity overarched by the wide lateral crest separating it from the supra-cranial hollow. Its lower part spreads outwardly on a broad ledge extending from the lateral occipital border forward upon the upper surface of the zygomatic root. This ledge resembles the long extension backward of the zygomatic root in the bear, and in like manner it projects over the auditory archway and the contiguous processes.

From the fractured condition of the specimen, I am unable to ascertain the position of the squamous suture, and this may be said also of other sutures. The temporal surface as formed by the squamosal plate and the neighboring portion of the parietal is pierced with a number of large vascular foramina. The occipital surface is broad, and it slopes inwardly from above to the occipital foramen.

The large condyles project strongly backward and downward, and are not in the least degree sessile, but well defined from the occipital surface by a deep groove. Their articular surface is broad, being within a fourth as great as the depth, and the flexure near its middle appears less pronounced than usual. The articular surfaces are not prolonged below on the basi-occipital, and the condyles in this position are separated by a deep notch twenty lines from each other.

The basilar process is broad and thick, and moderately tapering. Its under surface is transversely convex, especially anteriorly. On each side of the middle it presents a broad rough eminence for muscular attachment.

The relative positions of the paramastoid and mastoid processes, the auditory archway, and the post-glenoid tubercle are nearly the same as in feline animals, but here the resemblance ceases.

The paramastoid process is a comparatively slight roughened eminence, situated just above and external to the position of the fore part of the contiguous condyle. It is separated from the mastoid process by an archway directed inward and forward to the space usually occupied by a tympanic bone, but which in the specimen is filled with the matrix of the fossil.

The mastoid process, though much broader and longer than the paramastoid, does not project so much downward as the occipital condyle. It is semicircular below and roughened, and is compressed from without inwardly. Its outer surface presents a median fossa at the base.

The auditory archway expands outwardly in a funnel, and below is partially contracted by a short ledge, a process of the tympanic, projecting from the mastoid process.

The root of the zygoma is of great strength, and has, projecting downward from it, a post-glenoid tubercle of extraordinary size. The process is $2\frac{1}{2}$ inches in width, and projects externally in a rounded knob. Its lower part forms a slightly irregular flat surface, just above which, the tubercle is $1\frac{1}{2}$ inches thick. Its inner extremity slopes upward and inward.

The glenoid articulation is transverse, and its surface straight in this direction. Upon the post glenoid tubercle the surface is vertical until it curves forward and upward to the anterior edge of the zygomatic root. Its forward extension is about equal to that downward. The glenoid articulation is evidently adapted especially to a hinge-like motion, though not so restricted as that of carnivores.

Measurements derived from the cranial specimen of Uintatherium are as follows:

	Inches.
Breadth of the cranium at the outer part of the post-glenoid tubercles	10
Breadth of the cranium at the mastoid processes	$7\frac{1}{4}$
Width of the basi-occipital in front of the occipital condyles	$2\frac{3}{4}$
Breadth of the occipital condyles together	6
Breadth of each condyle	$2\frac{1}{4}$
Depth of each condyle	3
Distance between the condyles or breadth of the occipital foramen	$2\frac{1}{4}$
Length of the temporal fossa fore and aft	7
Breadth of cranium between the temporal fossæ where deepest, about	$4\frac{1}{4}$
Depth of cranium from bottom of supra-cranial basin to basi-occipital, about	$4\frac{1}{2}$

The lower-jaw specimens of Uintatherium are represented in Fig. 11, Plate XXV, and Figs. 32, 33, Plate XXVII. Both contain the last molar, and the better-preserved one also contains the fangs of the preceding molars and the last premolar. The space occupied by the molars is 4 inches, which appears small in relation with the size of the animal. The space which was occupied by the second molar is nearly as broad as the last molar. The crown of this measures 1½ inches. The space which was occupied by the first molar is little more than three-fourths of an inch, thus showing a great difference in the size of the first, compared with that of the succeeding molars.

The body of the jaw is of robust proportions. Its depth beneath the fore part of the last molar is 3¼ inches; its thickness just above the rounded base is nearly 1½ inches. A strong obtuse ridge sweeps from the root of the coronoid process downward and forward along the base of the jaw beneath the position of the molars.

Back of the position of the latter, the jaw bears more resemblance to the corresponding portion in the great felines than it does to that of ordinary pachyderms.

The coronoid process is a broad curved plate rising immediately in advance of the condyle, as in the lion. As in the latter, likewise, it is impressed externally with a deep masseteric fossa extending below on the body of the bone, but becoming more abruptly shallow approaching the base.

The entrance to the dental canal is nearly on a line with the alveolar border, 2¼ inches above the base of the jaw.

The condyle is a transverse convexity 2½ inches in breadth, and rather more than an inch in width at the middle. It is narrowest internally, the reverse of the condition in the lion.

The breadth of the jaw back of the molars is estimated to be about 5 inches; the breadth of the coronoid process at base is about 3 inches.

The specimen of a mutilated atlas, represented in Fig. 2, Plate XXVI, and Fig. 34, Plate XXVII, supposed to belong to Uintatherium, was found by the writer on the buttes west of Dry Creek Cañon. It accords in size with the cranium of Uintatherium above described, and fits the occipital condyles as well as the bone of one individual might be expected to adapt itself to that of another.

The atlas is very unlike that of any ordinary familiar animal. While it is much smaller than that of a mastodon, it includes a canal of even greater capacity. Unlike that of the animal just named, it is quite circular, and about 3½ inches in diameter. The portion occupied by the spinal cord is absolutely larger than in the mastodon, and it is but slightly defined, from the portion for the pivot of the axis, by slight tubercular elevations for the transverse ligament.

The atlas is proportionately longer than in the mastodon, but is of less width. The inferior arch beneath is nearly flat, and without a hypapophysis, and on each side presents a superficial, rough prominence for muscular attachment. The neural arch is comparatively long and narrow, and appears to be devoid of a protuberance.

The articular concavities for the occipital condyles are deeper and more strongly sloping than in the mastodon. They are separated below by a deep notch at the fore part of the inferior arch. Above, they are removed from each other double the distance.

The articular facets for the axis are ovoidal, slightly concave, and incline at an angle of nearly 45°. They are separated below for a couple of inches by the thick back border of the inferior arch of the atlas. Above, they are separated by the long semicircular edge of the neural arch.

The inferior arch of the atlas supports a facet for the odontoid process of the axis, which is distinct from the articular facets on each side of the latter.

The transverse processes are unlike those of the elephant and mastodon, and are more like those in ordinary ruminants, &c. The ends are broken off, but they appear as broad, thick plates, extending fore and aft, though not the entire length in either direction.

The canal for the vertebral artery perforates the transverse process fore and aft from the back half of the upper to the anterior part of the lower surface. As a groove, it then turns upward in advance of the root of the transverse process, and is directed inward to a canal perforating the neural arch anteriorly above the position of the articular concavities for the occipital condyles.

Approximative measurements of the atlas are as follows:

	Inches.
Breadth between the outer edges of the anterior articular concavities	6¼
Depth of the atlas posteriorly from above downward	4¾
Breadth between the outer edges of the posterior articular facets	6½
Fore and aft extent of the inferior arch at the middle	2
Length of the atlas laterally	5
Diameter of spinal foramen from above downward	3¾
Diameter of spinal foramen transversely	3½
Breadth fore and aft of transverse processes	3½

The humerus of Uintatherium, of which the anterior view of a specimen is given in Fig. 3, Plate XXVI, is very unlike that of any other familiar animal. In its peculiarity of form it presents no very evident relationship with that of the larger pachyderms, odd or even toed, the proboscideans, or the ruminants. It is shorter, in proportion with its breadth, than in the elephant. The shaft is narrowest and most nearly cylindroid at the union of the upper two-thirds with the lower third. The upper part is prominently produced outwardly to support a long triangular deltoid tract, the point of which reaches below the middle of the bone. The deltoid surface looks outwardly and backward, and is nearly flat, except below where it is depressed. The back of the shaft presents a broad, nearly flat surface, dividing near the middle in two portions, of which one extends nearly straight downward, while the other portion winds outward and forward below the deltoid tract to the front of the distal extremity.

The surface of the shaft internally to the deltoid tract is wide and sloping inwardly. It is slightly depressed on the deltoid expansion, but elsewhere is nearly flat transversely, and it narrows downward in its extension to the internal epicondyle.

The outer or greater tuberosity of the humerus and the contiguous portions of the head and deltoid tract are destroyed in the specimen. The inner side of the head of the bone presents a broad depressed tract rising on the shaft below in a triangular protuberance, which answers to the ordinary internal tuberosity of the humerus. From the apex of the angular protuberance, a ridge descends the shaft defining the inner or anterior aspect of the bone from the posterior.

The head is most convex from before backward, and in this direction it looks as if, in the complete condition, it had not been greater than the transverse diameter.

The external epicondyle is thick and prominent, but is of comparatively little vertical extent. Its summit forms a thick, rough eminence, extending an inch externally to the capitulum and several inches in width above it. Its outer face presents a broad crescentoid surface directed obliquely outward and downward. It is rough and pierced with vascular foramina, and is divided into several facets for the attachment of the extensors of the forearm and the external lateral ligament.

The internal epicondyle is a comparatively low, thick, and rough prominence, defined from the trochlea by a wide, pitted groove. Its upper part is destroyed in the specimen. Its back part barely projects posterior to the position of the trochlea.

Above the distal articulation, where the bone is expanded to form the outer epicondyle, it is depressed into a broad and unusually deep concave fossa.

The olecranon fossa is broad and moderately deep, but is not much extended by the protrusion backward of the epicondyles.

The distal articulation of the humerus presents a well-rounded capitulum on the outer condyle and a broad trochlea extending from it on the inner condyle. The capitulum is convex and narrows behind on a ridge separating the posterior prominence of the outer epicondyle from the trochlea. The trochlear groove is directed obliquely from the fossa in front of the outer epicondyle downward and inward, then backward, upward, and outward to the olecranon fossa.

The measurements of the bone are as follows:

	Inches.
Length of the humerus internally	20¼
Width transversely of the head	4¾
Width of shaft at the middle from the lower part of the deltoid tract to the postero-internal border	4¼
Thickness of shaft at middle of same position	2⅛
Circumference of narrowest part of shaft	9¼
Diameter transversely of narrowest part of shaft	2¼
Diameter antero-posteriorly of narrowest part of shaft	3
Breadth at the epicondyles	7½
Breadth of distal articulation	5¼

The mutilated upper extremity of the femur, represented in Fig. 4, Plate XXVI, was found by Dr. Corson, on the buttes west of Dry Creek Cañon, a dozen miles from the former specimens. It is suspected to pertain to Uintatherium, though it would appear to have belonged to a larger animal, and perhaps a different species, than the one to which the cranium and humerus are referred. The specimen has about the same size and form as the corresponding part in the elephant, but the great trochanter is destroyed. The length of the fragment is about 11 inches. The head is 5 inches in diameter, but its surface is too much mutilated in the specimen to determine whether it possessed a pit for the attachment of a round ligament, or whether it is absent as in recent proboscideans. The outer border of the shaft below the position of the great trochanter is 2¾ inches thick. From the appearance of the specimen, the femur in its entire condition has evidently resembled that of the elephant more than it does that of the perissodactyles.

The mutilated distal end of a femur, represented in Fig. 5, is also supposed to belong to Uintatherium, though it did not pertain to the same individual as the preceding specimen. It was found in the same locality, but at a distance from the former, and was derived from a different stratum, as it has an adherent friable sandstone matrix, while the other has an adherent indurated clay matrix. It is considerably smaller than the corresponding part of the femur of the elephant, and is very different in anatomical character. It is proportionately less thick. The shaft above the articulation, on the front and at the sides, presents a continuous transverse convexity, without any depression whatever above the position of the trochlea. The posterior surface in the same direction, between the position of the low epicondyles, is concave.

The loss of part of the outer condyle prevents a comparison of breadth with the inner one, but this is more prominent posteriorly than the former. The trochlea is shallow and but feebly prominent anteriorly in comparison with that in the elephant. Its articular surface is continuous with that of the inner condyle, and also that of the outer one, so far as it is preserved in the specimen, without the slightest definition. The intercondyloid notch commences at the bottom of the trochlea and gradually widens backward and upward with a curve outward.

The length of the fragment of the femur is $6\frac{1}{2}$ inches. The breadth between the epicondyles is about $5\frac{1}{2}$ inches; the thickness of the inner condyle and trochlea together is 5 inches, and the depth of the trochlea along its groove is $2\frac{1}{2}$ inches.

Several large tarsal bones, found together on the buttes to the west side of Dry Creek Cañon, may perhaps belong to Uintatherium. They consist of a calcaneum, astragalus, and cuboid bone of the left foot, and fit well enough together to have belonged to the same individual. In form and proportions, though somewhat peculiar, they more nearly resemble those of the mastodon and elephant than of other known animals.

The calcaneum, of which an upper view, half-size, is given in Fig. 6, Plate XXVI, is remarkable for its short robust character. The tuber calcis, in comparison with that in the ordinary proboscidians, is very short. The breadth of the tuber exceeds its length, and the depth exceeds the breadth. The thickened extremity narrows below and is continuous with the thick longitudinal plantar ridge. The upper part of the tuber inclines nearly straight backward from the articulation. Its outer surface forms part of an irregular plane with the fore part of the bone.

The sustentaculum is thick and three-fourths the length of the bone. The groove beneath for the flexor-tendons is well marked. The articular surface it supports for the astragalus, is larger than that on the body of the bone. The groove separating the articular surfaces for the astragalus nearly occupies the middle of the bone. Both surfaces are flat in front, but convex backward behind.

No articular surface exists for the fibula. At the fore part of the bone there is a small articular facet for the cuboid. The remaining portion of the front surface forms a deep and wide irregular plane.

The astragalus, of which upper and lower views are given half-size in Figs. 7, 8, Plate XXVI, resembles that of the ordinary proboscideans. The bone is irregularly square, with nearly equal horizontal diameters, and of less thickness than these.

The upper articular face has nearly the shape of that in a mastodon, but is rather more depressed posteriorly. The fibular extension holds about the same proportion to the tibial surface as in the animal named.

The calcanean articular surfaces are the reverse in their comparative size to what they are in the mastodon, the inner one being the larger. Both are also more concave fore and aft than in that animal.

The navicular articular surface is proportionately deeper in comparison with its width than in mastodon, and is well defined outwardly from the cuboid articular facet.

The cuboid is triangular in outline, with rounded angles, and with the thickness more than half the breadth or depth. Proximally it presents a double articular facet, of which the division for the astragalus is larger than that for the calcaneum. The former division is continuous with a narrow facet on the inner side for the navicular. Distally the bone also presents a double articular facet, the divisions forming an obtuse angle.

The measurements of the tarsal bones are as follows:

Calcaneum.

	Lines.
Length of calcaneum	42
Breadth at fore part	39
Depth at fore part externally	31
Length of tuber calcis from the outer articular facet above	20
Breadth transversely of the outer articular facet for the astragalus	14½

	Lines.
Breadth fore and aft of the outer articular facet for the astragalus	23
Breadth transversely of the inner articular facet for the astragalus	18
Breadth fore and aft of the inner articular facet for the astragalus.	24
Breadth transversely of the articular facet for the cuboid	14
Breadth vertically of the articular facet for the cuboid	10

Astragalus.

	Lines.
Greatest breadth fore and aft of the astragalus at inner side	50
Greatest breadth transversely of the astragalus	52
Greatest thickness of astragalus	32
Breadth of tibial articular surface at middle transversely	38
Breadth of tibial articular surface at middle fore and aft	32
Breadth of articular facet for scaphoid	40
Depth of articular facet for scaphoid	28

Cuboid.

	Lines.
Depth of the cuboid	25
Breadth of the cuboid inferiorly	25
Length of the cuboid at center	15

The canine tooth, originally described and referred to a carnivore with the name of *Uintamastix atrox*, is represented in Figs. 1 to 3, Plate XXV. The specimen is broken into two pieces, is mutilated at the point, and has lost apparently several inches of the base. In its perfect state the tooth approximated a foot in length, of which it now retains about three-fourths. It is saber-like in general form—long, laterally compressed cylindroid, and moderately curved. It appears more curved at the base, and from this position, also, has a somewhat outward deflexion, so that the tooth in its course curved forward and downward with an outward divergence. Laterally from the base it gradually tapers to the point; fore and aft it gradually narrows to near the lower third, when it becomes slightly expanded before tapering, so as to assume the shape of a lance-head. This likeness is rendered more striking internally by the surface being concavely impressed in front and behind the axis extending toward the trenchant borders of the lance-head extremity. Externally, it is impressed in like manner to a less extent posteriorly, but not anteriorly. Above the lance-head extremity of the tooth it is obtusely rounded in front and behind, and in this position is elliptical in transverse section, as represented by the outline, Fig. 5. A section near the middle of the lance-head extremity has the form represented in Fig. 4.

The tooth, so far as the specimen extends, appears to have been invested with thin enamel throughout. Externally, it reaches to the broken edge of the base, and, internally, appears to have been lost from the corresponding position by erosion. Externally, it is longitudinally rugose, and the rugosity appears to be greater toward the point, and, to some extent, is divergent toward the trenchant borders. Internally, the rugosity of the enamel is less marked, and toward the point it is worn off for several inches along the axis and near the borders from the attrition of an opposing lower tooth. The extent of attrition would apparently indicate large lower canines.

At the broken base of the specimen the borders of the exposed pulp cavity are nearly 4 lines thick. The fore and aft diameter of the tooth 2 inches below the broken base is a little under 2 inches; the thickness is 13 lines. The breadth of the tooth just before expanding in the lance-head extremity is 1½ inches. The widest part of the latter appears to have been a couple of lines greater.

The tusk above described, though apparently according in form with those of *Dinoceras mirabilis*, as described by Professor Marsh, exhibits different proportions, having less breadth and greater thickness. Thus Professor Marsh gives as the diameters of the tusks of *D. mirabilis* 64 millimeters breadth, and 25 millimeters thickness. The tusk above described has a breadth of 50 millimeters, and a thickness of 28 millimeters.

From the description of the skull of Dinoceras given by Professor Marsh, as before intimated, I have been led to view the large tusks above described, and originally referred to a carnivore with the name of Uintamastix, as really pertaining to Uintatherium, and perhaps to the same species as that indicated by the cranial specimen referred to *U. robustum*.

The molar tooth of Uintatherium, represented in Figs. 13, 14, found with the large tusk, has the same form and constitution as the upper molars first referred to the genus, except that it is considerably smaller, and has no tubercle behind the summit of the conjunction of the lobes of the crown. Proportionately, also, the basal ridge is much better developed at the inner part of the crown, where it is continuous with the stronger ridge in front and behind. The antero-posterior diameter of this tooth is 11½ lines, and its transverse diameter is estimated at 13½ lines.

The tooth I supposed to be an upper premolar of *U. robustum*; if, however, it is a true molar, its comparatively small size, and the absence of the characteristic tubercle on the posterior slope of the conjunction of the lobes of the crown, as existing in the species just named, would indicate that it probably belonged to a different one. Found in association with the canines referred to *Uintamastix atrox*, it may pertain to the same animal.

Order *Rodentia*.

Small quadrupeds with clawed toes. Teeth consisting of two long curved incisors in each jaw; no canines, and the molars separated from the former by a wide interval.

PARAMYS.

An interesting peculiar extinct genus of gnawers of the sciurine family is indicated by a number of specimens, consisting of fragments of lower jaws with teeth, which were discovered by Dr. Carter, in the summer of 1871, in the Tertiary formation in the vicinity of Fort Bridger.

As in the squirrels and marmots, the lower molars are four in number, and are inserted each by two fangs. They are nearly of the same size, but are proportionately narrower than in the animals just mentioned, as the fore and aft diameter exceeds the transverse, while in most sciurine animals the reverse condition usually exists.

The crowns are short, square, tuberculate, and enameled. The arrangements and proportionate size of the tubercles at the four corners of the crown, including a concave surface, are the same as in the squirrels.

The lower jaw is proportionately shorter and deeper than in most known rodents, the reduction in length being mainly due to a less development of that part of the bone in advance of the molars. To compensate for the difference in length and to make room to accommodate the incisors, these teeth reach farther back than usual. In squirrels and marmots their posterior extremity reaches a short distance behind and beneath the last molar. In Paramys it reached further backward, upward, and externally to a level with the crown of the last molar.

The jaw in advance of the molars is not only short compared with the usual condition in most known rodents, but the acute edge of the hiatus between the molars and incisors is almost on a level with the alveoli of the teeth, instead of forming a deep concave notch, so conspicuous a feature in the lower jaw of the gnawers generally.

In sciurine and most other rodents the ridge defining the masseteric fossa extends far forward on the side of the jaw to a position beneath the second or even the first molar tooth. In the rabbits the defining ridge is comparatively far back, extending only to the position of the interval of the last two molars. In Paramys it holds an intermediate position, extending as far forward as the position of the third molar, where it forms a conspicuous angular prominence, as in the marmots.

The mental foramen, much higher in relative position than usual in rodents, is situated in advance of the molars a short distance below the edge of the hiatus separating the latter from the incisor.

Paramys delicatus.

The largest species of Paramys was, perhaps, about a fourth less in size than the Maryland marmot, though its series of molar teeth is nearly equal in size, measuring three-fourths of an inch in length. It is represented by two specimens sent to me by Dr. Carter, consisting of portions of the right and left sides of the lower jaw, containing most of the molars and portions of the incisors. One of them is represented in Fig. 23, Plate VI, of the natural size. The triturating surfaces of the molars of both specimens, magnified three diameters, are represented in Figs. 24, 25.

In one of the specimens, Fig. 23, two mental foramina exist, one in the position, previously indicated, in advance of the molars, a short distance below the edge of the jaw; the other is situated lower down below the position of the first molar. In the other specimen the foramen exists in the latter position, and as the jaw is broken in advance, it cannot be determined whether a second existed, which is, however, probable, as it is the usual and normal position of one. A prominent tubercle is formed at the angle of convergence of the ridges which define the masseteric fossa.

Paramys delicatior.

A second species is indicated by a specimen consisting of the greater portion of the left ramus of a lower jaw, represented in Fig. 26, Plate VI. It retains the second molar tooth, the triturating surface of which, magnified three diameters, is represented in Fig. 27 of the same plate. The molar series has measured about $7\frac{1}{4}$ lines in length, and the animal was about the size of our common gray rabbit.

Since writing the above, I have received from Dr. Carter several additional specimens which I suspect belong to the same species. One of them, an intermediate lower molar, is represented in Fig. 16, Plate XXVII. It sufficiently resembles the tooth of Fig. 27, Plate VI, originally referred to *P. delicatior*, to pertain to the same species, though it is slightly larger.

The other specimen, apparently from the same individual, consists of a pair of upper molars represented in Figs. 17, 18, Plate XXVII, magnified three diameters. They have nearly the form and construction of those of the Sciurides.

The fore and aft diameter of the lower molar is 1.8 lines. The fore and aft diameter of the upper molars is 1.8 lines, and the transverse diameter is 2 lines.

Paramys delicatissimus.

A third and still smaller species of Paramys is indicated by a specimen consisting of the greater portion of the right ramus of a lower jaw containing all the molars, and a second specimen consisting of a small fragment of another lower jaw containing the second molar. The first specimen of the natural size is represented in Fig. 28, Plate VI. A view of the triturating surfaces of the molars, magnified three diameters, is given in Fig. 29. The molar series measures ½ an inch in length, and the animal was about the size of the common gray squirrel.

Comparative measurements are as follows:

	P. delicatus.	P. delicatior.	P. delicatissimus.
	Lines.	*Lines.*	*Lines.*
Length of lower molar series	9	7¼	6
Length of hiatus in advance of lower molar series	3¼	......	3
Depth of jaw below the second molar	6	5	4
Fore and aft diameter of incisor	2¼	2	1½
Transverse diameter of incisor	1½	1½	1½
Fore and aft diameter of second molar	2¼	1½	1½
Transverse diameter of second molar	1¼	1½	1¼

MYSOPS.

Mysops minimus.

A small rodent, intermediate in size to the common mouse and the brown rat, is indicated by a specimen discovered by Dr. Carter at Grizzly Buttes and sent to the author last summer. The specimen consists of the median portion of the right ramus of a lower jaw containing the last two molars, the fangs of the others, and part of the incisor. It is represented in Fig. 31, Plate VI, magnified two diameters.

The jaw in its form, proportions, and construction, and the number of teeth and their relative position, agree with the conditions in Paramys, but the form of the molars is sufficiently different to refer the specimen to a different genus, for which the above name has been proposed.

The molar teeth, as in Paramys, are four in number, inserted each by a pair of fangs. The crowns are quadrate and invested with enamel. The triturating surface, instead of being constructed like that of the squirrels, is more like that of the rats, as seen in Fig. 32, Plate VI, in which the last two molars of the specimen are represented magnified eight diameters. The crown of the third molar exhibits two transverse lobes, or ridges, joined by an intermediate narrow ridge, and the inner extremities of the lobes include a trilateral tubercle. The enamel being worn away from the prominences of the crown leave exposed a pair of transversely ellipsoidal dentinal surfaces joined by a narrow isthmus. Upon the summit of the internal tubercle a small islet of dentine also appears.

The last molar exhibits three transverse ridges or lobes, of which the anterior is the thickest, the middle one the thinnest, and the posterior the shortest. The anterior lobe is worn so as to exhibit a transversely elliptical surface of dentine bordered with enamel. The middle ridge of the crown appears sigmoid and is unworn. The posterior lobe presents an exposed islet of dentine on the inner half of its length.

The anterior molar of Mysops, like the last one, is more elongate fore and aft than the two succeeding molars, but it is proportionately of less size than in the rats, and has not three fangs as in these animals.

The length of the molar series is $\frac{1}{4}$ of an inch. The first and fourth molars are about $\frac{1}{2}$ of a line fore and aft; the intermediate ones about $\frac{1}{2}$. The incisor measures about $\frac{1}{2}$ of a line fore and aft by $\frac{2}{3}$ transversely. The depth of the lower jaw below the second molar is $2\frac{3}{4}$ lines. The length of the hiatus in advance of the molars is $1\frac{1}{4}$ lines.

MYSOPS FRATERNUS.

Since writing the foregoing I have received another specimen, which may belong to Mysops. It was found by a Shoshone Indian, and given to Dr. Carter. It consists of a portion of the right ramus of the lower jaw, represented in Fig. 14, Plate XXVII. It contains the last three molars, the triturating surfaces of which are represented in Fig. 15, magnified eight diameters.

The jaw is proportionately deep and short, compared with that of the rat. The masseteric fossa is deep, and defined by a rectangle, the apex of which reaches as far forward as the position of the third molar tooth. The border of the jaw at the hiatus in advance of the molars extends nearly on a level from their alveoli to that of the incisor.

The molar teeth, though having the same general constitution as the corresponding ones in the jaw-fragment of *Mysops minimus*, above described, appear sufficiently distinct to pertain to another species, and I have therefore distinguished it as such with the name of *M. fraternus*.

In the jaw-specimens of both species the molars are worn nearly to the same extent. In comparing the corresponding teeth, it will be seen that the third molar in *M. fraternus* has a greater breadth fore and aft, and the last molar is of more uniform width transversely. In both teeth the intermediate conical lobe, occupying the inner part of the crown, is proportionately more robust in *M. fraternus*.

The depth of the jaw below the third molar is 2.6 lines; the breadth of each of the three back molars fore and aft is about eight-tenths of a line; the space occupied by the four molars is a little over 3 lines.

SCIURAVUS.

In the American Journal of Science for July, 1871, Professor Marsh has described an extinct genus of rodents from remains found at Grizzly Buttes, under the above name, and refers them to two species with the names of *Sciuravus nitidus* and *S. undans*. The former, described from an upper-jaw fragment with three molars, was about the size of the brown rat. The latter, indicated by a lower-jaw fragment with the incisors and the anterior three molars, was a somewhat larger animal.

While we have not the means of determining whether Paramys is absolutely distinct from Sciuravus, we have the opportunity of examining a specimen belonging to a different genus from the former, and which we suspect pertains to the latter. The specimen in question consists of a fragment of the left side of the lower jaw, containing the third molar, the alveolus behind, and part of that in front. It belonged to an animal but little larger than the rat. The fossil was found at Grizzly Buttes by Dr. Carter. The only remaining tooth it contains is represented in Fig. 30, Plate VI, magnified eight diameters.

The tooth is about a line in breadth, and, together with the alveolus back of it, occupies a space of 2¼ lines. The crown of the tooth is quadrate, broader than wide, and is composed of four principal conical lobes, as in the squirrels, and as in its associate Paramys. The sculpture and connection of the lobes is different, as may be conveniently observed by comparing Fig. 30 with Fig. 27, representing a tooth of the same side of Paramys. It is especially to be noticed that in the latter the back pair of lobes include, between them and the anterior lobes, a broad hollow, and the former are connected behind by an acute ridge, which forms the posterior border of the crown. The broad hollow of the latter is closed externally by a festoon-like ridge connecting the outer lobes at their base.

In the supposed tooth of Sciuravus (Fig. 30) the broad hollow of the crown so conspicuous in Paramys and Sciurus is not evident. The posterior lobes are conjoined by a transverse ridge, and are bounded behind by a thick ridge descending inwardly from the postero-external lobe. The transverse valley of the crown is occupied by a pair of ridges diverging from the postero-external lobe to those in advance.

Order *Carnivora*.

PATRIOFELIS.

PATRIOFELIS ULTA.

A carnivorous animal, rather larger than our common American panther, and about the size of the jaguar, to which the above name has been given, is indicated by remains in the Bridger Tertiary formation. The specimens from which it was originally described in the Proceedings of the Academy of Natural Sciences for March, 1870, were obtained near Fort Bridger, Wyoming, during Professor Hayden's exploration of 1869. They consist of portions of both rami of a lower jaw, unfortunately with most of the teeth lost or mutilated. The right ramus is represented, one-half the natural size, in Fig. 10, Plate II.

The jaw of Patriofelis contains a series of five molar teeth immediately succeeding the canine tooth without conspicuous interval, as in some of the viverrine and musteline animals. The molar teeth are all inserted by a pair of fangs, and none of them appear to be of the purely tubercular kind. The first of the series is smallest, and the third the largest; the fourth was intermediate in size to the latter and the last one, which little exceeded the second.

The crown of the last molar in the specimen appears as if it had been composed of an anterior pointed, or perhaps trenchant, lobe, and a large posterior heel.

The crown of the penultimate molar appears to have been nearly of the same character. In the crown of the antepenultimate molar the posterior heel forms a median acute ridge from which the sides slope toward the bottom. The outer slope, nearly twice the depth of the inner, is bounded behind by a ridge descending from the summit of the heel. The inner slope is bordered by a basal ridge curving downward and forward from the summit.

The canines, as indicated by portions of the alveoli, are large and powerful teeth, as in feline animals. The alveoli are about $\frac{1}{2}$ inch in diameter.

The jaw has nearly the same form as in the panther, but is proportionately shorter, and beneath the molar teeth of greater depth, in this respect resembling more the condition in the striped hyena. The condyle has the same form and relative position as in ordinary carnivora, but is thicker or of greater extent on its articular surface fore and aft than in the panther. Its comparative breadth is undeterminate, from its being broken at both ends in the specimen.

The back portion of the jaw is proportionately narrower than in the panther; and the coronoid process, which appears to have had the same form as in this, is likewise narrower. The masseteric fossa is not so deep as in ordinary carnivora. Extending from the coronoid downward, a little below the level of the condyle, it becomes, rather abruptly shallower, and from this position gradually lessens in depth toward the base, from which it is not abruptly defined by a narrow ridge, as in the ordinary carnivora.

The symphysis is strong, and the rami approaching it thick, as in the panther. A group of seven mental foramina occupy a position at the side of the symphysis. The largest of them, as in the panther, is situated outside the back part of the canine alveolus.

From the absence of the characteristic portions of the teeth, the exact relationship of Patriofelis is not clear. It is perhaps intermediate to the feline and canine animals.

Measurements from the lower jaw of *Patriofelis ulta* are as follows.

	Inches.	Lines.
Estimated length of jaw	6	0
Distance from back of condyle to canine alveolus	5	4
Distance from back of condyle to back of last molar	2	3
Space occupied by the molar series	3	0
Breadth of coronoid at base	1	7
Depth of jaw below penultimate molar	1	4
Depth of jaw below back of last molar	1	6

Measurements of the molar teeth, estimated from their fangs and alveoli, are as follows:

	Lines.
Breadth of first molar tooth	5
Breadth of second molar tooth	7
Breadth of third molar tooth	8½
Breadth of fourth molar tooth	7
Breadth of fifth molar tooth	8

Fig. 20, Plate VII, represents a tooth discovered by Dr. Carter near Fort Bridger. It appears not to belong to the lower jaw of Patriofelis, but perhaps belongs to the upper jaw. The crown is composed of a large conical lobe with a broad heel, the sides of which slope from a median ridge. The breadth of the crown is 8½ lines; its thickness 5 lines.

SINOPA.

SINOPA RAPAX.

A lower-jaw fragment, containing two teeth and portions of two others, represented in Fig. 44, Plate VI, appears to indicate an extinct genus related to the canine family. The specimen was discovered by Dr. Carter in the vicinity of Fort Bridger, and was by him presented to the writer. It belonged to an animal about the size of the gray fox.

The specimen is insufficient to ascertain with any certainty the exact relationship of the animal to which it belonged, but the character of the teeth leads me to view it as having held an intermediate position to the existing genus Canis and the extinct one Hyænodon.

The teeth preserved entire in the specimen appear to correspond with the last premolar and the first or sectorial molar of the fox, and the remains of two teeth behind would be of the second and third molars. The last premolar is larger than the molars. Its crown is as wide, but is longer than that of the tooth retained behind it. The form of the crown is more like that in

Hyænodon than in the fox. It is proportionately longer and narrower than in the latter, and the accessory cusp at the back border of the principal one in the fox is nearly obsolete in the fossil. The heel of the crown is better developed than in the fox. It forms a median acute ridge, and slopes off on each side to the rounded base of the crown.

The first molar, as before intimated, is smaller than the last premolar. It is as wide as the second molar, but not so thick, and is slightly wider than the last molar. It is proportionately better developed in its relation with the succeeding molars than in Hyænodon. Its crown is intermediate in form and in the development of its parts to that in the fox on the one hand and the raccoon and badger on the other. The fore part of the crown, consisting of rather more than one-half, corresponds with the sectorial portion of the same tooth in the fox, but accords more in shape and the relative position and development of its points with the homologous portion in the raccoon and badger. The heel of the crown is bordered by a horseshoe-shaped ridge inclosing a cup-like concavity.

The heel of the second molar, the only portion of the crown retained in the specimen, is stouter than in the first molar, but has the same shape. The width of the crown is about equal to that of the tooth in advance, but has been slightly thicker.

The last molar is a two-fanged tooth like those in advance, but is not quite so wide, and a small portion of the back of the crown indicates it to have been of less thickness.

The base of the jaw-fragment is broken away in the specimen. The portion preserved presents nothing peculiar.

Measurements of the fossil are as follows:

	Lines.
Space occupied by the last premolar and molars	15
Space occupied by the molars	11
Breadth of crown of last premolar	4
Length of crown of last premolar at middle	$3\frac{1}{4}$
Breadth of crown of first molar	4
Length of crown of first molar at principal cusp	$2\frac{3}{4}$
Thickness of crown of first molar at heel	2
Breadth of crown of second molar	4
Thickness of crown of second molar at heel	$2\frac{1}{4}$
Breadth of crown of last molar	$3\frac{1}{2}$

The name *Sinopa*, applied to the extinct genus, according to Professor Hayden, is aboriginal, and is applied by the Blackfeet Indians to a small fox

While the original notice of *Sinopa rapax* was in print, in the Proceedings of the Academy of Natural Sciences of Philadelphia, Professor Marsh published a description, in the American Journal of Science for 1871, of some remains of a carnivore from the vicinity of Fort Bridger, under the name of *Vulpavus palustris*. It is characterized from several upper molars which accord in size sufficiently to pertain to the same animal as that above described. Further researches may prove the two animals to be the same.

SINOPA EXIMIA.

A jaw-fragment, discovered by Dr. Carter at Grizzly Buttes, and represented in Fig. 45, Plate VI, belongs to a smaller carnivore than the preceding. It was probably allied to the former, and may perhaps pertain to a smaller species of the same genus, of which I have some doubt, though, in the absence of more confirmatory evidence, I have considered it as such.

The specimen contains two teeth, which sufficiently resemble those retained in the jaw-fragment referred to *Sinopa rapax*, as to render it probable they are the corresponding ones, though the contiguity of the symphysis leads me to suspect that they may be the last two premolars. As seen in the figure, the back of the symphysis is just below the position in advance of the first tooth of the specimen. The teeth in shape are nearly like those in *Sinopa rapax*, but the proportions are reversed. The crowns of the two teeth have the same length, but the hinder one is wider and thicker.

The measurements of the specimen are as follows:

	Lines.
Depth of the jaw below the teeth	4¾
Space occupied by the two teeth	4½
Width of the crown of the first tooth	2
Length of the crown of the first tooth	2¼
Width of the crown of the second tooth	2½
Length of the crown of the second tooth	2¼

UINTACYON.

UINTACYON EDAX.

An interesting fossil, recently received from Dr. Carter and discovered by him in the Bridger beds, consists of the greater portion of the right ramus of a lower jaw, represented in Fig. 6, Plate XXVII. The specimen indicates a carnivorous animal, probably marsupial, and of a hitherto unknown genus, for which the above name has been proposed.

The jaw contained a series of eight molar teeth and a canine separated from the former by a small hiatus. Of the molar teeth, the specimen retains part of the first molar and the succeeding molar, represented in Figs. 7, 8, and the intermediate three premolars represented in Figs. 9, 10.

The jaw-fragment agrees in its form and proportions with the corresponding part in the existing fox. The teeth also, so far as they are preserved, are nearly like those of the latter animal.

The canine tooth was equally well developed as in the latter, and the first premolar is inserted by a single fang. The second premolar likewise resembles the corresponding tooth of the fox.

The third premolar is peculiar, and perhaps anomalous. It resembles more the form of an upper premolar than the usual form of lower premolars. It has three fangs, of which two are inserted in a line with those of the contiguous teeth, while the third fang is external. The crown is a four-sided pyramid with projecting basal angles, of which the postero-internal one is the most prominent.

The fourth premolar is like the last one of the fox, but is proportionately thicker. The fifth premolar is lost, and, like the preceding tooth, was inserted by a pair of fangs.

The first molar has lost the fore part of its crown, and this appears not to have been proportionately so well developed as in the fox. The crown of the second molar is nearly of the same form as in the latter.

The last molar is a small tooth, as in the fox, and is also inserted by a single fang.

The measurements of the specimen are as follows:

	Lines.
Depth of lower jaw at second premolar	4.8
Depth of lower jaw at first molar	4.8
Distance from fore part of canine to back of last molar	18.0
Length of the molar series	14.8
Length of the premolar series	8.8
Length of the true molar series	6.0
Breadth of second premolar	1.7
Breadth of third premolar	1.5
Breadth of the fourth premolar	1.6
Breadth of the fifth premolar	2.4
Breadth of the first molar	3.2
Breadth of second molar	2.0
Breadth of third molar	1.0

The main peculiarity of the fossil is the presence of an eighth tooth to the molar series. The one in excess of the usual number, without other consideration than convenience, I have viewed as a premolar. From its anomalous, or at least unusual, form, the fourth of the series of the premolars may be regarded as the additional tooth. Without it, the jaw would indicate a small canine animal, or at least a species of a closely allied genus. The animal was about half the size of the common fox.

UINTACYON VORAX.

Perhaps a larger species of the genus just named is indicated by the jaw-fragment represented in Fig. 11, Plate XXVII. The specimen was obtained on Henry's Fork of Green River, during Professor Hayden's expedition of 1870.

The jaw-fragment agrees in form with the corresponding part of the jaw-specimen of *U. edax*, but from its proportions belonged to an animal twice the size. It contains the penultimate molar, the heel of the one in advance, and the alveolus of the last molar. The teeth agree in their proportions with those of *U. edax*, and the penultimate molar, represented in Figs. 12, 13, sufficiently resembles that of the latter to belong to the same genus. The breadth of the penultimate molar is 2¾ lines.

Order *Insectivora*.

OMOMYS.

OMOMYS CARTERI.

The first mammalian fossil described from the Bridger Tertiary beds consists of the fragment of a lower jaw with teeth, discovered by Dr. Carter on Twin Butte, about one mile from Fort Bridger. The specimen is represented in Figs. 13, 14, Plate XXIX, of "The Extinct Mammalian Fauna of Dakota and Nebraska," published as the seventh volume of the Journal of the Academy of Natural Sciences of Philadelphia for 1869, and is described on page 408 of that work.

The jaw-specimen was accompanied with fragments of the cranium, for the most part too much broken to determine anything from them. They would appear to indicate a skull about the size of that of the common weasel, but with weaker jaws.

A fragment of the cranium retains a straight linear sagittal crest about 14 lines in length to its bifurcation at the forehead. The temporal surfaces appear to be full and convex, as in the weasel.

An occipital condyle resembles those of the latter, and measures about 4 lines in its longer diameter.

The ramus of the lower jaw, compared with that of the weasel, is more slender and delicate in its proportions. In the specimen both extremities are broken, but a portion of the symphysis is still retained.

The jaw below the molars is of nearly uniform depth, and measures about 2 lines. The base is slightly convex fore and aft, but makes a concave turn toward the angle. The masseteric fossa below is well marked. A small mental foramen occupies a position below the antepenultimate premolar.

In the earlier description of the specimen, I remarked that seven molar teeth, in an unbroken series, appear to have occupied the side of the jaw. In the actual condition of the fossil there are four teeth, consisting of the anterior two molars and the two premolars in advance. In front of these there are two empty sockets and parts of two others, and behind them there are the imperfect alveoli for a third molar. The sockets at the front of the jaw I at first supposed were intended for two additional two-fanged premolars. They fill up the interval between the retained teeth and the edge of the symphysis so closely that, from this fact and their relative size, I now suspect that they may have been occupied by a single-fanged premolar, a small canine, and two incisors. Assuming that such was the case, without any certainty in the matter, the number of molar teeth in the series would be six, of which three were premolars and three true molars. In this view the teeth retained in the specimen consist of the second and third premolars and the first and second molars. Their constitution would appear to indicate an insectivorous animal which, perhaps, was marsupial in character.

The teeth successively decrease in prominence or height from the second premolar to the second molar. They resemble in constitution the corresponding teeth of the opossum.

The crown of the premolars is laterally compressed conical, thicker behind than in front, and is embraced by a basal ridge. The crown of the second premolar, more prominent than in any of the other teeth, is triangular, longer than broad, and sharp-pointed. Its anterior slope is slightly convex and acute; its posterior slope is longer, slightly concave, and wide. The basal ridge forms an excavated heel behind, a more elevated ledge in front, and a pair of festoons both internally and externally. The inner side of the crown is defined from the back border by an acute ridge.

The crown of the last premolar has the same construction as that in advance, but is shorter and wider. The heel is slightly wider and more excavated, but the fore part of the basal ridge is not so prominent. The ridge defining the inner side from the posterior border is slightly more advanced and prominent, and the surfaces it separates are more concave.

The crowns of the true molars are nearly alike in form and size, though the first is in a trifling degree more prominent and wider. They have the same general constitution as those of shrews, of the hedgehog, the galeopithecus, and the opossum. Each is composed of two divisions, of which the posterior is the larger. The anterior division consists of a small, outer demiconoidal lobe, with a V-like summit joining by its arms a pair of inner and smaller pyramidal lobes. The posterior division consists of an outer lobe like that in advance, but larger, and joining it by one of the arms of its V-like summit, while the other arm joins a small pyramidal lobe at the inner corner of the crown. The outer part of the base of the crown is embraced by a basal cingulum nearly half its depth.

The space occupied by the teeth, in the view that there were two incisors, a canine, and six molars, is $7\frac{1}{4}$ lines. The last two premolars and the succeeding two molars occupy a space of 4.6 lines.

PALÆACODON.

PALÆACODON VERUS.

Two small fossil specimens, discovered the previous summer by Dr. Carter at Lodge-Pole Trail, Wyoming, indicate an insectivorous animal, or, perhaps, a marsupial allied to the opossum. One of the specimens consists of an upper-jaw fragment containing a molar, which appears to be the penultimate one of the series; the other is an isolated tooth, perhaps the last upper premolar or first molar.

The jaw-fragment is the portion which forms the anterior abutment of the zygoma. In advance of the tooth it retains are the remains of the alveoli of two others, and behind it the remains of another.

The molar of the jaw-fragment is represented in Fig. 46, Plate VI, magnified four diameters. The crown bears some resemblance with that of the molars of the opossum, but is less narrowed internally, and is therefore more quadrate or less triangular in form. The constitution is similar, but the outer lobes are proportionately better developed and the median ones are much reduced in size. A basal ridge nearly embraces the crown, but is nearly obsolete internally, and is best developed posteriorly, where it forms a wide festoon.

The isolated tooth is a diminished representative of the one in the jaw-fragment, and probably held the position of the third in advance of it. It may, perhaps, represent a smaller species. The specimens indicate an animal but little more than half the size of the opossum. How it is related with Omomys the paucity of material prevents a positive determination. The size of the teeth indicates a larger animal than *Omomys Carteri*.

In the American Journal of Science for 1871, Professor Marsh has described a tooth, from Grizzly Buttes, which he likens to the premolars of some insectivora, and refers it to a species with the name of *Triacodon fallax*.

He remarks that the species was probably about two-thirds of the size of the opossum, which dimensions would be too great for the animal we have named *Palæacodon verus*.

The sizes of the teeth referred to the latter are as follows:

	Lines.
Space occupied by the penultimate and antepenultimate molars	4
Breadth of penultimate molar	2
Width of penultimate molar	2¼
Breadth of last premolar	1½
Width of last premolar	1¾

WASHAKIUS.

WASHAKIUS INSIGNIS.

A jaw-fragment of a small animal recently sent to me by Dr. Carter is represented in Fig. 3, Plate XXVII, magnified three diameters. The specimen was found in the Bridger beds by a Shoshone Indian and given to Dr. Carter. It is quite different in appearance from any similar fossil from the same formation submitted to my inspection, and appears to indicate a different genus from those described in the preceding pages. I am uncertain as to its ordinal affinities, but suspect it to have pertained to an insectivorous animal, perhaps one of the many which have been indicated by Professor Marsh from fossils of the Bridger beds.

The jaw-fragment contains the last two molars, the triturating surfaces of which, considerably worn, are represented in Fig. 4, Plate XXVII, magnified eight diameters.

The portion of jaw is of moderate depth and stout in proportion. The base is thick and rounded. The masseteric depression is well marked, and is defined at its lower part in front by a strong ridge descending from the fore part of the coronoid process and ending in a conspicuous angular tubercle.

The teeth resemble most nearly those of Microsyops. They are inserted with a pair of fangs; but in the last molar the posterior fang is a connate pair extended backward.

The crown of the antepenultimate molar is quadrate with rounded corners, and is composed of four lobes. The postero-external lobe is largest, and is crescentoid conical. The postero-internal lobe is smallest and conical, and is joined at the summit by the back arm of the postero-external lobe. The anterior pair of lobes are connate, and are joined about their middle by the fore arm of the postero-external lobe. A deep angular valley occupies the inner part of the crown between the anterior and postero-internal lobes, and bounded externally by the postero-external lobe. A basal ridge incloses the outer part of the crown, but is interrupted in the most prominent part of the postero-external lobe.

The crown of the last molar, at its anterior two divisions, is composed on the same plan as that of the molar in advance, but it is prolonged backward so as to form an additional lobe.

The measurements of the specimen are as follows:

	Lines.
Depth of lower jaw below the last molar	2.1
Space occupied by the last two molars	2.4
Breadth of second molar	1.2
Breadth of last molar	1.4

The genus I have named in commemoration of Washakie, chief of the Shoshone Indians, with whom I had the pleasure of meeting during my visit to Fort Bridger. He has always been distinguished for his high character, and for his friendliness to the white race.

ELOTHERIUM.

In the American Journal of Science of 1871 Professor Marsh has described a molar tooth, from Henry's Fork of Green River, which he attributes to a suilline pachyderm with the name of *Elotherium lentus*. The specimen, he observes, indicates a species about half the size of *E. Mortoni*, the remains of which are found in the Miocene Tertiary deposit of the Mauvaises Terres of White River, Dakota.

Among the collections of fossils from the Bridger beds I have seen no remains which could be ascertained to belong to this genus. Figs. 28 and 29, Plate VII, represent two views of an incisor tooth which looks as if it might pertain to *E. Mortoni*. The specimen was found by Mr. Pierce, of Denver, twenty miles southeast of Cheyenne City, Wyoming.

REPTILIA.

The Bridger Tertiary formation, in comparison with the earlier Tertiaries of White River, Dakota, and of the Niobrara River, Nebraska, is remarkable for the variety as well as the number of its reptilian remains. Amid the multitude of fossils which have been collected in the latter localities nearly all belong to mammals; and though the remains of turtles are abundant, they appear all to be referable at most to a single species for each locality. No fragment of a crocodilian, lacertian, or serpent has yet been discovered either in the Mauvaises Terres of White River, Dakota, nor in the sands of the Niobrara River, Nebraska. From the Bridger beds there have been collected many remains of different species of crocodiles, turtles, lacertians, and serpents.

Order *Crocodilia*.

Body lizard-like in form, with four short limbs and feet, and a long, powerful tail. With long jaws, provided with a single row of teeth inserted in distinct sockets. Skin protected by bony plates.

CROCODILUS.

The Bridger Tertiary formation contains numerous remains of crocodiles. Many collected by Professor Hayden's party in 1870, and others obtained by Drs. Corson and Carter during the same and the succeeding year, have been submitted to the inspection of the writer. The specimens were found in various localities in the vicinity of Fort Bridger, as Little Sandy River, Big Sandy River, Green River, Black's Fork of the same, Church Buttes, Grizzly Buttes, &c. The specimens examined indicate several species, though from their generally detached and imperfect condition we have not been able to collocate them so as distinctly and clearly to establish the species. Some of the specimens we have referred to two named species. Professor Marsh subsequently named five species from remains obtained in the same localities during his exploration of 1870. Professor Cope has more recently named four additional species. It is probable that when the fossils are more carefully studied, the number of species to which they have been referred will be reduced.

CROCODILUS APTUS.

This species was originally named in 1869 from a fossil preserved in the Geological Cabinet of the General Land-Office in Washington. The specimen was obtained by Colonel John H. Knight, United States Army, near South Bitter Creek, Wyoming. Though consisting of a detached vertebra, it especially attracted my attention from having previously seen no remains of crocodiles in the large collections of fossils from the Tertiary formations of the west.

The vertebra represented in Fig. 2, Plate VIII, belongs to the cervical series, and resembles, both in size and form, the sixth or seventh of the Mississippi alligator. The bone appears to have been of mature age, and seems thoroughly petrified. It has lost the greater part of its neural arch and dependent processes, but is otherwise well preserved. From portions of adherent matrix, it has been imbedded in a soft rock similar to that adherent to some of the bones from other localities above mentioned.

The body of the bone in its axis is 16 lines long; its height and breadth in front are 14 lines. The hypopophysis, directed obliquely downward and forward, as in the alligator, is about 5 lines long. Back of the process the body is less prominently carinated than in the latter animal.

CROCODILUS ELLIOTTI.

The species thus named was originally designated from a specimen obtained, during Professor Hayden's exploration of 1870, at the junction of the Big Sandy and Green Rivers. It consists of an upper-jaw fragment containing two teeth and portions of two others, and is represented in Fig. 4, Plate VIII. It appears to be the anterior portion of the left maxillary, containing the fourth and fifth maxillary teeth and the fangs of the two succeeding ones. The shape of the jaw-fragment is nearly like that of the corresponding portion of the upper jaw in the mugger (*Crocodilus palustris*) of India, but is more rugose on its exterior surface, and the palatine surface is more vaulted. The teeth retained in the specimen have their crowns only partially protruded. They are proportionately more robust, or shorter and less pointed, than in the mugger. Strong ridges define the inner from the outer surfaces of the crown, which exhibits no indication of fluting, but the enamel is finely and closely wrinkled longitudinally.

The space occupied by the teeth, from the fourth to the seventh inclusive, is 35 lines. The entire length of the fifth or largest maxillary tooth is estimated at about $2\frac{1}{2}$ inches. The protruded portion measures externally $\frac{3}{4}$ of an inch in length, and its diameter at base fore and aft is $7\frac{1}{2}$ lines, and transversely $6\frac{1}{4}$ lines.

Fig. 6, Plate VIII, represents a large portion of the upper part of a skull, which has been attributed, but with no certainty, to the same species as the foregoing. The specimen, in a number of scattered fragments without teeth, was discovered, by Henry W. Elliott, on Little Sandy River, during Professor Hayden's exploration of 1870.

The fossil indicates a form of skull very different from that of our alligator, and is that of a true crocodile. It approached in form more that of the mugger of India or of the Nile crocodile than that of the American crocodile, ($C.\ americanus.$)

The cranium above is remarkably flat; from its lateral borders defined by the squamosals and post-frontals, and from the occipital border to the face in advance of the orbits, it forms a nearly uniform plane with no depression of the forehead nor eversion of the orbital margins. This uniform flatness is also extended along the middle of the face to the muzzle. This and the alveolar borders of the face are about as convex as in the mugger.

The sides of the muzzle are deeply notched at the conjunction of the premaxillaries and maxillaries, and the bottom of the notch exhibits a conspicuous recess for the accommodation of the large canine-like tooth of the mandible. A second and less conspicuous notch, as usual in the true crocodiles, occupies a position about the middle of the maxillaries.

The lateral borders of the cranium are less angular or more rounded approaching the orbits than in the mugger and the American crocodile. The superior temporal orifices are subrotund and nearly as wide transversely as fore and aft. The intervening parietal surface is broad and deeply pitted. The temporal surfaces of the parietal form a pair of deeply concave recesses.

The anterior orbital border, as constituted by the prefrontals and lachrymals, is depressed or slopes inwardly toward the orbits.

The nasal process of the frontal is much prolonged, extending 2 inches in advance of the position of the ant-orbital margins. The prefrontals are proportionately long and narrow compared with those in the mugger. Their length is about 4 inches; their breadth, where widest, is 14 lines.

The nasals are broad and flat at the back part. They are proportionately of greater breadth than in the mugger. Their estimated length is $9\frac{1}{2}$ inches; their breadth together in advance of the lachrymals is about $2\frac{1}{4}$ inches.

The fore part of the face, or the muzzle, has the same form as in the mugger and other true crocodiles, but is proportionately less thick than in the one specifically mentioned. The nasal orifice holds a more advanced position than usual, so that the alveolar border in front is barely more than half the extent it is in the mugger, nor is it perforated as in the latter and other true crocodiles. The upper surface of the skull is everywhere exceedingly rugose, with reticular ridges inclosing deep pits, and in some positions is deeply scored by vascular grooves.

Four teeth occupied the sides of the premaxillaries, forming an unbroken row. The intermediate pair are the larger and of nearly equal size; the others are also nearly of equal size. The first tooth did not occupy the fore part of the premaxillary as usual, in the true crocodiles, but is over an inch from the position of the symphysis, close to the second tooth. A large recess occupies the fore part of the palatine surface of the premaxillary, for the accommodation of the first mandibular tooth, as usual in the crocodiles, but it is closed or does not communicate by a perforation with the upper surface of the premaxillary border. The recess holds a position internal to the first premaxillary tooth. Smaller conical recesses occupy the intervals internally of the succeeding three teeth.

The maxillary appears to have accommodated fourteen or fifteen teeth, of which the fifth one was the largest, as in other crocodiles. The fourth, in comparison with the fifth one, was proportionately larger than in the mugger, and the sixth was not much less in size.

The depth of the socket of the fifth maxillary tooth is full 2 inches; its fore and aft diameter about $\frac{3}{4}$ inch. The depth of the fourth socket is 20 lines; its diameter 8 lines.

The premaxillary teeth, in comparison with those of the mugger, appear to

have been proportionately about as large. The anterior series of maxillary teeth were rather larger, and the posterior series smaller.

Detached portions of both quadrates accompany the other portions of the skull. They are somewhat peculiar in several anatomical points. The anterior surface is unequally divided by a conspicuous ridge, descending to within an inch of the articular surface for the mandible. The grooved or trochlear condition of the latter surface is much more decided than in the mugger or the American crocodile.

Measurements taken from the specimen above described are as follows:

	Inches.
Length from occipital border to end of muzzle	20
Breadth of cranium at occipital border between prominent angles of squamosals	7
Breadth of cranium at postorbital angles	$5\frac{1}{4}$
Breadth of cranium between temporal orifices	1
Breadth of forehead between orbits	$1\frac{1}{2}$
Breadth of temporal orifices	$1\frac{1}{2}$
Fore and aft diameter of the same	$1\frac{3}{8}$
Length of parietal	$2\frac{1}{4}$
Length of frontal	$5\frac{1}{2}$
Breadth of frontal where it joins the post-frontals	$2\frac{3}{4}$
Fore and aft diameter of the orbits	$2\frac{3}{4}$
Length of face in advance of the orbits	$13\frac{1}{2}$
Breadth of face outside the fifth maxillary teeth	$6\frac{3}{4}$
Breadth of muzzle as formed by premaxillaries	5
Breadth of muzzle at notch back of the latter	4
Length of premaxillaries	6
Breadth of nasal orifice	$2\frac{1}{2}$
Fore and aft diameter of the same	$2\frac{1}{4}$
Thickness of premaxillaries in advance of the same	$\frac{3}{4}$
Estimated length of entire alveolar border	$14\frac{1}{2}$
Space occupied by the anterior five maxillary teeth	$3\frac{3}{4}$
Space occupied by the posterior five maxillary teeth	3
Breadth of articular surface of quadrate for the mandible	$2\frac{1}{4}$

A detached basi-occipital, obtained near Little Sandy River, may, perhaps, belong to the same species as the preceding. The occipital condyle has nearly as great a vertical as a transverse diameter, the former measuring 15 lines, the latter 17 lines.

The last summer Dr. Joseph K. Corson sent, as a gift to the museum of the Academy of Natural Sciences of Philadelphia, a specimen consisting nearly of the whole of the lower jaw of a large crocodile. He discovered the fossil imbedded in a green sandstone in the vicinity of Fort Bridger. In removing it from its matrix it was much broken, and most of the teeth were destroyed.

The left ramus in a restored condition is represented in Fig. 8, Plate VIII. one-half the natural size.

The lower jaw belonged to a larger animal than the cranial specimen from Little Sandy River, and probably pertained to a different species. The form of the jaw is much like that of the mugger, but is of more robust proportions. The rami, in their dentary portions, are much thicker in proportion to their depth, and the symphysis is of greater extent, in this respect presenting a greater resemblance to the condition in the American crocodile.

The dentary portions of the rami the greater part of their length are as thick and thicker than the depth. Half way between the symphysis and the median enlargement of the dentary portion of the ramus the thickness is over 2 inches, while the depth is $\frac{1}{4}$ of an inch less. In the position of the enlargement just mentioned, the thickness is 2 inches and 2 lines, while the depth is only 2 lines more. The symphysis has measured about $4\frac{1}{4}$ inches fore and aft, and but slightly more than this transversely opposite the position of the large canine-like teeth.

The splenial bone, as if to give greater strength to the ponderous jaw, extends close up to the symphysis. The outer portion of the jaw in the position occupied by the teeth, is more rounded than in the mugger. The back portion of the jaw in form and constitution appears to agree with that in the mugger. The outer surface of the jaw, strongly foveated back of the large oval foramen, presents the usual vascular grooved and perforated appearance in advance.

About eighteen teeth occupied each ramus of the jaw, but all are broken from the specimen except one. Some of the broken and detached teeth accompany the jaw. They appear to have been comparatively robust, short, and blunt, conical in form, and but feebly curved. The enameled crown is rugose and longitudinally grooved, but not properly fluted; the narrow grooves separating wider convex and rugose longitudinal ridges. They sufficiently differ from those in the jaw-specimen referred to *Crocodilus Elliotti* to pertain to a different species.

The end of the symphysis of the jaw or of the chin is broken away, so that nothing can be ascertained in regard to the first pair of teeth of the two rami. A large tooth, canine-like in its relative position and size, as usual in the crocodiles, was number four in the series. The socket, occupied by green-sand matrix, is about 10 lines in diameter. The expansion of the symphysis in the position of this socket indicates its canine-like teeth to have been accommodated when at rest in a recess of the upper jaw at the junction of the premaxillaries and maxillaries.

Succeeding the canine tooth alveolus, there are the remains and sockets of five comparatively small teeth. Then followed several of the largest teeth accommodated by the second expansion of the jaw. The socket for the eleventh tooth is about the size of that of the canine tooth. In the left ramus it retains the tooth, the apex of which alone had protruded. After this tooth there followed a series of five others which successively decreased in size.

Measurements of the lower jaw are as follows:

	Inches.	Lines.
Length of rami of lower jaw	20	6
Width of lower jaw outside the glenoid articulations	12	0
Width of lower jaw a short distance in front of the glenoid articulations	13	0
Greatest width of symphysis	4	3
Width of jaw at second enlargement, below the eleventh tooth	6	0
Depth of jaw at oval foramen	3	10
Depth of last tooth	2	8
Depth of eleventh tooth	2	1
Thickness below the eleventh tooth	2	2
Depth of ramus near symphysis	1	9
Thickness of ramus near symphysis	2	2
Extent of symphysis fore and aft	4	3
Breadth of glenoid articulation	2	7
Length of hook-like process back of glenoid articulation	2	5
Space occupied by the teeth	11	0
Length of oval foramen	2	8
Width of oval foramen	0	11

Fig. 1, Plate VIII, represents the body apparently of a first lumbar vertebra; and Fig. 5 of the same plate, the proximal extremity of a left femur, large enough to belong to the same animal as the cranium above described. The two specimens were found together by Professor Hayden's party near Little Sandy River. They present no decided peculiarity distinguishing them from the corresponding part in the living crocodiles. The shaft of the femur contains a medullary cavity larger than usual in the latter, and in the specimen it is filled with chalcedony.

The measurements of the specimens are as follows:

	Lines.
Length of body of first lumbar vertebra beneath	20
Depth of body anteriorly	18
Width of body anteriorly	18
Width of head of the femur	26
Diameter of shaft below the inner process	12½

Fig. 3, Plate VIII, represents a specimen of a first caudal vertebra of a smaller species of crocodile than those indicated by the preceding specimens. It was obtained by Professor Hayden's party near Little Sandy River. The length of the body with its double ball is 21½ lines. Several other vertebræ from Black's Fork of Green River, and from near Church Buttes, Wyoming, from their size and conformation, would appear to belong to the same species.

Order *Chelonia*

No other Tertiary deposit in North America has yielded such an abundance of remains of different species and genera of turtles as the Bridger beds. The fossils represent a large proportion of fresh-water and paludal forms; the others pertain to land tortoises. Fragments of turtle-shells are the most frequent of the vertebrate fossils met with, strewed on the bare tops and sides of the buttes or among the *débris* at their base. Entire shells are comparatively rare, and if they have been complete as fossils, they soon undergo disintegration after exposure on the buttes. Most of them have been much crushed, while embedded, under pressure of the superincumbent strata, and now when exposed from the softening of the matrix they readily fall to pieces.

The greater quantity of the turtle remains are referable to a species of fresh-water terrapin of the genus Emys, the shells of which present sufficient variety as to have at first misled me in referring them to several different species. The next most abundant remains are those of one or two species of soft-shelled turtles of the genus Trionyx, and after these the remains of a large land-tortoise. Besides the species and genera described in the succeeding pages, Professor Cope has recently indicated a number of others from the same formation.

TESTUDO.

TESTUDO CORSONI.

Among the many remains of turtles from the Bridger Tertiary beds are those apparently of a large land-tortoise. Small and for the most part uncharacteristic fragments of the shell were obtained by Dr. Carter in 1869 and during Professor Hayden's exploration of 1870, but it was not until I received the specimen represented in Fig. 7, Plate XV, that I recognized the character of the species to which they pertained.

The last-mentioned specimen was discovered by Dr. Corson at Grizzly

Buttes. It consists of the anterior extremity of the under shield or plastron, consisting of the fore part of the episternals and the end of the entosternal. The specimen might be supposed to belong to an Emys, but its resemblance in form with the corresponding part in living species of Testudo leads me to place it with this genus. The episternals project together rather abruptly into a long, thick, and broad spade-like process, nearly straight at the front border, but slightly notched at the middle. The projection behind is defined by the outer extremities of deep grooves defining the gular and humeral scute impressions. Its lower surface is strongly convex; the upper surface slopes forward to the acute border of the process.

Back of the gular surface above, the plastron is deeply concave, but is not excavated beneath the former as in the gopher, (*Testudo carolina*.)

The end of the entosternal plate is impressed by the contiguous ends of the gular scutes. The episternal process is about 2 inches long; its breadth at base is $5\frac{1}{4}$ inches. The extremity of the process is $3\frac{3}{4}$ inches in width. The thickest part of the episterna measures $1\frac{1}{2}$ inch.

The species I have named in honor of its discoverer, Dr. Joseph K. Corson, United States Army, who to a love of his profession adds a special interest in the promotion of the natural sciences.

During my recent trip to Fort Bridger I was so fortunate as to obtain a number of additional specimens referable to *Testudo Corsoni*. Some of them had been previously collected by Drs. Corson and Carter, and others were found during our explorations of the buttes near Fort Bridger, and those of Dry Creek ten miles from the former.

One of the best preserved specimens consists of a nearly complete ventral shield or plastron, represented in Fig. 2, Plate XXX. This was discovered by Dr. Corson at Grizzly Buttes, and presented by him to the Academy. In the complete condition it has measured upward of 2 feet in length, and is estimated to have been about 16 inches in breadth to its sutural conjunction with the upper shell.

In its form and proportions it resembles that of the living *Testudo radiata* of Madagascar more than it does that of the great Galapagos tortoise.

The lobes of the plastron are of nearly equal length and breadth. The prolonged extremity or spade-like process of the anterior lobe is lost in the specimen. The posterior lobe terminates in a deep, wide, angular notch included by two angular processes.

The fore part of the anterior lobe is slightly bent upward and nearly straight transversely. The plastron from the position of the pectoral scute impression backward becomes gradually and deeply concave. The deeper part of the concavity is defined on the posterior lobe of the plastron by a narrow flat ledge laterally, which widens behind on the angular processes terminating the plastron. The sternal bridges are moderately convex, and are wide fore and aft.

The anatomical structure of the osseous plastron and the relative position and number of its scutes are the same as in modern species of Testudo.

The entosternal bone is subpyriform and wider than long. Its fore extremity reaches just in advance of the ends of the gular scute impressions, and its back border reaches the groove defining the humeral and pectoral scute impressions.

All the grooves defining the scute impressions are well marked, being deep and wide. The proportions of the scute impressions are nearly the same as in recent testudines.

The pectoral scute impression is longer at both extremities than intermediately. The groove defining it in front, commencing externally just in advance of the bottom of the axilla, curves backward and inward, and then turns forward and inward to the position of the back suture of the entosternal plate.

The measurements of the specimen are as follows:

	Inches.
Estimated length in median line to bottom of poststernal notch	24
Estimated length on each side to ends of poststernal processes	26¼
Estimated width	20
Estimated length of anterior lobe of plastron in median line	8
Length of posterior lobe of plastron in median line	5
Length of posterior lobe of plastron laterally	7
Width of anterior lobe at bottom of axillæ	12
Width of posterior lobe at bottom of inguinal fossæ	13
Width at bottom of anterior prolongation of plastron	5
Width at ends of poststernal angular processes	7½
Depth of poststernal notch	2½
Width fore and aft of sternal bridge	9¾
Length of entosternal plate	4½
Width of entosternal plate	5¾
Length of hyosternal plate in median line	6¼
Length of hyposternal plate in median line	5¾
Length of xiphisternal plate in median line	4¾
Length of pectoral scute impressions in median line	1¾
Length of pectoral scute impressions where narrowest	1
Length of abdominal scute impressions in median line	9
Length of femoral scute impressions in median line	3¾

	Inches.
Length of caudal scute impressions in median line	2¼
Thickness of plastron at base of anterior prolongation	1¾
Thickness of anterior lobe laterally near bottom of axilla	1
Thickness of posterior lobe near bottom of inguinal fossa	1½
Thickness of plastron near the center	¾

During a day's excursion to Dry Creek Buttes, ten miles from Fort Bridger, Mrs. Anna Carter, the wife of Dr. Carter, who accompanied us, discovered a large turtle partially imbedded in a green sandstone on the top of a butte. The upper shield had been destroyed by recent exposure, but the nearly complete plastron was obtained by removing the cast of the shell above it. The sutural connections of the bones are somewhat obscured by the firm adherence of particles of sand. It retains the anterior spade-like process nearly entire, and this is represented in Fig. 4, Plate XXX.

The specimen presents some differences from the former, which, however, I have not regarded as specific, though they may be so. The spade-like prolongation of the plastron is more abrupt and considerably longer than in the fragment upon which the species was originally founded. The fore part of the anterior lobe of the plastron approaching the lateral border along the groove defining the gular and humeral scute impressions is much more convex than in either of the preceding specimens. From the position of the entosternum backward, the plastron becomes concave, as in the former specimen, but the concavity is comparatively shallow. The poststernal notch is also of less depth than in the previous specimen, but otherwise the plastron is sufficiently like the latter to pertain to the same species.

The measurements of the specimen are as follows:

	Inches.
Length of plastron in median line	25
Length of plastron on each side	26½
Width of plastron at middle, estimated at about	20
Length of anterior lobe	9
Length of posterior lobe at middle	6
Length of posterior lobe to ends of angular processes	8½
Width of anterior lobe at base	12
Width of posterior lobe at base	11½
Length of episternal prolongation	2½
Width of episternal prolongation at base	5¼
Width of episternal prolongation near end	4¾
Breadth of sternal bridges fore and aft	9
Breadth at ends of poststernal angular processes	7½
Depth of poststernal notch	1¾
Length of entosternal bone	4¾
Breadth of entosternal bone	5¼

In some low buttes on the road to Carter Station, about three miles from Fort Bridger, Dr. Carter found a large turtle, which I viewed as pertaining to *Testudo Corsoni*. As it lay partially exposed it measured about 2 feet 4 inches in length, and approximated 2 feet in breadth. It was so much broken that in the attempt to remove it, it fell into a multitude of fragments.

In Dry Creek Cañon we discovered another turtle, which I viewed as *T. Corsoni*. The shell was in great part decomposed, but the rock which had occupied the interior still preserved its form. From this cast we estimated the shell to measure 28 inches long, 20 inches broad, and 14 inches high.

Another specimen of a large turtle, discovered by Dr. Corson on the buttes of Dry Creek, consisted of fragments of a plastron with a few marginal plates of the carapace. The plastron, of which we have been enabled to restore the greater part of the anterior lobe, presents peculiarity enough to pertain to a distinct species from *T. Corsoni*. It was about the size and proportions of the plastron attributed to the latter, but the episternals are neither so abruptly nor so much prolonged as in the former specimens, and the front part, as represented in Fig. 3, Plate XXX, is decidedly notched. The under surface of the extremity of the anterior lobe is flatter.

The bony construction of the plastron, so far as preserved, is the same as in the former specimens, and the entosternal is nearly of the same size and shape.

The scute impressions are also the same as in the former specimens, except that the pectoral scute impressions are nearly twice as long.

Fragments from the back lobe of the plastron retaining the bottom of the poststernal notch indicate this to be more acute than in the former specimens.

The measurements of the specimen are as follows:

	Inches.
Length of anterior lobe of the plastron	8
Breadth at base	10½
Length of episternal prolongation	1⅞
Breadth of episternal prolongation at base	5½
Breadth of episternal prolongation near the extremity	4¼
Length of entosternal plate	4⅞
Breadth of entosternal plate	5½
Length of gular scute impressions	3¼
Length of humeral scute impressions	4¼
Length of pectoral scute impressions	3
Length of pectoral scute impressions where least	2

Portions of the shell of another specimen, apparently referable to *Testudo Corsoni*, were discovered by Dr. Corson on Dry Creek Buttes. Several of the fragments so far recompose one side of the back lobe of the plastron as to determine its identity with that of *T. Corsoni*. It is especially interesting from its being accompanied by a number of fragments of the upper shell, which being reunited compose the middle portion, as represented in Fig. 1, Plate XXX. This specimen tends to confirm what I have latterly suspected, namely, that the specimens formerly described and represented in Plate XI, under the name of *Emys Carteri*, really belong to *Testudo Corsoni*. The specimens originally referred to the former, though much more complete than the one upon which the latter was founded, completely misled me. The spade-like process of the plastron was not simply broken off, but, while imbedded in its matrix, was crushed or squeezed off in such a manner as to leave but little trace of its true character. The accompanying portion of the carapace exhibited the costal plates with strong costal capitula as in living species of Emys. This emydoid character with others are probably sufficient indications that the specimens would properly be referable to a genus distinct from either Testudo or Emys, and is probably the same as that recently proposed by Professor Cope, under the name of Hadrianus.

The specimens originally referred to *Emys Carteri*, but now viewed as pertaining to *Testudo Corsoni*, were discovered by Dr. Carter in the buttes near Fort Bridger. They consist of the greater part of a mutilated plastron with the ends broken off, and the anterior median portion of the carapace.

The plastron represented in Fig. 1, Plate XI, resembles, in its size, form, and proportions, the nearly complete specimen above described and represented in Fig. 2, Plate XXX. It is not so concave posteriorly, but otherwise presents nothing peculiar.

The portion of the carapace represented in Fig. 2, Plate XI, consists of the nuchal and anterior three vertebral plates with fragments of the contiguous costal plates.

The anterior border of the fragment is slightly emarginate. The vertebral region is flat, and slopes forward from the anterior half of the first vertebral plate. The nuchal plate is nearly as long as wide, and its antero-lateral borders are moderately convergent.

The first vertebral plate is clavate in outline with the broad end behind. The anterior narrow end dips into an emargination of the nuchal plate. Its widest part is less than a fourth of its length in advance of its posterior border. The second vertebral plate presents the usual hexagonal coffin-like outline, but in a reversed position, its broadest part being about one-fifth of its length in advance of its back border. The third vertebral plate is oblong quadrate, with the fore and lateral borders convex, and the back one nearly straight.

The sutures defining the first costal plate depart from the anterior narrow end and the posterior widest part of the first vertebral plate. The scute impressions of the carapace are well defined by deep grooves.

The nuchal scute impression is flat, and widens anteriorly. The first marginal scute impression is wider than long.

The first vertebral scute area is longer than broad, and is purse-like in outline.

The second vertebral scute area is also longer than broad, and is quadrate, with the lateral borders nearly parallel.

The fragment of the carapace from its front border to the back border of the third vertebral plate measures 13¼ inches.

Other measurements of the carapace are as follows:

	Lines.
Length of nuchal plate	56
Breadth of nuchal plate in front	44
Breadth of nuchal plate where widest	60
Length of first vertebral plate	48
Breadth of first vertebral plate in front	9
Breadth of first vertebral plate where widest	30
Breadth of first vertebral plate at back border	14
Length of second vertebral plate	27
Breadth of second vertebral plate where widest	26
Breadth of second vertebral plate at back border	16
Length of third vertebral plate	29
Breadth of third vertebral plate at middle	22
Breadth of third vertebral plate at back border	17
Length of nuchal scute impression	21
Breadth of nuchal scute impression in front	11
Breadth of nuchal scute impression behind	6
Length of first marginal scute impression	26
Breadth of first marginal scute impression behind	38
Length of first vertebral scute impression	67
Breadth of first vertebral scute impression in front	32
Breadth of first vertebral scute impression near middle	52
Breadth of first vertebral scute impression at back border	43
Length of second vertebral scute impression	58
Breadth of second vertebral scute impression at middle	48

The accompanying plastron measured, in its complete condition, upward of 2 feet in length and about 1½ feet in breadth.

Other measurements of the specimen are as follows:

	Lines
Width of anterior lobe of plastron at base	108
Width of posterior lobe of plastron at base	120
Breadth of sternal bridges fore and aft	114
Length of entosternal plate	56
Breadth of entosternal plate	63
Length of hyosternals in median line of plastron	60
Length of hyposternals in median line of plastron	61
Length of humeral scute impressions	48
Length of pectoral scute impressions	26
Length of abdominal scute impressions	82
Length of femoral scute impressions	47

The portion of a carapace represented in Fig. 1, Plate XXX, and previously referred to as tending to confirm the impression that *Emys Carteri* was the same as *Testudo Corsoni*, retains most of the vertebral plates with contiguous fragments of the costal plates.

The anterior three vertebral plates, corresponding with those which are retained in the specimen originally referred to *Emys Carteri*, have the same form, but are wider. The succeeding two plates have the same form as the second vertebral plate in a reversed position. The sixth vertebral plate is too much broken to ascertain its exact form, but it would appear to be nearly the same as those in advance. The seventh plate is hexagonal, with the breadth more than twice the length; and the eighth plate has the same form, but is not so broad.

The length of the fragment of the carapace from the anterior broken end of the first vertebral plate to the back border of the eighth plate is 16 inches.

Other measurements are as follows:

	Lines
Length of first vertebral plate, estimated	40
Breadth of first vertebral plate in front	14
Breadth of first vertebral plate where widest	34
Breadth of first vertebral plate at back border	17
Length of second vertebral plate	27
Breadth of second vertebral plate where widest	28
Breadth of second vertebral plate at back border	18
Length of third vertebral plate	28
Breadth of third vertebral plate at middle	23
Length of fourth vertebral plate	26
Breadth of fourth vertebral plate in front	29
Length of fifth vertebral plate	24
Breadth of fifth vertebral plate in front	27
Length of sixth vertebral plate	20

	Lines.
Length of seventh vertebral plate	13
Breadth of seventh vertebral plate	28
Length of eighth vertebral plate	13
Breadth of eighth vertebral plate	24

The costal capitula of *Testudo Corsoni* appear in the specimens as robust conical eminences, with a broad, expanding base, and are proportionately better developed than in living species of Testudo, and even many of the species of Emys.

Figs. 2, 3, Plate XXIX, represent the upper extremity of a humerus, and Fig. 4 the lower extremity of a femur, which were found in association with the fragment of a carapace last described, and may reasonably be supposed to pertain to the same animal. Both fragments resemble the corresponding parts of a modern Testudo.

The head of the humerus has an inner trochlear extension, as in recent species of Testudo. Independent of this process, the transverse diameter of the head is nearly as great as the fore and aft diameter. In the specimen it presents a discoidal, flat surface, but this is evidently accidental.

The measurements of the specimens are as follows:

	Lines.
Breadth of humerus between tuberosities	29
Breadth between outer tuberosity and inner extension of the head	32
Breadth of the head with its inner trochlea	20
Fore and aft diameter of the head	17
Breadth of the distal end of the femur	24

EMYS.

EMYS WYOMINGENSIS.

Of the many remains of turtles from the Bridger Tertiary deposits I have had an opportunity of examining, most of them appear to me to belong to a species of Emys, which presents so much variation in anatomical details that the first specimens brought to my notice were viewed as pertaining to no less than four distinct species. These were named *Emys wyomingensis, E. Stevensonianus, E. Jeanesi,* and *E. Haydeni.* A subsequent examination of additional specimens, collected by Dr. J. Van A. Carter and Dr. Joseph K. Corson, United States Army, and presented by them to the Academy of Natural Sciences of Philadelphia, has led me to regard all those indicated under the above names as really pertaining to a single species. I admit that I may be wrong in this determination, but if such is the case, it would appear that almost every specimen presents characters to distinguish a species.

Regarding all the specimens under consideration as pertaining to a single species, this would retain the original name of *Emys wyomingensis*.

The composition of the shell so far as relates to the attachment of the carapace and plastron, the number of bones or plates and the number and relation of the corneous scutes, is the same as in living species of the genus Emys.

In the mature condition, the shell of *Emys wyomingensis* is upward of a foot in length, with about three-fourths the same measurement in breadth.

To what degree the shell varies in form, that is to say in relation of length and breadth with the height, and in outline, cannot be determined from the material at command, on account of the imperfection of the specimens, or their distortion from the original condition, due to pressure or to a crushing force applied to them while inbedded in the strata from which they were obtained.

The elements of composition, especially the vertebral plates and scutes, differ more or less in different specimens, both in form and in the relation of length to the breadth. While the length of the vertebral scutes in general exceeds the breadth, especially in the case of those intermediate, in some specimens even to the extent of being a third greater, it nevertheless varies so much that in some instances it barely exceeds the breadth. The vertebral plates vary in the same manner in different specimens, nor does this variation always accord with that of the same character in the vertebral scutes, that is to say the elongation of the scutes is not always accompanied in a proportionate degree with elongation of the plates.

1. *Emys wyomingensis* was originally described from an isolated episternal bone, sent to the writer by Dr. Carter. It was the first of the remains of turtles from the Bridger Tertiary deposits, which could be referred to the genus. It is represented in Fig. 5, Plate IX, and exhibits the usual form of that in living species, but further presents the appearance of being impressed by a narrow intergular scute. The presence of the latter I suspect to be accidental or anomalous, though it may be normal, and may really indicate that the fossil belongs to a species distinct from those which I am now disposed to view as the same. The front of the specimen is truncated and slightly notched at the outer part.

2, 3. *Emys Stevensonianus* is the name originally given to a supposed species founded on the specimens represented in Figs. 2, 4, Plate IX. These were collected by Dr. Carter in the vicinity of Fort Bridger, and sent to the Smithsonian Institution, whence I obtained them for examination. The specimens consist of portion of a carapace, Fig. 2, and portions of two plastrons, Figs. 3, 4. Slight difference in the corresponding portion of the latter specimens with that attributed to *E. wyomingensis*, but especially the absence of any evidence of an intergular scute, led to their being referred to another species.

The sternal specimen, represented in Fig. 3, accompanied the portion of a carapace, represented in Fig. 2, and from its appearance was assumed to have belonged to the same individual.

In the sternal specimen just indicated, the entosternal plate is lozenge-shaped in outline, but constricted at the middle of its posterior part. Its length is equal to the breadth, but the evidence from the isolated episternal, first referred to *E. wyomingensis*, is that its entosternal was wider than long.

The divisions of the plastron and its impress by scutes, as seen in the more perfect specimen, appear to agree pretty closely with the arrangement observed in ordinary living emydes.

The second sternal specimen, represented in Fig. 4, was supposed to pertain to the same species as the former one, though exhibiting differences which rather approached it nearer to that first referred to *E. wyomingensis*, except that it exhibited no trace of the existence of an intergular scute. The entosternal bone is wider than long, and without conspicuous constriction at its posterior part. The anterior truncated border of the episternum is conspicuously notched at its outer part.

In all the sternal specimens indicated, the gular and humeral scutes have doubled over the edge and extended upon the upper surface in the same manner as in living emydes.

In the carapaceal specimen, Fig. 2, the vertebral plates, consisting of the series from the first to the eighth, inclusive, successively decrease in length except that the third is a little longer than the second, and the fourth and fifth are nearly equal. To the fourth inclusive, the length much exceeds the breadth, but they successively diminish in this proportion. The fifth is but slightly longer than wide, and the remaining plates are much wider than long. The second and fifth are of the same width, and in this respect exceed the first and intermediate ones, which are likewise of nearly uniform breadth. The sixth plate is the widest of the series; the others successively diminish.

They exhibit the usual forms, the first being oblong with the borders convex outwardly, the others to the fifth being wide coffin-shaped, and the remaining ones are more regularly hexagonal.

The second and third vertebral scute-spaces are quadrate with the lateral defining-grooves strongly double sigmoid. The second space is broader than long, but the third is the reverse.

4. It was the nearly complete shell, represented in Plate X, which was attributed to a different species from the former specimens, under the name of *Emys Jeanesi*. This fine fossil was obtained near Fort Bridger, Wyoming, during Professor Hayden's exploration of 1870. It is considerably distorted from pressure, the right side being crushed inwardly so as to be nearly vertical. The shell, completely petrified like all its associate fossils, is filled with a greenish-gray sandstone. Its prominence or convexity in the original condition was perhaps not greater than in some of the ordinary living emydes, but it is apparently more prominent, from the lateral pressure to which the shell has been subjected.

The carapace is oval in outline with the borders moderately deflected, acute, and without conspicuous indentations, except that it is slightly notched in the position of the nuchal plate. The plastron has the same form and degree of development in relation with the carapace as in living species of the genus. It is truncated in front, and notched behind.

Although the sutures of the shell are conspicuously visible, the bones or plates are all closely united, and the specimen appears to have been nearly or quite in the adult condition. No lines of successive growth are visible on the plates, which are everywhere smooth. The position or boundaries of the scutes are indicated by deeply marked grooves.

Ten vertebral plates appear to constitute the series, the connection of the last two in the specimen being destroyed. In form and proportions they bear a near likeness to those in emydes in general. They are rather wider proportionately than those in the specimen first referred to *E. Stevensonianus*, but otherwise are sufficiently alike to pertain to the same species.

As usual, the first vertebral plate is longest; then follows the third. The second, fourth, and fifth are nearly equal. The others, to the eighth, successively diminish. The second vertebral plate is as wide at its fore part as it is long, but the succeeding two plates are considerably longer than wide.

The fifth is as wide as it is long, and the remaining plates are considerably wider than long.

The costal plates have about the same form as in recent species of the genus, but the first one is of greater proportionate breadth. Besides the nuchal plate, it articulates with four marginal plates. The remaining costal plates are of nearly uniform width as in recent emydes.

The second costal plate articulates with the fourth and fifth marginals; the third, with the fifth and the anterior angle of the sixth marginals; the fourth, with the sixth marginal alone; the fifth, with the sixth and seventh marginals; the sixth, with the seventh and eighth marginals; the seventh, with the eighth to the tenth marginals inclusive, and the eighth with nearly the whole of the tenth and the angle of the eleventh marginals.

The marginal plates have nearly the same form and proportions as in recent emydes.

The nuchal plate also has nearly the form and proportions as in the latter. The pygal plate, likewise, has the same form, but is proportionately smaller.

The vertebral scute-tracts have nearly the same form as in living species of Emys, but the intervening ones are longer than wide. They are proportionately somewhat narrower than in the specimen first referred to *E. Stevensonianus*.

The first vertebral scute at its fore part extends outwardly nearly to the line between the first and second marginal scutes, and in this position is widest.

The last vertebral scute, at its posterior border, crosses the last vertebral plate a short distance back of the middle. In recent species of Emys it impresses the pygal plate.

The costal scutes resemble those of ordinary emydes, and as in these impress the marginal plates at their conjunction with the corresponding scutes.

The nuchal scute is comparatively short and wide. The specimen being imperfect at the back part prevents us from ascertaining positively whether there existed a pair of pygal scutes as in living emydes, but an apparent curve upon the bone renders it probable that two also belong to the extinct species.

The marginal scutes resemble those of recent emydes, but the anterior are wider than high, and the posterior, including the pygal scutes, are higher than wide.

The fore part of the plastron has a half-oval outline slightly projecting, and truncated at the extremity as in ordinary emydes. The back part likewise has the same form as in the latter, and is also notched at the extremity.

The pedicles are less elevated than in most recent emydes, and are rather wider to the acute border of the carapace.

The constitution of the plastron is so nearly like that of ordinary living emydes as hardly to need special description.

The entosternal plate is nearly lozenge-shaped, and is widest transversely.

The humeral scutes at their posterior border barely cross the posterior extremity of the entosternal bone.

The pectoral and abdominal scutes extend outwardly to conjoin the marginal scutes upon the marginal bones. In ordinary recent species of Emys the marginal scutes extend upon the hyosternal and hyposternal plates to join the pectoral and abdominal scutes.

The axillary and inguinal scutes are large, and impress each a marginal and a sternal plate.

The length of the carapace in a curved line is within half an inch of a foot and a quarter; its breadth, in the same manner, 11 inches; in a straight line it is little over a foot in length and about 10 inches in breadth. The plastron is less than a foot in length, and its pedicles measure, fore and aft, $4\frac{3}{4}$ inches.

5. The specimen originally referred to *Emys Haydeni* is represented in Fig. 6, Plate IX. It consists of a portion of the carapace attached to a mass of indurated clay, and was obtained near Fort Bridger, Wyoming, during Professor Hayden's exploration of 1870. Since the specimen was figured, additional portions of the shell have been found which allow the restoration of the fore part of the carapace. It belonged to a larger individual than the specimen first attributed to *E. Jeanesi*, and from the appearance of the marginal border of several of the costal plates to a less mature one.

The form of the carapace in front and its constitution in detail are very similar to the corresponding portion in the former specimen attributed to *E. Jeanesi*. The proportions of the vertebral plates is more nearly as in the latter than in the specimen attributed to *E. Stevensonianus*.

An apparently important difference between the fossil under examination and the one attributed to *E. Jeanesi* is the less uniformity of width of the intermediate costal plates. These alternately become wider and narrower toward their outer extremities, whereas in the specimen referred to *E. Jeanesi* they are nearly uniform.

As peculiarities of the fossil, the fourth vertebral plate is octagonal, and the fifth one in consequence quadrate.

The second and third vertebral scute-tracts are much longer than wide, and proportionately much longer than in the former specimens. The anterior division of the second vertebral scute forms three sides of a square; and the posterior groove defining the third scute crosses the sixth vertebral plate instead of the fifth as in the other specimens.

The peculiarities indicated in the fossil under examination I regard as being of an individual character and in some degree anomalous.

A fragment of the fore part of the plastron accompanying the specimen referred to *E. Haydeni*, and apparently belonging to the same individual, resembles the corresponding part in the specimens previously described, but is not notched at its anterior truncated border.

6. Another specimen, referable to *Emys wyomingensis*, consists of a nearly complete shell except the posterior third of the carapace. It was discovered by Dr. Carter in the bluffs of the Cottonwood, seven miles from Millersville, in the vicinity of Fort Bridger, Wyoming, and presented by him to the Academy of Natural Sciences of Philadelphia. It is occupied in the interior with a greenish-gray sandstone, including indurated clay pebbles. In form and size it approaches closely the specimen first referred to *Emys Jeanesi*. In the form and proportions of its vertebral scute impressions it more nearly resembles the specimen originally referred to *E. Haydeni*. The intermediate ones are, however, more strongly double sigmoid at their lateral borders; the fore part of the second vertebral scute is less square; and the anterior border of the third is strongly bowed forward instead of being nearly straight.

An accidental fracture of the specimen across the posterior third exposes to view the lateral supports of the carapace ascending from the plastron. These are much wider than in any of the living emydes, and approach in their proportions those of the living fresh-water turtle Batagur, of India.

7. A seventh specimen of *E. wyomingensis* consists of an intermediate portion of a carapace and nearly the whole of the sternum. It was obtained by Dr. Carter in the vicinity of Fort Bridger, and presented by him to the Academy of Natural Sciences of Philadelphia.

The vertebral plates of the carapace are in general of proportionately greater breadth in comparison with the length than in the former specimens, and in this respect most nearly approach the one which was referred to *E. Hay-*

deni. The vertebral scute impressions likewise most nearly resemble those of the latter but are proportionately broader, and the posterior border of the third vertebral scute crosses, as usual, the fifth vertebral plate.

The interior of the carapaceal specimen being freed from matrix, exhibits the costal plates with strong, well-developed costal capitula.

The plastron is flat; rather more strongly notched at its posterior extremity than in the former specimens in which it is preserved.

The thickness of the costal plates ranges from 2 to $4\frac{1}{2}$ lines. The thickness of the hyposternal plates internally ranges from $5\frac{1}{2}$ to $8\frac{1}{2}$ lines.

8. An eighth specimen, consisting of the greater part of a plastron with fragments of the carapace, was obtained by Dr. Carter near Lodge-Pole Trail, thirteen miles southeast of Fort Bridger, and presented to the Academy of Natural Sciences.

9. A fragment of a carapace, from Grizzly Buttes, presented to the Academy of Natural Sciences by Dr. Joseph K. Corson, United States Army, has the intermediate scute impressions much longer than the width, not more so, however, proportionately, than in the nearly complete specimen numbered as the sixth.

10. A similar fragment of an apparently young specimen, presented by Dr. Carter, has the second vertebral scute impression nearly equal in length and breadth; and the third one is but little longer than the breadth. Their lateral grooved borders are strongly double-sigmoid.

11. Part of a carapace and plastron of a still younger specimen, obtained by Dr. Carter near Lodge-Pole Trail, twelve miles southeast of Fort Bridger, nearly agrees in the form and proportions of its corresponding vertebral scute impressions with that last described. The second is nearly of equal length and breadth; the third and fourth are wider than the length. In its details of structure it accords sufficiently with the older and more complete specimens to render it probable that it pertained to the same species, except that the carapace is obtusely carinated its entire length. The entosternal bone is more rounded at its fore part than in previous specimens, and its length is about equal to the breadth.

12. A fragment of a plastron of another young individual, from the same locality and gentleman as the preceding, nearly agrees with the corresponding part. The entosternal is a little longer than broad, and is pyriform, with lateral projecting angles.

13. A specimen, apparently of a still younger individual of the same species, presented to the writer by Dr. Carter, was about the size of the palm of the hand. It consists of small portions of the carapace and more than half the plastron. The carapace is carinated as in specimen No. 11, and otherwise agrees with this in its details. The plastron has the same form as in the more complete and older specimens previously indicated, but the entosternal is more pyriform, considerably longer than wide, and the posterior defining groove of the pectoral scute crosses its middle.

If it is admitted that the specimens Nos. 11 and 13 belong to *Emys wyomingensis*, it would appear that the carinated condition of the carapace is a juvenile character, disappearing with growth. It would also appear that during growth the breadth of the entosternal plate became proportionately greater in relation with its length

None of the specimens viewed as young ones exhibit upon the surface lines of growth, except the sternal one, No. 12, in which they are feebly marked. A distal fragment of several posterior costal plates of specimen No. 11, in the immature appearance of its border, clearly proves its youthfulness.

Besides the thirteen characteristic specimens of *E. wyomingensis* which have been described or mentioned, fragments of many others are contained in the collections I have had the opportunity of examining. From their comparative frequency, this appears to have been the most abundant of the freshwater turtles of the Bridger Tertiary epoch.

14. Since writing the foregoing, I have had the opportunity of examining another specimen of *Emys wyomingensis* in the possession of Dr. Hiram Corson, which was sent to him from Fort Bridger by his son, Dr. Joseph K. Corson. The specimen consists of a nearly complete shell except the posterior fourth of the carapace. It is a little smaller than the fourth-described specimen, represented in Plate X, and is crushed and distorted nearly in a similar manner.

The most striking peculiarities of this, which may be distinguished as the fourteenth specimen, are the unusual depth and width of the scutal grooves of the carapace and the proportionate shortness and breadth of the costal scute areas.

The intermediate vertebral plates to the first and fifth are absolutely longer and narrower than in the rather larger fourth-described specimen. The costal plates are shorter, and the second to the fourth, inclusive, are broader. The first and second vertebral scute areas are wider and the third one longer. The forms of the plates and scute areas indicated are nearly the same in both specimens.

The plastron is slightly convex in both directions, and its extremities for half the length are more parallel at the lateral borders than in the fourth-described specimen. The axillary and inguinal scute areas are longer and narrower than in the latter, and somewhat modified in form. The length of the plastron in the median line is 11 inches. Other measurements in detail are given in the annexed table under the head of specimen 14.

Comparative measurements of the specimens referred to *Emys wyomingensis* are as follows:

150

Specimen.	1.	2.	3.	4.	5.	6.	7.	8.	9.	10.	11.	12.	13.	14.
	Lines.	*Lines.*	*Lines.*	*Lines.*	*Lines.*	*Lines.*	*Lines.*	*Lines.*	*Lines.*	*Lines.*	*Lines.*	*Lines.*	*Lines.*	*Lines.*
Length of carapace in the curve				175										
Breadth of carapace in the curve				132		132								132
Length of carapace in a straight line				150		108								120
Breadth of carapace in a straight line				120		134	132							120
Length of plastron in median line				125			96							
Breadth of plastron to sutural union				92		84	36							
Length of plastron at fore part				40		38	34							
Width of anterior truncation	24		19	21	26	31	22				20			32
Width at post-episternal suture	48	50	51	53		32	27				12	30		30
Width at post-humeral groove		32	51	60	60	33	45				28	11		60
Length of plastron at back part				45		42	45					25		42
Width at anterior femoral groove				66		60	60	62		10		23		30
Width of anterior xiphisternal groove				30		60	36	56		12	20			28
Width of poststernal notch				14		6	20			10				11
Depth of poststernal notch					16		7			11			1	7
Fore and aft extent of sternal pedicles				57		22	57				30			26
Measurements of the plates of the carapace:														
Length of first vertebral plate		16		20	22	20	21				10½			12
Breadth of first vertebral plate		11		13	14	11	16			10			4	11½
Length of second vertebral plate		14		14	13½	15	16	12½		12	8		4	15
Breadth of second vertebral plate		12½		14	14	12	16	14½		12	8		4	15
Length of third vertebral plate		15		15	11	13	15	12		11	7		4½	14½
Breadth of third vertebral plate		11½		13	14	12	15	13		12	7			12
Length of fourth vertebral plate		13½		11	12½	11	11	12		11	4			15½
Breadth of fourth vertebral plate		11½		13	12½	12	11	13½		9	6			12
Length of fifth vertebral plate		13		14	12	12	11	15		10½	5			12
Breadth of fifth vertebral plate		12½		11	12½	12	15	13½		10½	6			12
Length of sixth vertebral plate		10½		12	12	12	11				8			14½
Breadth of sixth vertebral plate		14		15	12		16				6			11
Length of seventh vertebral plate		9		11	16		9				4			
Breadth of seventh vertebral plate		11		15			10				6			
Length of eighth vertebral plate		6½		10			11				5			
Breadth of eighth vertebral plate		11		15			10½				7			

Specimens.	1.	2.	3.	4.	5.	6.	7.	8.	9.	10.	11.	12.	13.	14.
	Lines.	Lines.	Lines.	Lines.	Lines.	Lines.	Lines.	Lines.	Lines.	Lines.	Lines.	Lines.	Lines.	Lines.
Measurements of the plates of the carapace:														
Length of nuchal plate				22	26	23	24				15		10	21
Breadth of nuchal plate				31	35	35	26				18		12	25
Length of first costal plate				44	46	30					25		12	29
Breadth of first costal plate				25	30	29	31				16		9	30
Length of second costal plate				49	50	41								16
Breadth of second costal plate internally				14	16	16	16			10	8		4	18
Breadth of second costal plate externally				14	18	17								14
Length of third costal plate				50	50	44								43
Breadth of third costal plate internally				14½	17	12	18		18	12	7			15
Breadth of third costal plate externally				14½	12	14	18				9			17
Length of first marginal plate				19	22	17	19				9			17
Breadth of first marginal plate at outer edge				17	22	16	20				9			
Length of third marginal plate to the bend				16		15					10			
Breadth of third marginal plate at the bend				15		17								
Length of sixth marginal plate to the bend				19		17								
Breadth of sixth marginal plate at the bend				19		15								
Length of eighth marginal plate				24										
Breadth of eighth marginal plate				16										
Length of tenth marginal plate				21										
Breadth of tenth marginal plate				19										
Measurements of the plates of the plastron:														
Length of entosternal plate		30	21	25		22	21	24			13½	12½	10	22
Breadth of entosternal plate		20	20	28		25	25	20			11½	10½	9	20
Length of episternal plate at inner border	26	12	13	12	14	17	12	11			7	7½	6	9
Breadth of episternal plate at posterior border	12	16	13	12		15	15	14			9	8	7	14
Length of hyosternal plate at anterior border	11	17		33		33	27	20			15		5	36
Breadth of hyosternal plate at middle				51		51	46						17	49
Length of hyposternal plate at inner border		26		45		41	38	42			21		13½	38
Breadth of hyposternal plate at middle				36		30	41	43					16	43
Length of xiphisternal plate at inner border				24		27	57	30					10½	45
Breadth of xiphisternal plate at anterior border				20		29	29				18		11½	39

Specimen.	1.	2.	3.	4.	5.	6.	7.	8.	9.	10.	11.	12.	13.	14.
	Lines.	Lines.	Lines.	Lines.	Lines.	Lines.	Lines.	Lines.	Lines.	Lines.	Lines.	Lines.	Lines.	Lines.
Measurements of the scutes of the carapace:														
Length of nuchal scute				5	7	6	5	5			3½		2½	
Breadth of nuchal scute				6	10	9	10½	7½			6		3	
Length of first vertebral scute				39	35	29	29	30			17½		12	30
Breadth of first vertebral scute				35	41	40	34	40			17½		12	31
Length of second vertebral scute		29		30	32	31	29	32		21	17		10	34
Breadth of second vertebral scute		30½		36	26	29	29	29		21	16½		12	33
Length of third vertebral scute		30		30	33	33	31	29	31	23½	16			33
Breadth of third vertebral scute		29		36	35	27	27	32	23	22	18			32
Length of fourth vertebral scute				32							14			
Breadth of fourth vertebral scute				28	23						20			30
Length of fifth vertebral scute				20										
Breadth of fifth vertebral scute				18										
Length of pygal scute				16										
Breadth of pygal scute				43			46	45			23		13	33
Length of first costal scute at middle				41	45	35	46	44			22			35
Breadth of first costal scute at middle				32	22	36								44
Length of second costal scute at middle				27	20	42	28				14			34
Breadth of second costal scute at middle				20	31	,31	39							34
Length of third costal scute at middle				25	31	39	39				14			45
Breadth of third costal scute at middle				24										
Length of fourth costal scute at middle				22	14	11		11			6		5	
Breadth of fourth costal scute at middle				15	13	11		19			10		6½	
Length of first marginal scute externally				12	11	14		12			7		4½	
Breadth of first marginal scute externally				16	15	13		18			9		5	
Length of second marginal scute externally				18	13	21								
Breadth of second marginal scute externally				13	19									
Length of sixth marginal scute externally				19	19									
Breadth of sixth marginal scute externally				19										
Length of eighth marginal scute externally				20										
Breadth of eighth marginal scute externally				18										
Length of eleventh marginal scute externally														
Breadth of eleventh marginal scute externally														

153

Specimens.	1.	2.	3.	4.	5.	6.	7.	8.	9.	10.	11.	12.	13.	14.
	Lines.	Lines.	Lines.	Lines.	Lines.	Lines.	Lines.	Lines.	Lines.	Lines.	Lines.	Lines.	Lines.	Lines.
Measurements of the scutes of the plastron:														
Length of gular scutes at inner border	18	17	19	19	21	19	16	17			10½	10½	6¼	16
Breadth of gular scutes in front	14		13	13	15	13	13½	13			8	7½	6	12
Length of humeral scutes internally		12	7	19		18	16	14			9	10	5	19
Breadth of humeral scutes at back border			25	20		24	29	31			17	16	12	25
Length of pectoral scutes internally		24		23		21	22	26			12	13	9	12
Breadth of pectoral scutes at middle				49		42	46						16	44
Length of abdominal scutes internally		28		30		25	32	31			15		9½	39
Breadth of abdominal scutes at middle				49		42	44	45					15	45
Length of femoral scutes internally				28		21	27	31					10	26
Breadth of femoral scutes at anterior border				31		31	32	31			19		12½	31
Length of caudal scutes internally				16		18	18						6	11
Breadth of caudal scutes at anterior border				21		21	21	36					10	23
Length of axillary scutes				13		19								11
Breadth of axillary scutes				11		9								10
Length of inguinal scutes				11		16								16
Breadth of inguinal scutes				11		6								11

The specimen No. 1 was that originally referred to *E. wyomingensis*; Nos. 2, 3, those referred to *E. Stevensonianus*; No. 4 to *E. Jeanesi*; and No. 5 to *E. Haydeni*. The others are, for the first time, described in this communication.

BAPTEMYS.

A peculiar and interesting genus of extinct emydiform turtles, apparently intermediate in its characters to the existing American genera Dermatemys and Staurotypus, is founded on remains in the Bridger Tertiary formation of Wyoming.

In shape and constitution, the shell of Baptemys (Plate XII) approaches most nearly that of Dermatemys, more especially the carapace, while the sternum partakes of the character of that of Staurotypus.

The carapace is oval in outline, apparently not wider behind than in front, and with the prominence or convexity about equal to half its breadth. The convexity is nearly uniform fore and aft, and laterally to the flexure of the marginal plates. The anterior border is barely everted and is thick and rounded. The imperfection of the fossils prevents a determination whether the posterior border departed from the general convexity of the back of the shell. The surface formed by the first and second marginal plates is feebly depressed.

A median carina or thick rounded ridge starts upon the sixth vertebral plate and extends backward.

Eleven vertebral plates enter into the constitution of the carapace. Those anteriorly are proportionately much longer than in emydes. They also appear proportionately of greater extent than in Dermatemys.

The first vertebral plate is oblong, somewhat narrowed behind, and with the sides convex. Those to the sixth inclusive are hexagonal coffin-shaped. From the fifth they rapidly decrease in length to the eighth inclusive, and then increase again to the last. The seventh is more uniformly hexagonal than the others. The ninth is quadrate and wider than long. The tenth is quadrate, widest behind, with the lateral borders convex and the back border concave.

The costal plates are like those of Dermatemys, and as in this widen outwardly more than in ordinary emydes in accordance with the greater convexity of the carapace.

The nuchal plate and marginal bones, so far as preserved, appear to be nearly as in Dermatemys.

The scute impressions of the carapace, as in the latter, are not defined by such deep grooves as are usually observed in emydes.

The vertebral scute impressions have the same form and general proportions as in Dermatemys. The first is wide, urn-like in outline, and is broader than long. The succeeding three are quadrate, with the length greatly exceeding the breadth, and with the usual lateral brace-like or double-sigmoid borders. The last impression narrows for a short distance and then diverges in the usual manner.

The costal scute impressions resemble those of emydes and extend further upon the marginal bones than in Dermatemys, nearly reaching the middle of their outer face at the sides of the carapace, as far back as they are preserved in the fossils, as well as in front.

The position of the nuchal scute is not preserved in the fossils, but the part immediately contiguous in one of them indicates that it had about the same proportions as in Dermatemys.

The marginal scute impressions about occupied the lower two-thirds of the outer aspect of the marginal plates. The line intervening to the first two marginal scutes is continuous with that between the first vertebral and the succeeding costal scute.

Considering the striking resemblance of the carapace of Baptemys to that of Dermatemys, it is not a little surprising to observe so much difference in the plastron, though this also is nearly alike in the scute impressions.

Compared with that of Dermatemys, the plastron is remarkably small, leaving proportionately much larger spaces in advance and behind the bridges for the movements of the animal. As before intimated, it is intermediate in character to that of the last-named genus and that of Staurotypus. The pedicles are intermediate in extent to what they are in the two genera just mentioned. The fore part of the plastron has nearly the same shape as in Dermatemys, but is widely emarginate at the extremity, and it is thick and rounded at the border instead of being acute as usual in emydes. The back part of the plastron is narrower than in Dermatemys, but less so than in Staurotypus. It terminates in a rounded extremity as seen in Fig. 2, Plate XII. In Dermatemys, it ends in a wide notch; in Staurotypus, in a point.

The entosternal bone is proportionately as large as in Dermatemys, and has nearly the same form. The same may be said to be the case with the episternals, (Fig. 6, Plate XV,) except that their anterior border is more concave.

The hyosternals and hyposternals have nearly the same extent. Their intervening suture crosses the sternum near the middle of the pedicles.

Dr. Gray, who established the genus Dermatemys, represents the South American species *D. Mawii*, with a pair of gular scutes. *D. Berardii*, of Mexico, is represented by Dumeril as possessing a single symmetrical gular scute, and this also is the case in two shells from Balize River, Yucatan, and Tabasco, Mexico, described by Professor Cope as pertaining to another species which he has named *D. abnormis*.

In Baptemys there is no trace of separation of gular scutes from the humeral scutes as indicated in Fig. 6, Plate XV. The grooves defining the latter from the pectoral scutes occupy nearly the same position as in Dermatemys, crossing nearly through the middle of the entosternal plate.

In Emys the gular and humeral scutes fold deeply upon the upper surface of the sternum, but in Baptemys, as is also the case in Dermatemys, the corresponding scutes fold only to the upper edge of the rounded border of the sternum.

The intervening grooves of the pectoral, abdominal, femoral, and caudal scutes nearly equally subdivide the sternum of Baptemys.

The pectoral and abdominal scutes extend upon the sternal pedicles, and are there separated from the marginal scutes by large intervening scutes, as in the sea-turtles and in Dermatemys. In the same position in *Dermatemys abnormis* there are four of these scutes. In one of the specimens I have had the opportunity of seeing there are four of these scutes on one side and three on the other; but in this case it appears evident that the reduction is not the usual condition in the species.

There are three scutes on the sternal bridge of Baptemys which successively increase in size. The first or axillary scute joins the fourth and fifth marginal scutes and the pectoral scute. The middle or submarginal scute is hexagonal, widest transversely, and it joins the fifth and sixth marginal scutes and the pectoral and abdominal scutes. The third or inguinal scute, nearly twice the extent of that in advance, is also hexagonal. It extends across the hyposternal upon the hyosternal plate, and joins the sixth and seventh marginal scutes and the abdominal scute, an outward prolongation of which to the inguinal notch separates it from the femoral scute.

The axillary fossa reaches as far back as the posterior third of the fourth marginal bone; the inguinal fossa extends forward nearly on a line with the posterior border of the sixth marginal bone.

The interior of the fossils being occupied by the rocky matrix, all the internal anatomical details are concealed from view.

Baptemys in the relatively smaller size of the plastron to the carapace, and in the presence of submarginal scutes to the sternal bridges, is more nearly related to the marine turtles than the genus Emys.

Baptemys appears also to have been nearly related with the equally ancient and extinct genus Pleurosternon, of the English Tertiary formation. In this the vertebral scute areas of the carapace are remarkable for their breadth, which considerably exceeds the length, whereas in Baptemys the intermediate vertebral scute areas are much longer than broad. The plastron in Pleurosternon is intermediate in its proportions to that of Emys and Baptemys, and has an additional pair of bones entering into its composition which do not exist in the latter genera. In Pleurosternon a pair of integular scutes intervene to the gular scutes; in Baptemys there appears to be no distinction of gular scutes from humeral scutes. In Pleurosternon, as in Baptemys, large accessory or submarginal scutes intervene to the comparatively large axillary and inguinal scutes.

Baptemys wyomingensis.

The species thus named, as well as the genus, was first characterized from a beautiful specimen of the turtle-shell, discovered by Mr. O. C. Smith, of Leverett, Massachusetts, while engaged in the service of the Union Pacific Railroad Company, near Fort Bridger, Wyoming Territory. The specimen was loaned to Professor Hayden, by whom it was sent to the writer for examination. It is represented in Plate XII, one-third the natural size.

The specimen consists of a shell which nearly retains its original form, but has lost the front marginal plates on one side, all those behind, most of those of the left, and the front part of the plastron. It is black, as is frequently the case with the fossils from the same locality; and it is filled in the interior with a gray sandstone mingled with coarse pebbles of indurated bluish clay.

In its perfect condition the shell has measured about a foot and a half in length, and in breadth about a foot. Following the curvature of the carapace fore and aft, it has measured about 20 inches in length, and the transverse arch from a level has been nearly as great.

The length of the plastron has been about 11½ inches; its breadth from its sutural junction with the carapace is 9 inches.

The sides of the plastron slope inwardly to a moderate degree. The pedicles are nearly on a level with the rest of the plastron, but are somewhat prominent in front and slope backward, and are concave approaching the inguinal fossæ. The rise of the shell appears mainly to commence in the marginal bones from the sternal pedicles; to what degree is uncertain, as this part of the fossil is somewhat crushed inwardly. The rise is greater anteriorly, and gradually appears to subside behind.

The fore and aft extent of the pedicles is 4¼ inches. The length of the anterior extension of the plastron has been about 3½ inches; its breadth at the bottom of the axillary fossæ is 5¼ inches. The length of the posterior extension of the plastron is a little more than 3½ inches, and its width at the bottom of the inguinal fossæ nearly 4½ inches.

The marginal bones appear more abruptly bent to join the sternal bridge than in Dermatemys, but the difference is partially due to the crushing inward of the under part of the shell in the fossil.

A second specimen of the shell of *Baptemys wyomingensis* was subsequently discovered during Professor Hayden's exploration of 1870 at Church Buttes, Wyoming. The shell is of a different color, and is filled with and partially imbedded in a different matrix from the former specimen. The bones are brown, and the matrix consists of a very hard sandstone. The specimen, though far less complete than the former, fortunately retains one-half of the anterior part of the plastron. Most of the carapace is lost or imbedded in the hard rock. The sternum on one side from its fore extremity to the commencement of the xiphisternal bone, together with the pedicle and its characteristic scute impressions, is well preserved.

The measurements of this second specimen indicate an individual of the same size as the former. Slight differences existing between corresponding parts of the two appear to be variations only of an individual character. In the second specimen the large inguinal scute passes just over the back edge of the hyosternal plate, while in the former one it extends upon it for half an inch.

Measurements derived mainly from the more complete specimen are as follows.

	Length.	Breadth.
	Lines.	*Lines.*
First vertebral plate	27	13
Second vertebral plate	22	14
Third vertebral plate	23	14
Fourth vertebral plate	22	14
Fifth vertebral plate	22	15
Sixth vertebral plate	17	16
Seventh vertebral plate	10	17
Eighth vertebral plate	7	12
Ninth vertebral plate	9	14
Tenth vertebral plate	13	16

	Length.	Width internally.	Width externally.
	Lines.	*Lines.*	*Lines.*
First costal plate	45	26	37
Second costal plate	60	24	26
Third costal plate	70	24	27
Fourth costal plate	73	22	24
Fifth costal plate	69	20	30
Sixth costal plate	60	17	22
Seventh costal plate	47	15	20

The nuchal plate fore and aft has been about $2\frac{3}{4}$ inches; its breadth about an inch greater.

The marginal bones, so far as preserved, appear to have nearly the proportions and aspects as in Dermatemys. Their vertical measurement is about 2 inches, and their width about the same.

The measurements of the scute impressions are as follows:

	Length.	Breadth.
	Lines.	*Lines.*
First vertebral scute impression	39	52
Second vertebral scute impression	50	30
Third vertebral scute impression	43	33
Fourth vertebral scute impression	38	32
First costal scute at middle	54	56
Second costal scute at middle	74	46
Third costal scute at middle	70	46

The episternals at their inner border measure 1 inch in length; at their posterior extremity 10 lines.

The entosternal bone is 2½ inches fore and aft, and 2¾ inches wide.

In the better preserved of the two specimens the plastron presents the irregularity of having the left hyposternal and xiphisternal near half an inch more produced forward than upon the right side, as seen in Fig. 2, Plate XII.

Measurements of the remaining sternal bones are as follows:

	Lines.
Length of right hyosternal internally	30
Length of left hyosternal internally	25
Breadth of hyosternals at middle	58
Length of hyposternals internally	34
Breadth of hyposternals at middle	52
Length of right xiphisternal internally	34
Length of left xiphisternal internally	40
Breadth of anterior border	24

Measurements of the scutes are as follows:

	Lines.
Length of gular-humeral scute internally	28
Breadth of gular-humeral scute posteriorly	26
Length of pectoral scute internally	18
Breadth of pectoral scute posteriorly	36
Length of abdominal scute internally	29
Breadth of abdominal scute posteriorly	28
Length of femoral scute internally	26
Breadth of femoral scute posteriorly	22
Length of caudal scute internally	27
Axillary scute obliquely from within outward	32
Axillary scute at posterior border	13
Middle scute of sternal bridge fore and aft	17
Middle scute of sternal bridge at middle, transversely	25
Inguinal scute fore and aft	32
Inguinal scute at anterior border	13
Inguinal scute at middle between prominent angles	29

BAENA.

By this name I have distinguished a remarkable genus of turtles, indicated by remains in the Bridger Tertiary beds. It partook of characters of the snappers or chelydroids, the terrapins or emydoids, and the sea-turtles or chelonioids. The specimens upon which the genus is founded consist of shells, which are mostly so much crushed and distorted as to render it somewhat uncertain as to their exact original and perfect form. They were apparently about as prominent as in our snapper, and had nearly the same outline of shape. The middle of the carapace is not depressed as in the latter, but is somewhat flattened, and forms a continuous convexity with the sides. The posterior extremity presents a deep emargination as in the snapper, and on each side is notched likewise as in the latter.

The plastron of Baena is emydoid in character, and in its degree of development in relation with the carapace approaches that of its associate genus Baptemys. As in this, large spaces exist between the extremities of the plastron and carapace, but comparatively of much less extent than in Chelydra. The pedicles of the plastron are immovably conjoined with the carapace. They are as wide relatively as in the emydoids, but are much longer. The two extremities of the plastron are nearly alike in shape, being tongue-like and feebly emarginate at the end.

The number, arrangement, and general form of the corneous scutes of the carapace appear to have been the same as in Emys and Chelydra. The plastron exhibits two pairs of gular scute areas, which, together with the other scute areas, made seven pairs to the plastron. In addition to these the pedicles exhibit a row of scute areas between the former and the marginal scute areas of the carapace, as in the sea-turtle, the snapper, Dermatemys, and Baptemys.

A feature which may be regarded as a character of Baena is the obliteration of the sutures, and the shell at maturity has the bones so co-ossified that their original boundaries cannot be traced.

The true ribs or costal arches, connate with the costal plates, are remarkably prominent in Baena, and the costal capitula are well developed. In several specimens, in which portions of the carapace are broken away, the mass of rock within exhibits deep concave grooves indicating the former position of the rib-arches.

The sustaining columns of the carapace, springing as processes from the hyosternal and hyposternal bones of the plastron, are of great comparative breadth, and subdivide the interior of the shell into three compartments as in the Batagur, a genus of fresh-water turtles now living in India.

Baena arenosa.

The species thus named was originally founded on a specimen consisting of a nearly complete turtle-shell discovered at the junction of the Big Sandy and Green Rivers, Wyoming, during Professor Hayden's exploration of 1870. The specimen is represented in Figs. 1, 2, Plate XIII.

The shell, besides appearing to be in some degree crushed downward or flattened, has lost the fore part and right border of the carapace. The plastron, less injured, has lost its anterior extremity.

The outline of the carapace appears to have been broadly oval; and the shell was apparently not more elevated than in our common snapper.

All the bones of the carapace and plastron are so intimately co-ossified that the position of the former sutures cannot be detected. The grooved boundaries of the scutal areas are, on the other hand, well marked.

The carapace corresponding with the position of the intermediate vertebral scutes is flattened and slightly depressed at the middle. It is most prominent along the lateral boundaries of the vertebral scutes. In the position of the last of the latter it is most prominent at the middle. No distinct carination exists, but a feeble and widely interrupted ridge occupies the median line of the carapace, scarcely noticeable were it not better developed in other specimens. The sides of the carapace slope evenly outward to the rounded flexure of the lateral marginal plates.

The posterior marginal plates are notched as in the snapper, and are slightly recurved at the prominent ends. Between the last pair of marginal bones a wide concave emargination exists, as in Chelydra, but of less depth.

The intermediate vertebral scute tracts are nearly square, and are as broad as, or a little broader than, long. The lateral grooves have the usual brace form. The groove between the second and third tracts is convex forward; the succeeding one much less so; and that between the fourth and fifth tracts is much produced forward with a mammiform outline.

The costal scute tracts are nearly like those of Emys and Chelydra. Their grooves are directed nearly parallel outwardly, except the extreme back and front ones.

The plastron appears quite flat and nearly on the same level with its pedicles, but this condition is evidently in some degree the result of accidental pressure from above. The posterior extremity is broad, linguiform, with the end slightly and concavely emarginate.

The pectoral scute impressions, as in the Chelydra, are larger than any others of the plastron. They extend outwardly on half the breadth of the sternal bridges. The anterior groove is directed outwardly on a level with the bottom of the axillary fossæ, and near its end turns abruptly and obliquely forward to the edge of the latter.

The abdominal scute impressions, shorter than those next in front and behind, extend upon the posterior half of the breadth of the sternal bridges. Their posterior groove is directed obliquely outward and backward to the bottom of the inguinal fossæ.

The femoral are considerably larger than the costal scute impressions, and defined from them by a sigmoid groove.

The bridges of the plastron present a row of four large scutal areas intervening between the pectoral and abdominal scute areas internally, and the marginal-scute areas of the carapace externally. The first and last of these may be regarded as homologues of the comparatively small axillary and inguinal scute areas of Emydes; the intermediate ones are superadded.

The axillary scute area, partially broken away in the specimen, appears to have had four borders, of which the anterior formed the outer boundary of the axilla, and the internal joined the pectoral scute area.

The second submarginal scute area, the smallest of the series, is quadrate, and internally joins the pectoral scute area. The succeeding submarginal area, larger than those in advance, is pentagonal, with the two shorter sides forming a projecting angle joining the pectoral and abdominal areas.

The inguinal scute area, larger than the others, has four borders, of which the internal joins the abdominal area, and the posterior bounds the greater part of the bottom of the inguinal space.

The surface of the carapace is somewhat irregular; that of the plastron is more regularly and minutely roughened or fretted in appearance.

A second nearly complete specimen of a shell of Baena was discovered by Dr. J. Van A. Carter at Church Buttes, on Black's Fork of Green River, three miles north of Fort Bridger, and was obligingly sent to the writer as a gift. The shell, like the former one, is considerably crushed, so as to render an exact determination of its original form uncertain. It approximated the other specimen both in shape and size, and, like it, has all the bones so completely co-ossified that their limits are obliterated.

This second specimen presents several differences from the former one, which led to its having been considered as pertaining to another species, to which the name of *B. affinis* was given. Additional specimens since obtained and exhibiting other variations have led to viewing all of them as belonging to a single species.

The carapace measures 13 inches in length following the curvature. Its anterior portion, preserved in the specimen on one side, has a rather obtuse border, and is not recurved. In front it is prominent, as far as seen in the specimen, corresponding with the position of what appears to be the outer portion of the nuchal scute area. The latter apparently is of great width, at least an inch at its conjunction with the first vertebral scute area.

The latter and the last of the series are prominent in the median line, where they form a thick, rounded ridge. A low interrupted ridge extends along the median line of the carapace, which is barely evident in the first-described specimen. The short divisions of the ridge are flanked by equally long fusiform elevations slightly divergent forward. In addition, the carapace is rather irregularly prominent along the position of the lateral grooves of the vertebral scute areas. The intermediate vertebral scute areas are proportionately narrower than in the first specimen. The second and third are slightly longer than wide; the fourth a little wider than long; and the first and last in width considerably exceed the length.

The plastron is preserved nearly complete, and is represented in Fig. 3, Plate XIII. It appears as if originally it had been less flat than in the former specimen, as, independently of fractures, it turns up more at the extremities as well as at the bridges.

The anterior extremity, which is lost in the former specimen, affords an opportunity of completing our knowledge of the plastron. It is shorter and narrower than the posterior extremity, but is nearly like it in shape. The free border is obtusely rounded, and is slightly more thickened and prominent at the divisions produced by the scute impressions. These do not mark the upper surface as in the Emydæ. The lower surface exhibits one of the most remarkable peculiarities of the genus, which is the possession of two pairs of gular scutes.

The first pair of gular scutes are comparatively small, and are defined posteriorly, in the usual manner, by oblique grooves diverging at an angle of 45°.

The second pair of gular scute impressions escaped my notice until I had seen several additional specimens. As this did not occur until after the drawing of Fig. 3 was made, they are not there represented. They are seen in Fig. 1, Plate XV, which was subsequently and more accurately drawn from the same specimen. They are rather larger than the first pair, and are defined posteriorly by a serpentine groove directed outwardly nearly from the same point as the grooves in advance.

The remaining scute areas of the plastron are nearly like those of the preceding specimen, except those covering the pedicles.

Only three scutes covered the latter in the second specimen, the one corresponding with the first submarginal scute area of the first specimen being deficient. In consequence of its absence, a modification of the outlines of the contiguous ones resulted. The posterior groove of the axillary scute, and the anterior groove of the area corresponding with the second submarginal scute in the first specimen, instead of being transverse are oblique and join each other at an angle externally. The posterior two scutal areas also differ from those of the first specimen in being separated by a groove directed obliquely outward and backward instead of nearly transversely.

The surface of the plastron exhibits the same minutely fretted appearance as in the former specimen.

In the perfect condition the two specimens of Baena which have been described differed but little in size. The length of the carapace in a straight line has approximated 13 inches, the breadth 9 or 10 inches. The length of the plastron is 11 inches; its breadth to its conjunction with the carapace about 8 inches.

A third and less perfect specimen of the shell of *Baena arenosa*, consisting of the central portion of the carapace and nearly the corresponding portion with the anterior extremity of the plastron, was found by Dr. Carter on Henry's Fork of Green River, and presented by him to the Academy of Natural Sciences.

This specimen had about the same size as the previous ones, and like them has all the bones completely co-ossified. The median ridge of the carapace is more distinct than in the other specimens, and its divisions appear more or less distinctly to mark the position of the vertebral plates, while the lateral diverging prominences also appear to mark the sides of these plates.

The intermediate vertebral scute areas are intermediate in proportions to those of the former specimens.

The surface of the plastron is smooth and exhibits no trace of the minutely fretted condition observed in the former specimens. The grooves defining the sternal scute impressions, the median groove as well the more transverse ones, are less regular in their course than in the other specimens.

The anterior extremity of the plastron, represented in Fig. 2, Plate XV, is flat, and exhibits the second pair of gular scute areas larger than in the former specimen in which they exist, while their more tortuous back groove starts from the median groove a half inch behind that in front. The rounded border is more prominent in the position of the gular scute impressions than in the former specimen.

A small part of the sternal bridges retained in the specimen shows a portion of the second submarginal scute area with an internal projecting angle intermediate in extent to that of the former specimens. On one side, also, a small portion of the first submarginal area is retained, and this appears to indicate that it was nearly of the size and shape of that in the first-described specimen of Baena.

Comparative measurements of the three described specimens of *Baena arenosa* are as follows:

	Lines.	Lines.	Lines.
Length of first vertebral scute at middle		22	
Breadth of first vertebral scute at middle		36	
Length of second vertebral scute at middle	36	35	35
Breadth of second vertebral scute at middle	36	30	32
Length of third vertebral scute at middle	35	35	34
Breadth of third vertebral scute at middle	37	31	32
Length of fourth vertebral scute at middle	29	28	
Breadth of fourth vertebral scute at middle	37	30	33
Length of fifth vertebral scute at middle	32	26	
Breadth of fifth vertebral scute at middle	48	40	
Width of first costal scute internally		34	
Width of second costal scute internally	37	36	38
Width of third costal scute internally	32	30	30
Width of fourth costal scute internally	21	17	
Length of anterior prolongation of the plastron		35	37
Breadth at base of anterior prolongation of the plastron	46	41	44
Length of posterior prolongation of the plastron	48	48	
Breadth at base of posterior prolongation of the plastron	50	52	52
Breadth of pedicles of plastron	62	58	56
Length approximately of pedicles of plastron	26	24	24
Length of gular scutes internally		10	16
Breadth of gular scutes at back border		14	15
Length of humeral scutes internally		26	21
Breadth of humeral scutes at back border	24	19	22
Length of pectoral scutes internally	32	24	22
Breadth of pectoral scutes at middle	41	45	50
Breadth of pectoral scutes at back border	31	31	33
Length of abdominal scutes internally	15	22	20
Breadth of abdominal scutes at middle	41	38	44
Breadth of abdominal scutes at back border	34	29	28
Length of femoral scutes internally	28	26	24
Breadth of femoral scutes at back border	21	22	19
Length of caudal scutes internally	21	22	
Fore and aft diameter of first scute of pedicle		18	
Fore and aft diameter of second scute of pedicle	16		
Fore and aft diameter of third scute of pedicle	20	24	
Fore and aft diameter of fourth scute of pedicle	22	20	

A fourth specimen referable to *Baena arenosa* consists of a small portion of the carapace with a large portion of the plastron, from Henry's Fork of Green River, found by Dr. Carter, and presented by him to the Academy of Natural Sciences of Philadelphia. This exhibits no peculiarity, excepting that the scutal grooves of the plastron are more irregular in their course than in the preceding specimens. The median groove of the plastron is especially tortuous, while in the other specimens it is nearly straight. A retained portion of one of the sternal bridges exhibits evidences that four scutes impressed them, arranged nearly as in the first-described specimen of Baena. The surface of the plastron is less smooth than in the previous specimen, but it does not present the fretted appearance of the two former ones.

A fifth specimen, apparently referable to *B. arenosa*, consists of the anterior extremity of a plastron, represented in Fig. 3, Plate XV. It was found at Grizzly Buttes by Dr. Joseph K. Corson, and by him presented to the Academy of Natural Sciences. It would appear from its size as if it had belonged to a larger individual than the preceding specimens. It is nearly flat, or in a trifling degree convex, and is smooth, or without any appearance of fretting. It exhibits the four gular scute areas of unequal extent.

Another specimen, consisting of the anterior extremity of a plastron, apparently of a young animal of the same species, is represented in Figs. 4, 5, Plate XV. The specimen was found at the junction of the Big Sandy and Green Rivers, Wyoming, during Professor Hayden's exploration of 1870.

In this specimen the sutures are visible, and the contiguous bones defined. The grooves defining the two pairs of gular scutes start all from the same point, which is near the center of the entosternum. The entosternal bone viewed below is pyriform, but in the reverse position to that ordinarily observed in emydes. Viewed above, it resembles that of the snapper, (*Chelydra*,) or that of the sea-turtle, (*Chelone*.) In front it is received between the episternals; behind, it forms two lateral barbs projecting obliquely outward between the episternals and the hyosternals, and a long, median, pointed process extending between the hyosternals. The episternals posteriorly are angular, and are there received into a notch of the hyosternals.

From the matrix of the first-described specimen of the shell of Baena I obtained a portion of the pelvis, which presents some anatomical points of importance.

Comparative measurements of the anterior extremity of the plastron, where this is present in the specimens, are as follows:

	First specimen.	Second specimen.	Third specimen.	Fourth specimen.	Sixth specimen, arch, young.
	Lines.	Lines.	Lines.	Lines.	Lines.
Length of plastron to anterior pectoral groove	....	34	36	37	....
Breadth of plastron at anterior pectoral groove	44	36	42	44	....
Breadth of plastron at anterior humeral groove	....	28	28	31	20
Breadth of plastron at middle gular groove	....	18	18	20	13

The pelvis is more expanded above than in Emys, and in this respect is more like that of the snapper. The sacrum represented in Fig. 9, Plate XVI, is intermediate in its proportions to that of the two genera just mentioned.

The length of the sacral vertebræ of Baena, independently of the wings or transverse processes, exceeds the breadth, the proportions in this respect according more with the condition in the terrapin than in the snapper. The second sacral vertebra is, however, larger than the first, as in the latter turtle, and the reverse of what it is in the former. The inferior surface of the bodies of the sacral vertebræ is half cylindroid, depressed at the sides in the first one, but scarcely so in the second.

The anterior articular surface of the first sacral centrum is moderately convex; the posterior articular surface of the second centrum is concave. In Emys the corresponding surfaces are flat, or nearly so; in Chelydra the anterior one is concave, the posterior convex, with lateral extensions nearly flat.

The proportionate length and robustness of the sacral alæ of Baena agree more nearly with the condition in the snapper than in the terrapin. In Emys the posterior alæ are comparatively feeble appendages, and they join the ends of the anterior alæ by means of a ligament. In Baena the posterior alæ are strong processes, as in the snapper, and likewise, as in this, join the ends of the alæ in advance by suture, but appear not to be prolonged to join the ilium.

The innominatum of Baena, as represented in Fig. 8, Plate XVI, is proportionately of more robust character than in Emys. The ilium in shape is more like that of this genus than that of the snapper, but is proportionately of much greater breadth, the wing being of nearly double the expanse.

The expanded extremity of the first sacral wing articulates with the anterior extremity of the crest of the ilium. In Emys it articulates with the latter midway to the two prominent extremities of the crest.

The acetabulum and commencements of the ischiatic and pubic rami present nothing peculiar from the condition observed in the snapper.

Measurements of the pelvic specimens are as follows:

	Lines.
Length of sacrum beneath the centra	9½
Length of first sacral centrum	4½
Breadth of first sacral centrum	4½
Length of second sacral centrum	4½
Breadth of second sacral centrum	4½
Breadth of sacrum at first pair of alæ	30
Length of first sacral alæ	13
Length of second sacral alæ	11
Length of innominatum	23
Breadth of crest of innominatum	18
Height of acetabulum	6
Breadth of acetabulum	9½

CHISTERNON.

CHISTERNON UNDATUM.

A large turtle-shell, discovered by Dr. Carter in a chain of buttes a few miles from Fort Bridger, and presented to the Academy of Natural Sciences of Philadelphia, was originally described by me under the name of *Baena undata*. A careful examination of the specimen has led me to view it as pertaining to a different and heretofore undescribed genus.

The specimen represented in Plate XIV, one-half of the diameter of nature, consists of the intermediate portion of a shell, with the extremities broken away nearly in the position of the broad columns which spring from the plastron to support the carapace. Though much fractured, it appears to have been but little so while it lay imbedded in the deposit from which it was derived, so that it now retains its orginal form. The upper shell is as much vaulted as in some of the living land-turtles. This form, together with the thick bone and strong, broad sternal supports, enabled it to sustain the great weight of superincumbent pressure which has crushed so many of its associates. The interior of the shell is occupied by a greenish-gray sandstone, from which I obtained a pair of sacral vertebræ.

The outline of the shell in its perfect condition was ovoid as in ordinary Emydes; narrower and more elevated in front, wider and more depressed behind. The fore part and sides of the carapace are uniformly convex, but the hind part appears to have had the margin somewhat recurved. Over the position of the vertebral scute areas the surface is flat and even and without a carina. The plastron is flat, but its bridges turn from their commencement upward and outward to the border of the carapace, which is elevated $2\frac{1}{2}$ inches above the level. The highest part of the shell is nearly 6 inches above the level of the plastron. The bones of the shell, especially those of the carapace, appear co-ossified, but not so completely as in *Baena arenosa*, for those of the plastron can be distinctly traced.

The intermediate vertebral scute areas have nearly the form and proportions of those of *Baena arenosa*, and are rather longer than wide. The costal scutes widened more outwardly than in that turtle, indicating a proportionately greater degree of prominence of the shell.

The lateral marginal scute areas are much like those in Emydes, but the groove defining them from the costal scute areas exhibits an unusually undulating course, not angular but serpentine or waving.

The number and relative position of the scute areas of the plastron and its bridges are the same as in Baena, but the median sternal groove defining them on the two sides is remarkable for its irregular serpentine course, repeatedly crossing the also somewhat irregular course of the median suture of the plastron.

The sutures of the plastron being visible, they reveal to us an unexpected peculiarity, the existence or absence of which cannot be determined in the shells of *Baena arenosa* from the total obliteration of the sutures.

The peculiarity in the plastron of Chisternon, to which the genus owes its name, is the presence of a large triangular bone, added to those which usually exist in turtles, on each side of the shell. This interealated or mesosternal bone commences at the center of the plastron and gradually widens outwardly to where it conjoins the marginal plates of the carapace at the intermediate half of the sternal bridge. The four sutures defining the mesosternal plates from those in front and behind cross the plastron obliquely. A similar bone exists in another extinct genus, the Pleurosternon, of the early Tertiary formation of England, but in this it has the shape of a parallelogram.

The sternal bridges of Chisternon present four large scutal areas nearly resembling those of *Baena arenosa*. They are not quite symmetrical on the two sides.

The axillary scute area is pentagonal, and is the smallest of the series. The anterior border is oblique, and bounds the axillary notch externally. Two outer borders form an obtuse angle and join the third and fourth marginal areas. The inner border joins the pectoral area.

The second submarginal area is second in size of the series. It is longer than broad, and nearly quadrate, but has its outer angles cut off. The inner border conjoins the pectoral area; the outer the fourth and fifth marginal areas.

The third submarginal area is but little larger than the axillary area. It joins the pectoral and abdominal areas internally, and the fifth and sixth marginal areas externally.

The inguinal area, the largest of the submarginal areas, is obliquely quadrate, longer than broad, and with the outer angles cut off. The posterior border bounds the inguinal notch; the inner border joins the abdominal area, and the outer border joins the sixth, seventh, and eighth marginal areas.

The inferior surface of the plastron is comparatively smooth. Striations cross the sutures, and elsewhere it presents a finely reticulo-vascular appearance.

The fractured condition of the shell affords us an opportunity of seeing the strong hyosternal and hyposternal columns which aid in sustaining the carapace. These columns are broad, vertical plates reaching far into the cavity of the shell and dividing it into three compartments, as in the Batagur of India.

The hyosternal columns are $2\frac{1}{4}$ inches wide from their inner concave border to the axilla. The aperture of the shell between them is a doorway 3 inches wide near the roof and $4\frac{1}{4}$ inches near the floor. The hyosternal columns, partially exposed in the specimen, appear to be co-extensive with the anterior supports.

The breadth of the shell of *Chisternon undatum*, between the lateral obtuse borders of the carapace, is 15 inches. The length of the shell, or of the carapace, in a straight line is estimated to have been about a foot and a half. The length of the plastron is estimated to have been about 14 inches; its breadth at the root of the posterior extremity is $5\frac{1}{2}$ inches; and at the root of the anterior extremity has been rather less. The sternal bridges measure 7 inches fore and aft, and their length to the outer edge of the carapace is 5 inches.

Other measurements of the shell are as follows:

	Inches.
Length of second vertebral scute area, estimated at	$4\frac{1}{4}$
Breadth of second vertebral scute area	$3\frac{1}{4}$
Length of third vertebral scute area	$4\frac{1}{8}$
Breadth of third vertebral scute area	$3\frac{1}{2}$
Length of fourth vertebral scute area	$3\frac{1}{2}$
Breadth of fourth vertebral scute area	$3\frac{1}{2}$
Breadth of second costal scute area, internally	$4\frac{3}{8}$
Breadth of second costal scute area, externally	$4\frac{7}{8}$
Breadth of third costal scute area, internally	$3\frac{5}{8}$
Breadth of third costal scute area, externally	$3\frac{3}{4}$
Height of sixth and seventh marginal scute areas	$2\frac{1}{2}$
Length of hyposternals internally	5
Breadth of hyposternals internally	$5\frac{3}{4}$
Breadth of hyosternals	$5\frac{3}{4}$
Breadth of plate intercalated between the hyosternals and hyposternals	6
Extent of the same plate at the base externally, fore and aft	$4\frac{3}{8}$
Breadth of groove between pectoral and abdominal areas where it joins the projecting angle of the third sterno-costal scute areas	$7\frac{1}{2}$
Breadth of plastron at anterior suture of xiphisternals	$4\frac{1}{2}$
Breadth of pectoral scutes to sterno-costal scutes	4 and $4\frac{1}{2}$
Length of abdominal scute internally	3
Length of femoral scute internally	$3\frac{3}{4}$
Axillary scute area, length at middle	$1\frac{3}{4}$
Axillary scute area, breadth at middle	$2\frac{1}{2}$
First submarginal scute area, length at middle	$2\frac{1}{2}$
First submarginal scute area, breadth at middle	$2\frac{3}{4}$
Second submarginal scute area, length at middle	2
Second submarginal scute area, breadth at middle	$2\frac{3}{4}$
Inguinal scute area, length at middle	$2\frac{3}{4}$
Inguinal scute area, breadth at middle	$2\frac{1}{4}$

The sacral vertebræ, represented in Figs. 11, 12, Plate XIX, are of proportionately greater length than in Baena. The first one is nearly as long as it is broad; and the second is half as long again as the former, and is equal in this respect to its breadth.

The anterior articulation of the first sacral centrum forms a decided cup-like depression, and not merely a transverse concavity like that in the snapper. The second sacral centrum is prolonged to an unusual degree beyond the neural arch. It ends in a flat, roughened articular surface, as if intended for the conjunction of another vertebra entering into the constitution of the sacrum. The neural arches of the sacral vertebræ are proportionately higher than in the snapper, and they appear to have articulated movably with each other by zygapophyses alone.

The diapophyses are about equally developed with those in the snapper. The neural arch is not co-ossified with the centrum; nor are the pleurapophyses co-ossified with either.

Measurements of the sacral vertebæ are as follows:

	Lines.
Length of the sacrum inferiorly	18
Length of first sacral centrum	7
Breadth of first sacral vertebra, with diapophyses	9
Length of second sacral centrum	11
Breadth of second sacral vertebra, with diapophyses	10½
Height of first sacral vertebra to end of spinous process	13
Height of anterior articulation of first sacral centrum	5
Breadth of anterior articulation of first sacral centrum	5

An isolated vertebra, from Henry's Fork of Green River, looks as if it might be the first sacral of *Chisternon undatum*. The body is little more than half the length of that of the last sacral above described, but its anterior articular surface agrees in size, form, and roughness with the posterior surface of the last sacral centrum just mentioned. The pleurapophyses have about the same degree of development as in the snapper.

Fig. 10, Plate XVI, represents a caudal vertebra, obtained by Dr. Carter near Lodge-Pole Trail. In construction it resembles the caudals of the snapper, the centrum, as in this, being opisthocoelian, or having a cup behind and a ball in front. The proportions of the vertebra accord best with the more anterior caudals of the snapper, but its transverse processes are as small as in the terminal caudals of the latter. Perhaps it may belong to Chisternon, but the opinion is conjectural. If the former isolated vertebra belongs to Chisternon, it is doubtful whether this second one does.

Fig. 7, Plate XVI, represents an isolated ilium of a turtle, found at Grizzly Buttes by Dr. Carter. It resembles in its form that of a snapper, but is more robust in proportion to its length. The inner surface at the upper extremity is flat and longitudinally striated, but is devoid of the fossa existing in the snapper. The length of the bone is 3¼ inches; the width of its upper end ¾ of an inch; the width at the lower end is 17 lines. From the form of the bone I would suspect that it belonged to Chisternon, rather than to either *Emys Carteri* or *Baptemys wyomingensis*.

HYBEMYS.

HYBEMYS ARENARIUS.

Two little specimens, obtained by Professor Hayden, in the Tertiary formation of Little Sandy Creek, Wyoming, appear to indicate a previously undescribed turtle, to which the above name was given. They consist of a detached marginal bone, and a fragment of a costal plate of a species about the size of the common spotted turtle, *Emys guttata*. The bones are unusually thick in proportion to their breadth, compared with those of ordinary recent Emydes. Their surface is smooth and strongly marked by the lines of separation of the scute areas. The costal ridge on the interior of the costal plate is scarcely perceptible; the costal capitulum is rather stouter than in Emydes.

The marginal plate represented in Fig. 9, Plate XV, is especially remarkable, and it is upon its peculiarity that the genus is inferred. It would appear to correspond with the ninth of the series, and has the same form as in the corresponding plate of ordinary Emydes. The outer portion of the upper surface, strongly defined by the groove of the costal scute, exhibits at its fore and back part a half-circular boss, occupying the middle of the marginal scute areas. As we may safely infer the other marginals to have the same construction, it follows that the margin of the carapace is ornamented with a circle of hemispherical bosses, each of which is crossed by the sutures of the marginal bones.

ANOSTEIRA.

ANOSTEIRA ORNATA.

Among the many remains of turtles from the Bridger Tertiary formation, submitted to my examination from time to time, by Dr. Carter and Professor Hayden, there were a few isolated plates of peculiar character which were described and referred to a genus and species under the above name. Subsequently Dr. Carter discovered many parts of a shell of the same species, which we have endeavored to collocate as represented in Figs. 1, 2, Plate XVI.

Anosteira is a remarkable genus, very unlike any other turtle, previously described, recent or extinct. The carapace and plastron, while being completely ossified as in Testudines, Emydes, &c., are ornamented in a manner only seen to the same degree in the soft-shelled turtles. True, we see something like ornamentation of the same kind in some of the Emydes, but in them the condition is comparatively feeble. The osseous shell also appears to be devoid of the usual outlines more or less strongly expressed of the investing scutes. A few of the plates exhibit obscure lines, but I am uncertain as to whether they accord with the areas of the scutes.

The outline of the carapace is broadly cordiform and somewhat resembles that of the ordinary sea-turtles, but is not acute posteriorly as in these, being obtuse as in the Emydes. The prominence of the carapace is moderate as in the less elevated forms of the latter. It is uniformly convex, except that it is acutely carinated in the median line posteriorly.

The margin of the carapace anteriorly is rather obtuse, but laterally and posteriorly is quite sharp. It is broadly and concavely notched in front; the first pair of marginal plates being the most prominent portions anteriorly. Antero-laterally it is slightly concave, and from this position posteriorly is uniformly convex.

The plastron with its bridges is flat, and is intermediate in its relative proportions with that of the snappers and Emydes. The bridges articulate with the carapace by gomphosis, as seen in Fig. 2. They join the marginal plates from the fifth to the eighth inclusive. The extremities of the plastron are both broken away in the specimen.

The vertebral plates of the carapace are narrow coffin-shaped. Those anterior are nearly level; those posterior are acutely carinated.

The costal plates within exhibit no costal elevation, but are quite level, as represented in Fig. 3. The costal capitula are unusually broad but thin.

The inner surface of the nuchal plate at the posterior border presents a pair of round articular processes for conjunction with the contiguous vertebra.

The upper surface of the carapace is ornate with rugosities. These are obsolete on the vertebral plates. On the costal plates they appear as longitudinal, undulating, and nearly parallel ridges crossing the plates. Internally they are feebly developed and become more strongly marked proceeding outwardly.

On the marginal, including the nuchal and pygal plates, the rugosities are finer, closer, more interrupted, and in part even granular.

Beneath, the rugosities of the marginal plates have a decidedly radiant appearance. The under surface of the marginals in advance of the axillary notches, and the corresponding surface of the nuchal plate, are smooth or devoid of the ornate rugosities.

The pygal plate and the contiguous marginals increase in thickness from their free acute edge inwardly, so as to be wedge-shaped in section. The base of the wedge, directed toward the cavity of the shell, is strongly grooved in the pygal plate, and gradually less so in the contiguous marginal plates. The groove contributes to the general cavity of the shell. Fig. 6 represents a fore and aft section of the pygal plate, exhibiting the groove on its inner part.

The plates of the plastron exhibit their ornate ridges arranged in a radiating manner, as seen in Fig. 2, but they are less prominent than those of the carapace.

The shell of the specimen of Anosteira, from which the above description was taken, in its entire condition, was about 5 inches in length in the median line and about 4½ inches in breadth.

Figs. 4, 5, represent two anterior marginal plates, showing that the species reaches a much greater size.

TRIONYX.

TRIONYX GUTTATUS.

One or more species of the soft-shelled turtles (*Trionyx*) are indicated by an abundance of fragments of shells which have come under my notice in the various collections of fossils from the Bridger beds. Anything like complete shells appear to be rare, as the best preserved which has yet been submitted to my examination is the portion of a carapace represented in Fig 1, Plate IX. The specimen, attached to a mass of sandstone, was obtained at Church Buttes, near Fort Bridger, during Professor Hayden's exploration of 1868

The osseous carapace in its entire condition is estimated to have been about a foot and a quarter in length, and, independently of the extension of the free ends of the ribs, has nearly reached that breadth. The bones range from three to four lines in thickness, except along the position of the costal ridges and near the thinner edges.

The carapace appears to have had the usual composition of seven vertebral plates, and eight pairs of costal plates back of the nuchal plate. It was moderately convex, and the posterior border in the specimen is deeply scolloped.

The vertebral plates in the specimen, consisting of part of the second, third, and fourth, are reversed coffin-shaped, and nearly twice as long as wide. Their anterior border is convex, and the posterior border concave.

The fifth vertebral plate is smaller than the preceding, and becomes earlier narrowed from the sides toward the back end.

The sixth plate is lozenge-shaped, about as long as it is wide, and occupies the space between the truncated angles of the sixth and seventh costal plates. The latter meet in the median line for more than half their width.

The seventh vertebral plate is a very small lozenge-shaped bone, with a crucial ridge on its surface, occupying an interval produced by the truncation of the contiguous angles of the seventh and eighth pairs of costal plates.

The costal plates, from the fourth to the sixth inclusive, are nearly of the same width internally, and they successively become more widened outwardly. The seventh costal plate is rather wider at the extremities than intermediately. The last costal plates are nearly as wide fore and aft as from within outwardly.

The surface of the carapace is sculptured for the most part with broad, rounded, and isolated concave pits resembling the impression of rain-drops on a soft surface. Only near the outer border of the costal plates, where these are preserved, do the pits become more or less confluent, usually in twos and threes. The reticular ridges bounding the pits are broad and low, and often as wide as the included pits.

Measurements of the specimen are as follows:

	Lines.
Length of third vertebral plate	23
Width of third vertebral plate in front	7
Width of third vertebral plate behind	12
Length of fourth vertebral plate	$20\frac{1}{2}$
Width of fourth vertebral plate in front	$7\frac{1}{2}$
Width of fourth vertebral plate behind	$10\frac{1}{4}$
Length of fifth vertebral plate	$19\frac{1}{2}$
Width of fifth vertebral plate in front	7
Width of fifth vertebral plate at middle	$8\frac{1}{2}$
Length of sixth vertebral plate	10
Width of sixth vertebral plate at anterior third	10
Length of seventh vertebral plate	$5\frac{1}{2}$
Width of seventh vertebral plate at middle	5
Width of fourth costal plate fore and aft at inner part	23
Width of fifth costal plate fore and aft at inner part	22
Width of sixth costal plate fore and aft at inner part	21
Width of seventh costal plate fore and aft at inner part	$19\frac{1}{2}$

	Lines
Width of eighth costal plate fore and aft at inner part	18
Length of sixth costal plate at middle	58
Length of seventh costal plate at middle	41
Length of eighth costal plate at middle	22

Many fragments, both of the carapace and plastron of soft-shelled turtles, collected during Professor Hayden's expedition of 1870, and subsequently by Drs. Carter and Corson at various localities in the vicinity of Fort Bridger, appear to be referable to the same species as the above.

A specimen consisting of the right half of a nuchal plate, with an attached piece of a first costal, derived from the same locality as the specimen above described, belonged to an animal about the same size. The width of the scabrous portion of the nuchal plate in its complete condition was about $6\frac{3}{4}$ inches; its fore and aft extent $1\frac{1}{2}$ inches. The sculpturing of the surface is more interrupted or broken than in the specimen specially referred to *Trionyx guttatus*. The reticular ridges are narrower and sharper, and exhibit a disposition to rise in points at their intersection.

A specimen consisting of an outer portion of an intermediate costal plate measures $3\frac{3}{4}$ inches wide, and is 5 lines thick. The reticulation of its surface is unbroken, but otherwise it resembles that of the nuchal plate just described.

TRIONYX UINTAENSIS.

During my stay at Fort Bridger, in a trip to Dry Creek, Major R. S. La Motte discovered the nearly complete carapace of a Trionyx, which he presented to the Academy of Natural Sciences of Philadelphia. The specimen is represented in Fig. 1, Plate XXIX, one-half the natural size. On first view I supposed it to belong to the same species as the former, but comparison of the specimen with that of Fig. 1 of Plate IV leads to the belief that it pertains to a different one.

The carapace is about $16\frac{1}{2}$ inches long and 16 inches broad, so that its proportions are reversed from those in our living *Trionyx muticus*. It is about as convex as in the latter, and appears to have been slightly depressed along the position of the vertebral plates, judging from that portion of the shell back of the fifth costal plates, as in advance of this the specimen has been crushed inwardly. The fore and back part of the carapace is truncated, as in *T. muticus*. The posterior truncation, slightly sinuous, extends the width of the last two pairs of costal plates. In *T. guttatus* the corresponding border occupied by the latter is convex, and exhibits three deep sinuosities—the middle one and the one on each side, as seen in Fig. 1, Plate IX.

Eight pairs of costal plates succeed the nuchal plate. The second, fifth, and sixth pairs expand considerably outward, more especially the last of these. The others are of more uniform breadth.

The specimen possesses only six vertebral plates. Of these, the first is the longest and widest. Its fore border is convex, and nearly in a line with the suture between the nuchal and first pair of costal plates. The lateral borders diverge to the back angles, which are truncated to join the second pair of costal plates.

The second and third vertebral plates are nearly equal in size, and are reversed coffin-shaped. The fourth plate is smaller, and oblong quadrate, with convex borders. The fifth plate is obverse coffin-shaped, shorter, but wider than the former. The sixth vertebral plate has not more than half its usual development. It is pentagonal shield-like, and is included between the angles of the fifth and sixth costal plates.

The posterior half of the sixth costal plates, and those succeeding them, unite in the median line by a tortuous suture.

The surface of the carapace presents a nearly uniform reticular aspect and the thickness of the bones is of the usual proportion.

The measurements of the specimen are as follows:

	Inches.
Space occupied by the six vertebral plates	10½
Breadth of nuchal plate	9½
Extent of nuchal plate in median line	2¼
Breadth together of seventh pair of costal plates	8¾
Breadth together of eighth pair of costal plates	3½

	Length.	Breadth.
	Lines.	Lines.
First vertebral plate	28	16
Second vertebral plate	22	15
Third vertebral plate	22	13
Fourth vertebral plate	20	10
Fifth vertebral plate	18	12
Sixth vertebral plate	11½	11

	Length.	Depth internally.	Depth externally.
	Inches.	Lines.	Lines.
First costal plate	6	29	32
Second costal plate	8	22	38
Third costal plate	8	21	26
Fourth costal plate	8	23	26
Fifth costal plate	8	22	33
Sixth costal plate	7	18	42
Seventh costal plate	4½	16	21

REMAINS OF TRIONYX OF UNDETERMINED SPECIES.

Small fragments of Trionyx shells, from the Bridger Tertiary strata, exhibiting a different kind of surface-marking or sculpture from that of the specimens referred to the preceding species, probably indicate others, or, perhaps, different genera.

A specimen found by Dr. Carter at Dry Creek, and represented in Fig. 11, Plate XVI, is an outer fragment of a costal plate. It is not pitted as in *Trionyx guttatus*, but is crossed obliquely by coarse ridges with the intervals occupied by a lattice of narrower ridges. Probably the specimen may belong to a species of Anosteira.

Another fragment of a costal plate, from Little Sandy Creek, is represented in Fig. 12 of the same plate. This specimen differs from the former in being crossed by widely separated ridges, with the intervals finely pitted.

Other specimens exhibit slight differences from the foregoing and from those of *Trionyx guttatus*, but are too imperfect to enable one to form any idea of their relationship.

Order *Lacertilia*.

The lizards have vertebræ with concavo-convex bodies, and have the teeth co-ossified with the jaws. The skin is furnished with horny or bony scales. True lizards, allied to the existing monitors, iguanas, and chameleons, appear to have been abundant and of varied character in the ancient Wyoming fauna. Few remains of these animals, described in the succeeding pages, have been submitted to my inspection, but Professor Marsh has indicated and briefly described twenty-one species of five extinct genera from fossils obtained by him from the Bridger beds.

SANIVA.

SANIVA ENSIDENS.

An extinct lizard, to which the above name has been applied, is indicated by some remains discovered during Professor Hayden's exploration of 1870, near Granger, Wyoming. The remains consist of portions of a skeleton, in a fragmentary condition, imbedded in an indurated ash-colored marl rock. The bones are black, and the hollows of the long bones, including the ribs and phalanges, are occupied with crystalline calcite.

The remains belong to a lacertian about the size of the existing monitor of the Nile, to which it appears to have been closely related. The bones indicate a robust body, a long tail, and limbs with long toes.

The vertebræ resemble those of the Nilotic monitor in form and proportions, and like them possess no zygosphenal articulation.

A pair of dorsal vertebræ are represented in Fig. 15, Plate XV. The body is $\frac{1}{2}$ an inch long inferiorly, and measures $\frac{3}{4}$ of an inch between the diapophyses. The ball and socket extremities are twice the breadth of the height. The ball measures 4 lines in breadth and 2 lines in height. The neural arch laterally at the zygapophyses is nearly 8 lines long.

An anterior caudal has the same length as the dorsals, but is narrower. The ball is of less width, but the same height. The hypopophyses for the chevron are quite prominent, and are situated a short distance in advance of the ball, as in the monitor.

A small detached tooth, imbedded in the same mass, in proximity to some small skull-fragments, presents the form and constitution of those of the monitors. It is represented in Fig. 35, Plate XXVII, magnified eight diameters. The length of the tooth is about $1\frac{1}{2}$ lines; its breadth, $\frac{3}{4}$ of a line; and its thickness, $\frac{1}{2}$ a line. It is compressed conical, feebly curved inwardly and backward, sharp-pointed, has abruptly impressed trenchant borders, and is smooth and shining. It is hollow, and has thick walls. The transverse section of the base is rhomboidally oval, with acute poles.

In breaking off portions of the rock containing the bones above described, there was exposed what appears to be the anterior extremity of a maxillary containing the remains of six teeth. The fragment is 4 lines long and $1\frac{1}{2}$ lines deep. The teeth are pleurodont in character, but appear different in form from the isolated tooth above indicated, and have more resemblance in shape to those of the iguana. The specimen appears so small in its proportions with the other bones, that it leads to the suspicion that it may not belong to the same skeleton.

No scales were found in association with the bones in the same mass of rock.

The name Saniwa, according to Professor Hayden, is used by one of the Indian tribes of the Upper Missouri for a rock-lizard.

SANIVA MAJOR.

Several fragments, discovered by Dr. Carter near the Lodge-Pole Trail, crossing Dry Creek, would appear to indicate a larger species of Saniva. The specimens are of a greenish hue and somewhat smooth or water-worn, and were derived from a green sandstone stratum.

One of the fossils consists of the distal extremity of a humerus, represented in Fig. 14, Plate XV. It resembles in general aspect the corresponding portion of the humerus of the monitor, but the shaft is proportionately more robust, and not so much narrowed toward the middle. It is occupied by a large medullary cavity with compact walls, as in the humerus of a bird. The internal epicondyle appears less prominent than in the monitor, in consequence of the less degree of contraction of the shaft. The external epicondyle does not reach upward to more than half the relative extent it does in the monitor, and it is not perforated. The ulnar eminence is prominent in front, but projects below to a less degree than the radial capitellum.

The breadth of the bone at the epicondyles is $\frac{3}{4}$ of an inch. The greater diameter or breadth of the shaft at the broken end is $3\frac{3}{4}$ lines; the short diameter is $2\frac{1}{2}$ lines.

Another of the fossils consists of a pair of dorsal vertebræ, represented in Figs. 36, 37, Plate XXVII. They agree in all respects with the two vertebræ referred to the former species, except that they are considerably larger.

The bodies of the vertebræ inferiorly measure $7\frac{3}{4}$ lines in length and $10\frac{1}{2}$ lines in width between the diapophyses. The ball measures 5 lines in breadth and 3 lines in height. The neural arch laterally at the zygapophyses is 10 lines in length.

GLYPTOSAURUS.

Professor Marsh has described, in the American Journal of Science and Arts for 1871, another genus of extinct lizards, under the above name, from remains obtained in the Bridger Tertiary. He observes that "the head was covered with large osseous shields symmetrically arranged and highly ornamented. Other parts of the body, especially the ventral region, were protected by rectangular, ornamented shields, united to each other by suture. The teeth are pleurodont, and are round with obtuse summits. The dorsal and caudal vertebræ have the same general form as those of Varanus, but show traces of a zygosphene articulation.

Professor Marsh indicates eight species, mainly founded on differences in the position, form, and ornamentation of the dermal osseous shields and the form of the teeth.

Dr. Carter has submitted to my examination a number of specimens collected by him at Grizzly Buttes, which in part or whole are attributable to the same genus, and mostly to the species named *Glyptosaurus ocellatus*.

Several of the dermal shields from the trunk of the body are represented in Figs. 13 to 15, Plate XVI, and several of the cranial shields in Figs. 16, 17, of the same plate, all magnified two diameters.

The dermal shields of the trunk are oblong quadrate, with the longer margins thick and roughened for sutural conjunction with one another. The extremities thin out for imbrication. The anterior extremity, which is overlapped by the shield in advance, extends a third or more of the length of the plate, and is smooth. The posterior two-thirds or less of the shields are ornamented on their free surface with rounded knobs or tubercles, closely arranged in more or less concentric rows.

The cranial shields are from four to six sided, and proportionately of greater thickness than the former. All their margins are roughened for sutural attachment together, and their free surface is ornamented in the same manner as the shields of the trunk.

Accompanying the specimens of dermal shields above described, there are several detached vertebræ. One of the specimens is a dorsal vertebra resembling those of Saniva, but somewhat smaller, and, like them, presents no zygosphene articulation. It may probably belong to that genus. The other specimen is an intermediate caudal vertebra of the same proportions of length and breadth as in Saniva, but the ball and socket articulation is as high as it is wide. It has no zygophene articulation, and the hypopophyses for the chevron are immediately beneath the ball of the body. The length of the latter inferiorly is $2\frac{3}{4}$ lines.

CHAMELEO.

CHAMELEO PRISTINUS.

A small fragment of a lower jaw with teeth, discovered by Dr. Carter in the Bridger Tertiary formation, is represented in Figs. 38, 39, Plate XXVII, magnified three diameters. In every respect it agrees with the corresponding part of the jaw of the living chameleons, but indicates a much larger species. In a space of 5 lines the alveolar border is occupied by eight teeth successively increasing in size from before backward.

The teeth are laterally compressed conical, with the borders in front and behind somewhat extended and acute, and at the base produced into a minute denticle. Externally the bases of the teeth are separated by perpendicular furrows descending on the face of the jaw to the position of a finely perforate horizontal line. Beneath the bases of the teeth internally there is a wider and more conspicuous horizontal and perforated groove. Below this, toward the rounded base of the jaw, the usual Meckelian groove is situated. The outer face of the jaw exhibits two vasculo-neural foramina. The depth of the lower jaw from the point of the last tooth of the specimen is $2\frac{1}{2}$ lines.

FISHES.

The remains of fishes in the Bridger beds are not so abundant as one might have supposed from the nature of their composition and the conditions of their origin. Nevertheless, it is probable that fishes were abundant in genera, species, and individuals in the great Uintah Lake and its tributaries, whose deposits form the Bridger beds. The same circumstances which removed the less coherent parts of the skeleton from the interior of the many turtles, and likewise scattered the bones of these and of the multitude of other reptiles and of mammals, no doubt served to destroy the more delicate structure of the fishes and to distribute their hard parts through the mud. It is probable that future explorations may lead to the discovery of some strata of the Bridger beds in which well-preserved forms of fishes may exist like those found in the shales of the deeper beds of Green River.

The remains of fishes from the Bridger beds, which, with few exceptions, were found by Dr. Carter and submitted to my examination, consist mainly of smoothly enameled ganoid scales, a few isolated specimens of vertebral centra, portions of spinous rays, and fragments of jaws with teeth. My means of comparison of these specimens with the skeletons of recent fishes are exceedingly meager, but they indicate forms which generally appear to be most nearly related with our mud-fishes, (*Amia*,) and the gars, (*Lepidosteus*.)

Professor Marsh (Proceedings of the Academy of Natural Sciences of Philadelphia, 1871, p. 105) has already noticed specimens from the same locality, which he refers to two species of Amia about the size of *A. calca*, and two species of Lepidosteus about the same size as the modern gar-pike.

AMIA.

AMIA (PROTAMIA) UINTAENSIS.

A number of specimens, discovered by Dr. Carter on the buttes about ten miles from Dry Creek Cañon, indicate a large fish related with the modern Amia, but exhibiting sufficient peculiarity to pertain to a different genus, for which the name of Protamia has been proposed.

Figs. 1, 2, Plate XXXII, represent one of the best-preserved specimens, a vertebral centrum from the fore part of the dorsal series. Its breadth is considerably greater in proportion with its length than in Amia; it is more prominent below; has a different transverse outline; has shorter parapophyses, which also spring from a higher position at the sides, and the bottom of the articular cones is situated considerably above the centre.

The centrum is nearly four times the width and three times the height of its length. It is slightly curved from side to side with the convexity directed forward. It is widest at the upper third, opposite the origin of the parapophyses, and is shortest at the sides intermediately.

The articular cones have their bottom considerably above the center, and are more minutely perforate for the notochord than in Amia.

The sides of the centrum are concave between the prominent articular margins, and slant in a nearly straight line to the ridges defining the narrow inferior surface. The latter is concave, and the lateral ridges are obtuse, and excavated in an oblong shallow fossa at their fore part.

The upper part of the centrum is transversely convex between the parapophyses. The articular fossæ for the contiguous neural arches, as in Amia, are in the form of the figure of 8, and their internal prominent borders form the lateral limits of the bottom of the neural canal.

The parapophyses are short, stout processes projecting above the middle of the centrum from its widest part, and on a line with the bottom of the articular cones.

The measurements of the specimen are as follows:

	Lines.
Length of centrum inferiorly	5.6
Height of centrum anteriorly	15.5
Width of centrum in line with the parapophyses	20.0

Another specimen from the same locality as the preceding is represented in Figs. 3, 4, Plate XXXII. It appears to be the centrum of an atlas, and may probably belong to the same species as the former, though, judging from its different aspect, to a different individual.

The centrum is transversely oval and slightly curved, with the convexity of the curve directed forward. Its breadth is two and a half times the height, and over five times the length.

The anterior surface is nearly flat and somewhat uneven, and just above the center is depressed into a concave pit about one-fifth the diameter of the centrum. The posterior surface presents the usual cone with its bottom just above the center.

The sides of the centrum are concave between the articular borders, and bear no trace of parapophyses. The lower part is more flat, and presents a shallow fossa on each side of a median concavity. The fossæ for the neural arch are quite prominent at their contiguous borders.

The measurements of the specimen are as follows:

	Lines.
Length of centrum inferiorly	4
Height of centrum anteriorly	14
Breadth of centrum at middle	22

A third specimen, from the same locality, less well preserved, resembles the first one, and belonged apparently to a somewhat smaller individual. Its parapophyses barely project beyond the sides, and are hollowed at the end. The ridges defining the inferior surface from the concave sides are barely excavated.

Its measurements are as follows:

	Lines.
Length of centrum inferiorly	5.5
Height of centrum anteriorly	15.5
Breadth of centrum at the upper third	18.25

A series of three specimens, with a portion of another, from the same locality as the preceding, appear to correspond with the anterior vertebræ of Amia from the second to the fifth inclusive. The fragment resembles the lateral half of the atlas above described, but is bi-concave. The other specimens resemble the first and third ones above described in the form of the centrum. In the third vertebra the parapophyses are high up, as in the first-described specimen. In the succeeding two they spring from near the middle of the sides of the centrum.

The measurements of the second cervical are as follows:

	Lines
Length of centrum inferiorly	4
Depth of centrum anteriorly	13
Breadth of centrum at middle	20

The measurements of the fifth vertebra are as follows:

	Lines.
Length of centrum inferiorly	3.6
Depth of centrum anteriorly	12.0
Breadth of centrum at middle, on line with parapophyses	17.0

A series of three posterior dorsal centra, from the same locality as the preceding specimens, perhaps belong to the same species, but, from their appearance, most probably to another individual. They are somewhat distorted from pressure, and appear in the original condition closely to have resembled corresponding vertebræ of *Amia calva*, but are nearly three times the breadth, and scarcely twice the length.

The three specimens together, represented in Fig. 5, measure 16 lines in length.

The anterior of the three presents the following measurements:

	Lines.
Length of centrum inferiorly	5
Depth of centrum anteriorly	11
Breadth of centrum inferiorly, opposite the diapophyses	14

A specimen from Dry Creek, consisting of a mutilated basi-occipital, about the size of that of the alligator-gar, differs considerably as well as from that of the mud-fish. It is represented in Figs. 6, 6ª, and may perhaps belong to Protamia.

The articular conical cup has its acute margin scolloped, as seen in Fig. 6. The deep median groove on the under part of the bone in Amia and Lepidosteus reaches the articular margin, but in the fossil, stops the fourth of an inch short of it. On each side of the bone at the articular margin corresponding with the lateral notch there is a conspicuous fossa not seen in the genera just named. In advance of the fossa on each side of the median groove there is a broad, slanting, flat surface, longitudinally ridged, of which there is likewise no exact counterpart in Lepidosteus, but appears to correspond with a smooth surface occupying the same position in Amia.

AMIA (PROTAMIA) MEDIA.

Figs. 7 to 9 represent a vertebral centrum, obtained at the junction of the Sandy and Green Rivers, during Professor Hayden's expedition of 1870. In its form and proportions it resembles a centrum from near the fore part of the dorsal series of *Amia calva*, but pertained to a species double the size. It presents several peculiarities which render it probable that it belongs to a related genus. The sides of the centrum are less contracted than in Amia, and the pair of ridges beneath are substituted by a pair of oval pits. The parapophyses project transversely just above the middle, and are very short.

The measurements of the specimen are as follows:

	Lines.
Length of centrum inferiorly	5.5
Height of centrum	10.2
Breadth of centrum	13.0

Figs. 10, 11 represent a vertebral centrum, found by Dr. Carter on Dry Creek. It resembles a centrum of *Amia calva* from the back of the dorsal series, but is double the size. It presents beneath a pair of grooved ridges, as in *A. calva*.

The specimen measures as follows:

	Lines.
Length of centrum inferiorly	4.0
Height of centrum	7.6
Breadth of centrum	8.6

AMIA (PROTAMIA) GRACILIS.

Figs. 23, 24, Plate XXXII, represent a vertebral centrum found by Dr. Carter, together with a number of ganoid scales, opposite the second crossing of Henry's Fork of Green River. The centrum has a different color from the scales, and clearly did not belong to the same fish. It is from near the middle of the dorsal series, and pertained to a smaller species than *Amia calva*. The two ridges beneath the centra of the latter are substituted by two oblong fossæ.

The measurements of the specimen are as follows:

	Lines.
Length of vertebral centrum inferiorly	1.8
Height of vertebral centrum	3.4
Breadth of vertebral centrum	3.8

HYPAMIA

Hypamia elegans.

Figs. 19 to 22, Plate XXXII, represent a vertebral centrum, found by Dr. Carter on Dry Creek. It is from near the middle of the dorsal series, and evidently indicates a genus distinct from but nearly related with Amia. As in this, the centrum is short in proportion with its breadth, and it presents sutural impressions for a contiguous pair of neural arches. The articular cups have their bottom central and minutely perforate. The sides below the parapophyses are concave, and converge to a median prominence, which is excavated into a pair of fossæ, separated only by a linear partition. The parapophyses are cylindroid and comparatively short.

The measurements of the specimen are as follows:

	Lines.
Length of centrum inferiorly	2.2
Depth of centrum anteriorly	6.5
Breadth of centrum anteriorly	7.6
Breadth of centrum, including parapophyses	8.5

The specimen indicates a species about one-third larger than *Amia calva*.

LEPIDOSTEUS.

Lepidosteus atrox.

During Professor Hayden's expedition of 1870, James Stevenson collected a number of remains of fishes at the junction of Big Sandy and Green Rivers, Wyoming. The specimens consist of isolated vertebral centra, ganoid scales, and portions of jaws with teeth, all of a black hue. Among them are several vertebræ indicating an extinct species of gar larger than the existing alligator-gar, *Lepidosteus ferox*.

Figs. 14, 15, Plate XXXII, represent the centrum of a vertebra from a position in advance of the middle of the dorsal series. The length of the centrum is not greater than the breadth. The extremities are hexahedral in outline. The under surface is flat, and ornate with longitudinal and somewhat reticular wrinkles. The sides beneath the parapophyses are impressed into a deep fossa. The neurapophyses are likewise impressed at the sides with a deep fossa, and a second deep pit occupies a position just behind and above the parapophyses. These appear rather narrower than in the alligator-gar, and are less anterior in position.

The measurements of the specimen are as follows:

	Lines.
Length of centrum inferiorly	8.6
Breadth of centrum posteriorly	8.6
Height of centrum posteriorly	6.0

Another specimen, consisting of a caudal centrum, perhaps belongs to the same species. It has about the same length as the preceding, and is hexagonal in outline at the ends. Its sides present a strong longitudinal ridge separating a deep fossa below from another above.

The measurements of the specimen are as follows:

	Lines.
Length of centrum inferiorly	8.4
Breadth of centrum posteriorly	4.4
Height of centrum posteriorly	4.8

LEPIDOSTEUS ———— ?

Accompanying the preceding specimens there are two vertebræ of another species of Lepidosteus. One of them is a posterior dorsal centrum, and is represented in Figs. 16, 17. It is about as long as the corresponding centra of the gar-pike, *Lepidosteus osseus*, but is proportionately broader and more robust. Its articular ends are hexahedral, with the upper and lower borders slightly emarginate. The lower surface of the centrum is nearly flat, nearly level, contracted at its anterior third, and deeply grooved along the middle. It is bounded by ridges defining it from the deeply impressed sides.

The other specimen is the centrum of a caudal nearly as long as the former, but narrower in proportion with its difference of position in the series.

The measurements of the specimens are as follows:

	Lines.
Length of dorsal centrum inferiorly	5.4
Length of caudal centrum inferiorly	5.0
Breadth of dorsal centrum posteriorly	5.4
Breadth of caudal centrum posteriorly	3.0
Height of dorsal centrum posteriorly	4.2
Height of caudal centrum posteriorly	3.2

These specimens may, perhaps, belong to one of the species indicated by Professor Marsh.

Figs. 27 to 30 represent several ganoid scales, accompanying the preceding specimens, which probably pertain to the smaller of the two Lepidostei. They are covered with perfectly smooth, shining ganoine without markings.

Fig. 25 represents a fragment of the fore part of the right ramus of the lower jaw accompanying the former specimens. Its construction is similar to the corresponding part in the alligator-gar, but is proportionately not so thick or robust near the symphysial end. The lower surface is reticulated with round meshes, and the ridges of the net are ornate with shining translucent tubercles.

The dental groove exhibits the remains of a row of large teeth, of which one retained exhibits the same character as those of the living gars. The outer edge of the groove was also furnished with minute teeth, but the inner edge exhibits no trace of these organs.

LEPIDOSTEUS SIMPLEX.

Some remains of a Lepidosteus, together with some fragments of a turtle-shell, were collected near Washakie Station, Wyoming, by James Stevenson, during Professor Hayden's exploration of 1870. The remains of the Lepidosteus consist of a mutilated basi-occipital and three succeeding vertebral centra, together with several small jaw-fragments and a number of large ganoid scales.

The basi-occipital and vertebral centra, represented in Fig. 18, Plate XXXII, resemble in form and proportions those of alligator-gar, but are smaller.

A tooth, represented in Fig. 26, contained in one of the jaw-fragments agrees in character with the larger teeth of living gars. The outer edge of the same jaw-fragment is furnished with smaller and more curved teeth of the same kind.

Figs. 31, 32 represent two lozenge-shaped scales of less breadth but thicker than those of the alligator-gar. The enamel surface is flat, smooth, and highly polished, and exhibits no markings except one or several minute puncta near the center.

Fig. 33 represents a similar scale, which appears to be traversed fore and aft by a canal communicating by a short cleft with the outer surface. The cleft is directed backward, and is protected by an angular elevation of the anterior border.

Fig. 34 represents another scale of a different form, probably from the median line of the back.

The measurements of the basi-occipital and vertebral centra are as follows:

	Lines.
Breadth of the articulation of the basi-occipital	10.0
Height from lower groove to edge of occipital foramen	5.0
Length of first vertebral centrum	4.0

	Lines.
Breadth of first vertebral centrum	10.0
Length of second vertebral centrum	4.4
Breadth of second vertebral centrum	7.6
Length of third vertebral centrum	5.0
Breadth of third vertebral centrum	6.8

A number of large ganoid scales, of the same character as the preceding, were collected in a sandstone stratum on Little Sandy Creek, during Professor Hayden's expedition of 1870. Several of these selected from the collection are represented in Figs. 35 to 38.

A number of similar scales were obtained by Dr. Carter in the vicinity of Fort Bridger. Figs. 39 to 42 represent several of the scales selected from the collection.

Fig. 43 represents a scale of a Lepidosteus found in association with the large saber-like canines described in the preceding pages, and supposed to belong to Uintatherium.

LEPIDOSTEUS NOTABILIS.

A vertebral centrum partially imbedded in a yellowish sandstone containing casts of shells was obtained near Washakie, Wyoming, during Professor Hayden's exploration of 1870.

The centrum is represented in Figs. 12, 13, and appears to indicate a fish related with Lepidosteus, but probably of a different genus. It pertains to an anterior dorsal, and is about the size of a corresponding centrum of the alligator-gar, but has the parapophyses much shorter. The centrum also differs in shape from those of the alligator-gar. The lower surface is broad and flat, and is marked with longitudinal curved and furcate ridges. The sides are perpendicular and depressed in a deep fossa beneath the parapophyses In the alligator-gar the sides slant outwardly from the lower surface.

The posterior end of the centrum of the fossil is four-sided, with the widest border above and convex, the shortest below and straight, and the lateral borders slanting with a slight sigmoid course.

The short parapophyses project from the upper part of the centrum nearly from the middle.

The measurements of the specimen are as follows:

	Lines
Length of centrum inferiorly	8
Height of centrum posteriorly	9
Breadth of centrum at upper part	10
Breadth of centrum at lower part	6
Breadth of centrum, including parapophyses	13

PIMELODUS.

PIMELODUS ANTIQUUS.

Among the fossil-fish remains of Professor Hayden's collection from the junction of the Big Sandy and Green Rivers, there are a number of fragments of pectoral spines and a few jaw-fragments of a species of cat-fish.

The pectoral spines, of which two fragments are represented in Figs. 44, 45, Plate XXXII, are like those of our living cat-fish. A fragment comprising about two-thirds of the symphysial portion of a dentary bone, Fig. 46, resembles the same in the recent cat-fish, and, as in it, was covered with a broad card-like surface of teeth. The breadth of the dentary surface near the symphysis is $3\frac{1}{2}$ lines. The pectoral spines have ranged from an inch to upward of 2 inches in length. The size of the specie was from a foot to 18 inches.

PHAREODUS.

PHAREODUS ACUTUS.

Accompanying the remains of gars and cat-fish, from the junction of the Big Sandy and Green Rivers, there are many fragments of jaw-bones and others with teeth, evidently not belonging to either of those genera of fishes. They also present sufficient peculiarity to render it probable that they may not belong in the same family with Amia, and therefore probably not to the closely allied genera supposed to be indicated by the vertebral specimens described in the preceding pages. The means of comparison at my command are too scanty to enable me to determine the affinities of the fish to which the fossils pertain.

Figs. 49, 50, Plate XXXII represent two of the best preserved and more characteristic of the specimens, consisting of fragments of dentary bones. These are proportionately deeper and stronger than in Amia. They support a single row of long teeth at the border, and possess no patch of smaller teeth internally such as exist in Amia. The teeth are cylindro-conical, with their somewhat thickened bases close together and firmly co-ossified with the jaw. Their shaft is straight and not curved as in Amia, but the sharp conical apex is bent inwardly.

Figs. 47, 48 represent fragments of premaxillaries. In these the teeth are of the same character as in the dentary bones, but are less bent at the tips.

Fig. 51 represents what I suppose to be a fragment of a maxillary of the same fish. It is provided with teeth as in Amia, Salmo, and some other genera.

Associated with the specimens of the character above described there are a number of others, consisting of small fragments of bones with close patches of short conical teeth, like the vomerine and other similar patches of teeth of Amia.

The dentary fragment of Fig. 49 contains the remains of a dozen teeth in the space of 11 lines. The specimen of Fig. 50 contained thirteen teeth within a space of 10 lines from the symphysis. Of the retained teeth the last is the longest, and measures nearly 3 lines. The others are about $2\frac{1}{4}$ lines in length. The premaxillary fragment of Fig. 48 contained seven teeth in a space of as many lines. The first tooth is the longest, and measures $2\frac{3}{4}$ lines. In the other fragment of Fig. 47, ten teeth occupied a space of $8\frac{1}{2}$ lines.

The genus supposed to be indicated by the specimens has been named from the light-house-like form of the teeth.

REMAINS OF FISHES FROM THE SHALES OF GREEN RIVER, WYOMING.

In Professor Hayden's Preliminary Report on the Geology of Wyoming for 1870, p. 142, the author remarks that soon after leaving Rock Springs Station, on the Union Pacific Railroad, the Green River group is seen on the bluff hills on either side of the road to the entrance of Bitter Creek into Green River. In the valley of the latter remarkable sections of strata are exposed to view. The group he calls the Green River shales, because the strata are composed of thin layers, varying in thickness from that of a knife-blade to several inches. The rocks all have a grayish-buff color on exposure, sometimes with bands of dark brown. These darker bands are saturated with a bituminous matter which renders them combustible.

About two miles west of Rock Springs Station there is an excavation on the railroad which has been called the Petrified Fish Cut, on account of the thousands of beautiful and perfect fossil-fishes which are found on the surface of the thin shales, sometimes a dozen or more on an area of a square foot. Remains of insects and aquatic plants are also found in the shales, and in one instance a well-preserved portion of a feather of a bird was discovered.

A large collection of fossil-fishes from the Petrified Fish Cut, obtained by Professor Hayden in 1870, was submitted to Professor Cope, who has described the different forms in the report above mentioned.

The first of the fossil-fishes of the Green River shales was discovered by the late Dr. John Evans as early as 1856, and was submitted to the examination of the writer. Several specimens, both of the buff-colored and dark bituminous shales, containing fossil-fishes, have been presented to me by Judge W. A. Carter and Dr. J. Van A. Carter, of Fort Bridger, Wyoming.

Professor Cope describes seven species, including one of those described by me, from the Green River shales. Two are named *Clupea humilis* and *C. pusilla*, and a third *Osteoglossum encaustum*. The others are referred to two extinct genera with the names of *Asineops squamifrons* and *A. viridensis*, *Erismatopterus Rickseckeri* and *E. levatus*.

CLUPEA.

CLUPEA HUMILIS.

The species was originally described in the Proceedings of the Academy of Natural Sciences of Philadelphia for 1856, page 256. It was indicated from a specimen consisting of an impression of a nearly complete fish in a piece of shale, which looks like one-half of a rounded, water-worn fragment. The fossil was found by Dr. Evans on Green River, and was stated by him to have been derived from the Tertiary rocks of that locality. The fish is represented in Fig. 1, Plate XVII, of the natural size. It has the ordinary form of living species of herring, and presents the characters of the genus.

This small herring in its total length has measured about $3\frac{3}{4}$ inches. The back is slightly arched, and the dorsal fin is situated just in advance of the middle. The ventral border is strongly arched, and is rather abruptly narrowed from the anus. The ventral fins are placed beneath the back of the middle of the dorsal fin. The head is pointed. The tail is deeply forked, and its pedicle is rather narrow.

The number of vertebræ appears to be about thirty-four, of which at least twenty are dorsal, the remainder caudal. The notochord appears to have extended continuously through the perforated vertebral bodies.

The depth of the body at the fore part of the dorsal fin is four and a half times less than the length. The length of the head slightly exceeds the depth of the body. The eyes are large.

The pectoral fins are destroyed, but their connection with the body was just below the position of the operculum. The ventral fins contain seven rays.

The dorsal fin appears to have had thirteen rays, of which the second was the longest, and from which the others gradually decreased. The anal fin contains fourteen rays. The caudal fin between its two extreme outer and longest rays, inclusive of these, appears to possess twenty rays.

The ventral carinated spines are twenty-five. Accessory ribs project from the vertebræ and ordinary ribs in the usual manner in the herrings.

Measurements of the specimen are as follows:

	Lines.
Total length of the fish	45
Length to commencement of the tail	34
Depth of body in front of dorsal fin	10
Depth at anus	6
Length of the head	$10\frac{1}{2}$
Depth of the head	$8\frac{1}{2}$
Length of the tail	10
Depth of pedicle of the tail	4
Distance from snout to commencement of dorsal fin	17
Distance from snout to anus	27

CLUPEA ALTA.

A slab of shale obtained from the so-called "Petrified Fish Cut," and submitted to my examination, contains ten herrings, in which the bones and scales are preserved, and stained of a dark-brown hue. The vertebræ, where broken, exhibit the position of a continuous notochord occupied by hyaline chalcedony that looks like the original substance of the latter itself. The most complete and largest of the fossil-fishes is represented of the natural size in Fig. 2, Plate XVII.

These fishes appear to belong to a different species of herring from the former, especially distinguished by the greater proportionate depth of the body and the more arched dorsal border. In most other essential characters the two appear to agree. It has the same number of vertebræ and of ventral carinated spines. The fins also, so far as can be determined, appear to contain the same number of rays.

The other specimens on the slab, though smaller, exhibit the same, or nearly the same, proportionate depth of the body.

The measurements of the specimen figured are as follows:

	Lines.
Total length of the fish	50
Length to commencement of tail	38
Depth of body in front of dorsal fin	$14\frac{1}{2}$

	Lines.
Depth at anus	9½
Length of head	12
Depth of head	10½
Length of tail	11
Depth of pedicle of tail	5
Distance from snout to commencement of dorsal fin	20½
Distance from snout to anus	31

DESCRIPTION OF REMAINS OF MAMMALS FROM THE TERTIARY FORMATION OF SWEETWATER RIVER, WYOMING.

A small collection of fossils, consisting of the remains of mammals, was obtained, during Professor Hayden's expedition of 1870, on Sweetwater River, eighteen miles west of Devil's Gate, Wyoming.

Professor Hayden, in his Preliminary Report of the Geological Survey of Wyoming, 1871, page 32, in relation to the locality whence the fossils were obtained, makes the following remarks: "Near Cloven Peak, fifteen miles west of Devil's Gate, there are some bluff-banks on the south side of the Sweetwater, about one hundred feet high, which indicate the existence of quite modern Tertiary beds, like those on the Niobrara River. They are composed of indurated sands and marls of a light-gray or cream color, and are in appearance precisely like those seen on the Laramie River, and many other places, which I have usually regarded as of the Pliocene age. Still farther to the westward are numerous exposures of these beds, which are weathered into the usual fortification-like forms, and scattered around their base are large numbers of remains of extinct mammals and turtles, apparently identical with those found on the Niobrara. They occur in the same beautiful state of preservation."

Professor Hayden's view of the age of the formation is confirmed by the zoölogical character of the fossils, which are nearly related with those from the Pliocene Tertiary sands of the Niobrara River, and are, without doubt, of a much more recent date than those of the Bridger beds.

The specimens submitted to my examination consist of fragments of jaws with teeth, portions of the larger limb-bones, small bones of the feet, and a few mutilated vertebræ. Most of them pertain to a species of Merycochœrus, an animal nearly related to Oreodon. A few apparently belonged to a smaller species, and several to a small equine animal. The others remain undetermined for want of ready means of comparison.

The fossils are all isolated specimens, which were picked up from the surface of the ground. Usually they are perfectly free from adherent matrix. They are white in appearance, and resemble recent bleached bones. They have lost their bone-cartilage, and are hard and brittle, though not friable. They are not in the least degree water-worn, and present no appearance of having been submitted to great pressure, as is so frequently the case with the ossils from the Cretaceous and Eocene formations of neighboring localities.

MAMMALIA.

Order *Ruminantia*.

MERYCOCHŒRUS.

MERYCOCHŒRUS RUSTICUS.

The genus above named was originally characterized from some remains, discovered by Professor Hayden during Lieutenant Warren's expedition of 1857, in a bed of dull, fine-grained grit, on the head-waters of the Niobrara River, near Fort Laramie, Nebraska.

Merycochœrus pertains to the same family as Oreodon, a genus characterized from a profusion of remains from the Miocene Tertiary deposit of the Mauvaises Terres of White River, Dakota. The general construction and form of the skull appear to be nearly the same, and such, also, is the case with the number, constitution, and relative position of the teeth. There are, however, certain peculiarities distinguishing the two genera.

The molar teeth of Oreodon have, comparatively with those of most genera of existing ruminants, short crowns as in the deer; and, as in this, at maturity they are all inserted alone by fangs. In Merycochœrus the crowns of the molars are proportionately longer, and in the mature condition of the animal, while the anterior ones were fully protruded, the posterior ones, though in functional position, were only partially protruded, and continued to advance as they were worn away. The difference between the two genera, Oreodon and Merycochœrus, in respect to the comparative length of the molar crowns, is like that existing between the molars of the deer and the ox, but not to the same degree. While the condition of the teeth of Oreodon corresponds to that of the deer, those of Merycochœrus rather hold an intermediate condition to those of the deer and the ox.

In Oreodon, when the last of the molar series was fully protruded so as to be inserted by the fangs alone, the anterior molars might still be in a condition to exhibit very conspicuously the anatomical characters of their triturating surfaces, as displayed in Plate VII of the Extinct Mammalian Fauna of

Dakota and Nebraska. In Merycochœrus, on the other hand, before the crown of the last molar was fully protruded, already the anatomical characters of the triturating surfaces of those in advance were, to some extent, destroyed; and in the case of the first true molar completely obliterated, in this state presenting simply a broad dentinal surface bordered with enamel. This condition is represented in the Fig. 3, of Plate X, of the same work just mentioned, though in this case the specimen belonged to an individual past maturity, and the last molar is fully protruded.

Another distinctive character in the teeth of Oreodon and Merycochœrus is expressed in the less degree of transverse symmetry of the crowns of the premolars in the latter. In Oreodon their various measurements are more uniform, and the summits of the principal constituent lobes of their crowns are nearly or quite median, and they nearly retained this relative position as the teeth were worn away. In Merycochœrus the length and fore and aft diameter of the crown exceed the transverse diameter except in the last upper one; and the summits of the lobes of the premolars, especially in the upper ones, are more or less in advance of the middle of the crowns, and they likewise retained this relative position as the teeth were worn.

In the original description of Merycochœrus, its distinction from Oreodon was mainly founded on peculiarities of the skull; the differences in the teeth above noted, especially those in the proportionate length of the crowns of the molars, and their relative mode of protrusion, were not recognized. This want of appreciation of the distinctive characters of the teeth of the two genera arose from the observations having been made on the jaw-specimens of Merycochœrus advanced beyond maturity, in which all the teeth were fully protruded, and in this condition did not strikingly differ from those of Oreodon.

A number of other fossils, discovered by Professor Hayden in the Pliocene sands of the Niobrara Valley, and described by the writer at the same time as those referred to Merycochœrus, from the difference in the proportionate length of the molar teeth in comparison with those of Oreodon, were referred to another genus with the name of Merychyus. In the same manner this was supposed to differ from the Merycochœrus; and though subsequently, in the preparation of the "Extinct Mammalian Fauna of Dakota and Nebraska," it was suspected that these two genera might prove to be the same, it was not until the present moment the suspicion appeared to be confirmed. From present observations and reflection, I am under the impression that Oreodon and Merycochœrus are two quite distinct though closely allied genera, of which the latter is geologically the later, and, perhaps, the successor by evolution from the former. Merychyus would appear to be the same as Merycochœrus, and the fossils which had been referred to it belong to the same geological horizon.

Of the specimens originally attributed to *Merychyus major* and *M. medius*, too little of the corresponding parts were preserved in such a condition as to enable us to make a comparison of the upper jaw and face with the same parts in Merycochœrus to ascertain how far they are like one another. The position of the infra-orbital foramen, which appears to be nearly or quite constant in a species, or varies but slightly in several species of a genus, in the jaw-specimen referred to *Merychyus major* is placed above the last premolar. It occupies the same position in *Merychyus elegans*; and in the upper jaw of a young animal, referred to *Merychyus medius*, it is placed above the last temporary premolar, therefore agreeing in position with that in the adults of the other two species. In *Merycochœrus proprius* the position of the foramen is further back, above the interval of the first and second molars, and this is, also, its position in the upper jaw of the Sweetwater species named *Merycochœrus rusticus*.

This difference of position is probably related with a difference in the shape of the face, which in Merycochœrus is rather abruptly narrowed in advance of the zygomata, as in the hog. The face of Merychyus I suspect rather to be more like that of Oreodon, narrowing gradually forward from the position of the orbits and zygomata, as in the peccary.

Admitting the three genera, Oreodon, Merycochœrus, and Merychyus, their distinctive characters, so far as ascertained from the materials at command, would appear to be as follows:

OREODON.—Molar teeth with short crowns, as in the deer; and, as in this, at maturity inserted by fangs. Anterior premolars straight, with the diameters nearly equal, and with their points median or nearly so. Face gradually convergent, conical. Infra-orbital arch narrow or of moderate depth; gradually declining upon the side of the face. Infra-orbital foramen small and situated above the third premolar. Nasal orifice nearly as wide as high, and situated immediately above the incisive alveolar border, as usual in most animals. Premaxillaries and maxillaries remaining distinct from one another. Incisive foramina of moderate size.

MERYCOCHŒRUS.—Crowns of the molars proportionately longer than in Oreodon, and protruding gradually as they were worn away; the anterior having their sculptured triturating surface obliterated before the posterior are fully protruded. Anterior premolars with the length and breadth exceeding the width, and the upper ones inclining posteriorly, and with their points in advance of the middle. Facial cone abruptly narrowed in advance of the orbits. Infra-orbital arches deep and rapidly declining on the face. Orbits smaller and more externally situated than in Oreodon. Infra-orbital foramen above the interval of the first and second molars. Nasal orifice situated far above the alveolar border, as in the tapir, and commencing below as an angular notch of the premaxillaries, which are firmly co-ossified together and with the maxillaries. Incisive foramen large.

MERYCHYUS.—Teeth as in Merycochœrus. Facial cone intermediate in character to the latter and Oreodon (?) Infra-orbital foramen situated above the last premolar, or in a position intermediate to that of Oreodon and Merycochœrus.

The more characteristic of the remains of Merycochœrus, from the Sweetwater River, consist of fragments of jaws with teeth from perhaps a half dozen individuals. One of the specimens consists of the greater part of an upper jaw, represented in Figs. 1, 2, Plate III, accompanied with a portion of the lower jaw, represented in Fig. 3 of the same plate.

The face of Merycochœrus, as indicated by the upper-jaw specimen just mentioned, would appear to differ in a remarkable manner from that of the closely allied genus Oreodon. In the species to which the name of *Merycochœrus rusticus* has been given, and which probably is the same as *Merychyus medius*, the face is narrowed in the same abrupt manner in advance of the orbits as in *Merycochœrus proprius*. It is, however, more convergent than in the latter, or is proportionately less widened at the extremity.

The relation of the orbits and zygomata to the fore part of the face in Oreodon is more like the condition in the peccary; in Merycochœrus more like that in the hog.

The side of the face in *M. rusticus* between the position of the orbit and the prominence produced by the canine, and above the alveolar ridge, is deeply concave, even more so proportionately than in the hog. In *M. proprius*, it is not depressed in this manner, so that the side of the face in the corresponding position is nearly vertical, and the large infra-orbital foramen opens forward on this vertical surface. In *M. rusticus*, the infra-orbital foramen is also large, and occupies a corresponding position, but is situated in the concavity of the side of the face, so that the surface of the alveolar border curves outwardly and downward from it.

The front of the snout or fore part of the upper jaw resembles in its construction the same part in the tapir more than that of Oréodon, but, as in the latter, it barely projects beyond the position of the canine alveoli. The premaxillaries are completely co-ossified with each other and with the maxillaries.

Viewed at the side, the fore part of the upper jaw is convex forward and downward, as in the tapir. Viewed in front, (Fig. 2, Plate III,) it presents a long slope, narrow above, widening below, depressed toward the median line, and bounded laterally by the convex curved prominences of the canine alveoli. About $1\frac{1}{2}$ inches above the alveolar margin the nasal orifice commences in an angular notch as in the tapir, but proportionately less narrow.

Behind the position of this nasal notch, bordered by thickened ridges ascending in a convergent manner from the canine alveoli, are the lateral concavities of the face before mentioned.

The upper part of the face being broken away, we can form no just idea of its character. If constructed as in Oreodon, by the conjunction of the maxillaries along the course of the nasals, it would appear to be exceedingly narrow, even less than half the width at the alveolar border. It would appear as if the construction might be somewhat similar to that in the tapir, so that the maxillaries bounded a large nasal aperture overhung by the nasals.

The infra-orbital arch is nearly twice the depth it is in Oreodon, and resembles in its proportions that of the hog. Its outer surface is nearly vertical or slopes slightly outward, and is nearly plane or slightly depressed. The anterior zygomatic root is an unusually prominent process of the maxillary. Its suture of conjunction with the malar descends nearly on a line with the anterior border of the orbit. The latter is smaller, and is situated more externally than in Oreodon.

The roof of the mouth is moderately concave, and the incisive foramen, apparently, is proportionately as large as in the tapir.

The lower jaw of Merycochœrus is like that of Oreodon, and, as in this and all living ruminants, has the rami united by suture.

The mental foramen, like the infra-orbital foramen, is proportionately larger than in Oreodon. Perhaps this difference in the size of the foramina, together with the other peculiarities of the face, may indicate that Merycochœrus was provided with large prehensile lips, or probably a short proboscis.

As in Oreodon, the dental series of the upper jaw consists of 3 incisors, 1 canine, 4 premolars, and 3 molars; of the lower jaw, 4 incisors, 1 canine, 3 premolars, and 3 molars.

In both jaws of Merycochœrus, as in Oreodon, the teeth form nearly closed rows. The largest interval is between the canine and first premolar of the upper jaw, to accommodate the lower canine, which in all the Oreodont family occupies a position behind the upper one.

A last upper incisor, retained in the upper-jaw specimen of *Merycochœrus rusticus*, resembles in its form and relative size to the others the corresponding tooth in *Merychyus elegans*.

In a fragment of a lower jaw represented in Fig. 5, Plate VII, and retaining most of the incisors, the lateral one is observed to be much larger in relation with the others than in Oreodon. Its crown, viewed in front, is nearly ovoid in outline. Its borders are acute and meet in a rounded point. The outer surface is convex. The inner surface, considerably shorter, is bounded by a basal ridge. The intervening incisors, about half the size of the outer one, successively but slightly decrease. Their crown is more truncate at the summit, and the internal basal ridge is stronger. The large lateral incisor is to be viewed as a modified canine in its relation with this tooth as present in animals usually.

The canines of Merycochœrus in all respects are like those of Oreodon. As in this genus, the lower ones are to be viewed as modified first premolars, assuming the form and function of canine teeth, but still holding in relation to the other teeth the ordinary position of the former.

The crowns of the premolars of *M. rusticus* in their earlier state are considerably longer proportionately than those of Oreodon, and by the time they became wholly protruded they were so much worn as to have the peculiar construction of their triturating surface obliterated.

The crowns of the upper premolars, except the last one, have a backward inclination, successively increasing from the third to the first. The points of these teeth occupy the anterior third of the crown in the earlier stage, and at a late period become so advanced as to appear to form the anterior corner of the crown. In Oreodon the corresponding teeth are nearly or quite straight, and the summit of the crown is median, and continues so as the teeth are worn away. The differences mentioned—that is to say, the backward inclination of the crowns of the premolars and the more advanced position of their points in Merycochœrus—would appear to be due to a comparative shortening of the face and a less consequent space for the development of the teeth.

The same differences which have been mentioned as existing between the premolars of *Merycochœrus rusticus* and Oreodon are also obvious in *Merychyus elegans*. The same may be said also of the third upper premolar in the fossil referred to *Merychyus major*, except that in this the crown of the tooth is proportionately not so long as in *Merycochœrus rusticus*, and was less worn when fully protruded.

In *Merycochœrus rusticus* the outer face of the upper premolars is convex longitudinally, but concave transversely; the lateral borders having a considerable degree of prominence. In *M. proprius* and *Merychyus major* they are likewise concave and bordered by a strong basal ridge which is absent in *Merycochœrus rusticus*. In *Merychyus elegans* the outer face of the upper premolars is convex transversely as well as longitudinally, and, as in the latter, is devoid of a basal ridge.

In a small fragment of an upper jaw of *M. rusticus*, containing the second and third premolars, represented in Figs. 3, 4, Plate VII, the crowns are comparatively but little worn and retain the characters of the triturating surface. These teeth are of less breadth in proportion to their thickness than in *M. proprius*, and in this respect are more like the corresponding teeth of Oreodon. Their outer part forms a strong curve from the ends of the fangs to the point of the crown, of which about one-fourth externally remains unprotruded, while it is fully protruded internally. The point of the crown is at the anterior third, and externally it appears to be continuous as part of the anterior projecting border of the crown. The inner portion of the crown exhibits three deep recesses inclosed by prominent loop-like folds. The posterior larger recess is separated from the anterior smaller pair by a ridge dividing the inner part of the outer or principal lobe of the crown. A basal ridge festoons the posterior internal loop of the third premolar, but does not exist in the second. The teeth are worn off in a slope on the postero-internal face of the principal lobe of the crown.

These teeth are sufficiently like the corresponding tooth in the jaw-specimen of *Merychyus major* to render it probable that this animal may belong to the same genus as the former.

The last upper premolar of *Merycochœrus rusticus* is like that of *M. proprius*.

The superior molars, the inferior premolars and molars, are so closely like those of *Merychyus elegans*, that they may be considered as their magnified representatives.

Fig. 1, Plate VII, represents a series of upper molars in a specimen in which the last one has not more than two-thirds protruded. A view of the outer part of this last molar is introduced in the representation of the upper jaw in Fig. 1, Plate III, so as to complete the series of upper molar teeth. In the first molar the anterior crescentic enamel pit is observed to be completely obliterated, and the posterior one nearly so. In the back two molars the inner faces of the internal lobes are decidedly concave longitudinally.

In Fig. 6, Plate VII, we have a presentation of the first and second upper molars of *Merycochœrus proprius* introduced for comparison. The specimen is from the head-waters of the Niobrara River, in the vicinity of Fort Laramie.

Fig. 2, of the same plate, represents the last premolar and the molar of the temporary series of *M. rusticus*. The molar is like those of the permanent set; the premolar resembles the former modified by having the anterior lobes, especially the inner one, proportionately less well developed.

In a small fragment of an upper jaw of another young animal, in which the temporary molars were retained and the first permanent molar had protruded, the maxillary presents a different appearance from that in the adult. The surfaces above and below the position of the ridge produced by the malar process are almost at a right angle to each other. The upper surface slopes forward and outward from the position occupied by the orbit, and upon it opens the infra-orbital foramen about half an inch within the ridge separating this surface from the lower one. In the progress of development from youth to age the angularity of the outer part of the maxillary became rounded, so that the surface assumed a convex instead of a nearly rectangular character.

Measurements of the jaws and teeth of *M. rusticus* are as follows:

	Lines.
Distance from upper incisors to back of last molar	62
Length of space occupied by upper series of molar teeth	53
Length of space occupied by upper premolars	22
Length of space occupied by upper molars	31
Breadth of upper jaw outside of canines	22
Breadth of upper jaw outside of second premolars	24
Breadth of upper jaw outside of second molars	30
Breadth of upper jaw at inner side of infra-orbital foramina	20
Breadth of face at lower margin of the orbits	55
Distance from lower incisors to back of last molar	62
Length of space occupied by lower series of molar teeth	48
Length of space occupied by lower premolars	17
Length of space occupied by lower molars	31
Depth of lower jaw at symphysis	29
Depth of lower jaw below last premolar	19
Depth of lower jaw below second molar	22
Width of condyle of lower jaw	19

	Antero-posterior.	Transverse.
	Lines.	*Lines.*
Diameter of upper canine	$4\frac{3}{4}$	$4\frac{3}{4}$
Diameter of lower canine	5	$4\frac{1}{4}$
Diameter of second upper premolar	$5\frac{1}{2}$	4
Diameter of third upper premolar	6	5
Diameter of fourth upper premolar	5	$6\frac{1}{2}$
Diameter of first upper molar	$8\frac{1}{2}$	$8\frac{1}{2}$
Diameter of second upper molar	11–12	10–$10\frac{1}{2}$
Diameter of third upper molar	11	$10\frac{1}{2}$
Diameter of first lower premolar	3	$2\frac{3}{4}$
Diameter of second lower premolar	6	$4\frac{1}{2}$
Diameter of third lower premolar	$6\frac{1}{2}$	5
Diameter of first lower molar	7	6
Diameter of second lower molar	10	$7\frac{1}{4}$
Diameter of third lower molar	$15\frac{1}{2}$	7

Of other bones referable to *Merycochœrus rusticus*, the collection contains the following:

Portions of several scapulæ. The glenoid cavity is oval, and measures 11 lines in its short diameter, and 15 lines in its long diameter, including the coracoid process.

A number of fragments, mostly distal extremities of tibiæ, of which one is represented in Fig. 9, Plate XX. The general construction is the same as in ordinary even-toed ungulates. The shaft approaching the articulation is three-sided, with the outer border subacute. This terminates in a triangular surface for junction with a fibula. The internal malleolus is comparatively long and pointed, and projects below the position of the anterior process of the tibia. The articular concavities are nearly of the same extent fore and aft, but the outer one is much the wider. The width of the end of the tibia in different specimens ranges from 13 to 14 lines, and the fore and aft diameter is $8\frac{1}{2}$ lines.

Of a number of specimens of the astragalus, one is represented in Fig. 10, Plate XX. It is about the size of that of the peccary, but is proportionately wider. The outer division of the trochlea is considerably larger than the inner one. The posterior articular surface for the calcaneum extends but little more than half the width of the bone. The length of the astragalus externally is 16 lines; its width at the lower tarsal articulation is $10\frac{1}{2}$ lines.

Of a number of specimens of the calcaneum, one is represented in Fig. 11. It is about the size of those of the peccary, but is more robust in its proportions. The tuber is a little shorter, but considerably thicker. A peculiarity of the bone is the absence of a sustentaculum tali, the usual articular surface of the latter being supported on a moderate expansion of the base of the tuber. The articular eminence for the fibula is but slightly prominent.

The length of the calcaneum is $2\frac{1}{2}$ inches; its width at the articulation below the tuber is 8 lines.

Another specimen of a calcaneum, interesting on account of its diseased condition, is represented in Fig. 15, Plate II.

MERYCOCHŒRUS —— sp?

Fig. 12, Plate XX, represents the distal end of the tibia, probably of a smaller species of Merycochœrus. The specimen was found with those above described. The transverse diameter of the articular end is 11 lines.

An astragalus resembling that above described probably belongs to the same animal as the latter. It is $10\frac{1}{2}$ lines long, and $5\frac{1}{2}$ lines wide.

Order *Solidungula*.

Associated with the remains of Merycochœrus, from the Sweetwater River, there are several bones of a small equine animal, probably a species of Hipparion. One of the specimens is an external cuneiform bone, of which an upper view is given in Fig. 13, Plate XX. Another specimen of the same bone has the navicular bone co-ossified with it. A third specimen consists of a first ungual phalanx 17 lines long at the side, and $14\frac{1}{2}$ lines wide at the upper extremity.

DESCRIPTION OF VERTEBRATE FOSSILS FROM THE TERTIARY FORMATION OF JOHN DAY'S RIVER, OREGON.

Through the Smithsonian Institution, at the suggestion of Professor S. F. Baird, a collection of fossils was submitted to the examination of the writer by Rev. Thomas Condon, of Dalles City, Oregon.

The fossils were discovered by Mr. Condon mainly in the valley of Bridge Creek, a tributary of John Day's River, one of the branches of the Columbia River. Some additional fossils from the same locality were also placed in my hands by Professor H. S. Osborn, of La Fayette College, Easton, Pennsylvania.

With the exception of a single turtle-bone, the fossils consist of remains of mammals. In general appearance and condition of preservation they resemble those of the Mauvaises Terres of White River, Dakota. They are nearly all specimens which have been found lying loose on the surface of the country, and are, therefore, more or less weathered, or much injured by exposure. A few of the fossils are imbedded in, and the cavities of others are filled with, a hard, compact, homogeneous rock of a bluish-gray hue. The rock appears to be an indurated marl, and contains abundance of lime. It bears a near resemblance to the matrix of the fossils of the Mauvaises Terres of Dakota, except that it is more compact and harder.

The zoological character of the fossils is such as to render it probable that the formation to which they belong is of contemporaneous age with the Tertiary deposit of the locality just named.

The greater number pertain to a species of Oreodon larger than any of those from White River, Dakota, and about the size of *Merycochærus proprius* of the Niobrara River, Nebraska.

A number of the fossils appear to belong to some of the same species as those of the Mauvaises Terres, as *Oreodon Culbertsoni*, *Agriochærus antiquus*, and *A. latifrons*, *Leptomeryx Evansi*, *Anchitherium Bairdi*, and *Rhinoceros occidentalis*.

The collections further contain remains of a second species of Rhinoceros, two species of Elotherium, &c., generally too scanty or imperfect to ascertain positively whether they pertain to species previously characterized

A descriptive list of the fossils is given below:

MAMMALIA.

Order *Ruminantia.*

OREODON.

OREODON CULBERTSONI.

This species, established on a multitude of remains from the Mauvaises Terres of White River, Dakota, is apparently indicated by some small fragments of upper and lower jaws with teeth, which are labeled "Big Bottom of John Day's River." One of the best-preserved and most characteristic specimens consists of a jaw-fragment containing the upper last premolar and the molars, the latter being represented in Fig. 12, Plate VII. In all respects it is like the corresponding part in *Oreodon Culbertsoni*, from White River. Other specimens show a slight variation in the size of the teeth.

OREODON SUPERBUS.

Nearly twenty-five years have elapsed since the first fossil remains of mammals from the Tertiary formations of the West were submitted to my examination. To the present time they have been coming to me in constant succession, so that I have had the opportunity of examining thousands of specimens, the collective weight of which would amount to several tons. From some of the first specimens brought from the Mauvaises Terres of White River, Dakota, after a few errors, I thought I had fixed upon well-marked characters distinguishing the extinct genus of hog-like ruminants, for which I proposed the name of Oreodon. Two species were described under the names of *O. Culbertsoni* and *O. gracilis*, mainly from a marked difference in size.

Several detached crania, differing from that of either of the species of Oreodon in the possession of large inflated ear-capsules, at first attributed to a peculiar genus with the name of Eucrotaphus, were subsequently referred to Agriochœrus, which had originally been described from jaws and teeth. Later this determination appeared to be confirmed by an almost complete skull in which the cranium agreed with the detached specimens.

Some small fragments, and finally a complete skull, appeared to indicate a third and larger species of Oreodon, to which the name of *O. major* was given.

It is especially remarkable for the great size of the ear-capsules compared with those of the other species, being proportionately quite as large as those in Agriochœrus.

Of the multitude of fragments of jaws with teeth, portions of skulls, and more or less complete skulls of Oreodon, which I have had the opportunity of examining, by far the greater number are referable to the species *O. Culbertsoni*, about a twentieth to *O. gracilis*, and one per centum to *O. major*. Specimens exhibit more or less variation, generally of a comparatively trifling character, but in some instances to such a degree as nearly to be distinctive enough for other species, and in some cases as nearly to remove the distinctions between the two species *O. Culbertsoni* and *O. gracilis*. Two specimens, presenting a greater extent of variation than usual, have been suspected to represent hybrids in the one case between *O. Culbertsoni* and *O. gracilis*, in the other case between the former and *O. major*. With the view that they may be specifically distinct, they have been named *O. affinis* and *O. hybridus*.

After a number of years, and after having seen many hundred specimens referable to *O. Culbertsoni*, to my utter astonishment one of the last ones received, consisting of the greater part of a skull, while agreeing in every other respect with the ordinary form of *O. Culbertsoni*, possesses ear-capsules as large as those of Agriochœrus. Looking upon this specimen as representing a species or an important variety, the name of *O. bullatus* was applied to it in allusion to its large inflated ear-capsules.

As the cranial portion of the skull of *O. bullatus* does not differ in size from the specimens originally referred to Eucrotaphus, we are now uncertain whether they pertained to *O. bullatus* or Agriochœrus. They correlate in size, construction, and form equally well with either.

Some remains from the Niobrara River, Nebraska, while clearly indicating members of the same family as Oreodon, appeared to me to belong to two different genera, to which the names of Merycochœrus and Merychyus were given. The recent discovery of additional remains of another species of Merycochœrus, on the Sweetwater River, Wyoming, while rendering the characters of the genus more obvious, rather tend to make the genus Merychyus doubtful.

The skull of Merycochœrus has the same general form and construction as that of Oreodon, and the teeth agree in number, relative position, and constitution. The crowns of the molar teeth in Oreodon are short and inserted by fangs, as in the deer. In Merycochœrus they are longer, and protrude more gradually as they are worn away. The face is more abruptly prolonged in front of the orbits; the infra-orbital arches are proportionately of much greater depth; and the infra-orbital foramina situated much further back. While the fore part of the upper jaw of Oreodon is constructed in the more ordinary manner of many animals—suilline pachyderms, carnivora, &c.—that of Merycochœrus is more like that of the tapir.

Merychyus, so far as known, is intermediate in character with Oreodon and Merycochœrus. Its molar teeth are like those of the latter; its face appears not to be so abruptly narrowed; and the infra-orbital foramina hold an intermediate position.

Another member of the oreodont family, from a formation probably of equivalent age to that which has yielded the remains of the Oreodons, has been named Leptauchenia. Its molar teeth agree in character with those of Merycochœrus and Merychyus, but are more strongly folded internally in the case of the lower ones, externally in the case of the upper ones. The face is more like that of Oreodon; has the infra-orbital foramina in the same relative position, but has large unossified spaces at the upper part of the face.

Oreodon superbus, the name which appears at the head of this chapter, was applied to a species, indicated more recently than any of the preceding, from specimens belonging to Mr. Condon's collection of Oregon fossils. The species exhibits characters which make it somewhat peculiar, and place it in a position intermediate to the White River Oreodons and the genus Merycochœrus. It is exemplified by a number of specimens, among which is the mutilated skull, represented, one-half the natural size, in Fig. 1, Plate I. Other specimens, consisting of detached mutilated crania, portions of others, and fragments of jaws and teeth, pertain to half a dozen or more individuals.

The skull of *Oreodon superbus* is about the size of that of *Merycochœrus proprius*. In form, proportions, and constitution, and in the number, relative position, and construction of the teeth, it nearly resembles the other known species of the genus from the Mauvaises Terres of White River, Dakota.

The cranium proper is a magnified likeness of that of *Oreodon Culbertsoni* or *O. major*, and more especially agrees with the latter in the possession of large inflated ear-capsules. It presents the same kind of variation in different specimens observed in *O. Culbertsoni*. In most of the specimens the temporal surfaces slope from the sagittal crest with a slight sigmoid curve. In one specimen the parietal surface is deeply depressed on each side of the sagittal crest. In another specimen a pair of well-marked grooves follow the course of the fore part of the squamous suture, one in front, the other behind it. In all the specimens the front groove is more or less distinct; in some of them the back groove is barely perceptible.

The auditory capsules are ovoidal, with the greater diameter fore and aft, and the length exceeding the width. They extend from the paramastoid process forward to the middle line of the glenoid articular surface, and project below the level of this for half their length.

The face of *Oreodon superbus* differs from that of the other species of the genus more than it does among these. It especially differs in the position of the infra-orbital foramen, and in the great proportionate depth of the infra-orbital arch. In the other known species of Oreodon the infra-orbital foramen occupies a position above the third premolar. In *O. superbus* it is placed above the last premolar, as in Merychyus. In Merycochœrus it is placed further back over the interval of the first and second true molars. The infra-orbital arch is proportionately as deep as in Merycochœrus, and like it presents a broad, nearly flat surface, extending forward below the position of the lachrymal fossa. The latter is relatively shallow. The forehead is more flat than is usual in *Oreodon Culbertsoni*. The anterior nasal orifice is like that in other species of the genus.

The teeth of *Oreodon superbus*, so far as we have had the opportunity of examining them, appear to agree in all respects with those of the other known species.

Fig. 16, Plate II, represents a fragment of the lower jaw, natural size, containing the premolars and the first molar. A view of the triturating surfaces of the premolars is given in Fig. 9, Plate VII.

Figs. 7, 8, Plate VII, represent a first molar, part of the second, and the last molar from a lower-jaw specimen.

Fig. 10, of the same plate, represents a facial specimen, with a view of the forehead, one-half the natural size.

Measurements obtained from several specimens of portions of skulls of *Oreodon superbus* are as follows:

Estimated length of skull, approximating 14 inches.

	Lines.	Lines.	Lines.
Breadth of forehead between orbits on line with supra-orbital foramina	50	....	50
Length of face from orbit to lateral nasal notch	54	....	....
Height of face on line with the second true molar	48	....	....
Depth of orbital entrance	21	....	....
Transverse diameter of the same	19	....	....
Depth of infra-orbital arch	18	21	....
Length of upper molar series, estimated at	70	....	....
Antero-posterior diameter of last upper molar	....	17½	....
Transverse diameter of last upper molar	....	15½	....
Breadth of nasals together	....	....	16

Measurements obtained from lower-jaw fragments detached and not pertaining to the preceding:

	Lines.	Lines.	Lines.	Lines.
Space occupied by the lower molar series	66	....	....	....
Space occupied by the lower premolars	25	....	30½	27½
Space occupied by the lower molars	40	41	....	....
Depth of jaw below back of last molar	....	33	....	....
Depth of jaw below fore part of last molar	27	26	....	....
Depth of jaw below middle of first molar	25	24	....	31
Depth of jaw below middle of second premolar	25	....	....	31
Antero-posterior diameter of first premolar	....	....	9	....
Transverse diameter of first premolar	....	....	3½	....
Antero-posterior diameter of second premolar	....	....	10	....
Transverse diameter of second premolar	....	....	4½	....
Antero-posterior diameter of third premolar	....	....	11	....
Transverse diameter of third premolar	....	....	6½	....
Antero-posterior diameter of first molar	....	10	....	....
Transverse diameter of first molar	....	7	....	....
Antero-posterior diameter of second molar	....	11½	....	....
Transverse diameter of second molar	....	8½	....	....
Antero-posterior diameter of third molar	....	12½	....	....
Transverse diameter of third molar	....	9	....	....
Transverse diameter crown of lower canine	....	....	....	11

LEPTOMERYX.

LEPTOMERYX EVANSI.

A small ruminant, related to the musks, was originally described under the above name, from remains discovered by Dr. John Evans and Professor Hayden in the Mauvaises Terres of White River, Dakota.

Two small fragments of jaws, the one containing a well-preserved upper molar and the other a lower molar, from John Day's River, agree in all respects with the corresponding parts of *Leptomeryx Evansi*.

AGRIOCHŒRUS.

AGRIOCHŒRUS ANTIQUUS. AGRIOCHŒRUS LATIFRONS.

The above genus and species were originally characterized from remains found in the Mauvaises Terres of Dakota. The genus is related with Oreodon, but exhibits peculiarities enough to regard it as pertaining to another family of extinct ruminating hog-like animals.

A small fragment of an upper jaw with portion of a molar, and a few fragments of detached molars from John Day's River, appear to indicate the presence of both the above-mentioned species.

Artiodactyla.

DICOTYLES.

DICOTYLES PRISTINUS.

An extinct animal about the size of and nearly allied to the living collared peccary, *Dicotyles torquatus*, is represented in the Condon collection by several detached lower molar teeth. These have nearly the size and constitution of those of the collared peccary, though considerably worn and therefore smoother than when in a younger condition. Independently of this smoothness, due to age, the constituent lobes of their crowns do not present the wrinkled condition observed in the living peccaries.

The last lower molar, represented in Fig. 14, Plate VII, has a five-lobed crown with a basal ridge in front and externally, and also postero-internally. The lobes of the crown are comparatively simple, or but slightly complicated by offsets or folds. The penultimate molar, represented in Fig. 13, Plate VII, has four principal lobes to the crown, arranged as in the recent peccary.

An upper molar, from a younger animal, perhaps, belonged to the same species. It is more square than in the recent peccary, and has the four constituent lobes of the crown comparatively smooth and devoid of wrinkles.

Measurements of the specimens referred to *Dicotyles pristinus* are as follows:

	Lines.
Antero-posterior diameter of last lower molar	9
Transverse diameter of last lower molar	5
Antero-posterior diameter of penultimate lower molar	7
Transverse diameter of penultimate lower molar	5¼
Antero-posterior diameter of first or second upper molar	6
Transverse diameter of first or second upper molar	5½

Professor O. C. Marsh has also described some remains of peccaries from the same locality, which he attributes to two species under the names of *Dicotyles hesperius* and *Platygonus Condoni*. The former is estimated as about half the bulk of the collared peccary, the latter as being about the size of the hog.

ELOTHERIUM.

ELOTHERIUM IMPERATOR.

Mr. Condon's collection of Oregon fossils contains portions of several teeth of large size, which are supposed to belong to a huge species of Elotherium, for which the above name is proposed.

One of the specimens, represented in Fig. 3, Plate II, is a portion of a large canine tooth, from Bridge Creek. In the perfect condition the tooth would appear to have measured upward of 7 inches in length. The crown has measured about 3½ inches long, with the diameter at base antero-posteriorly about 22 lines, and transversely about 20 lines. The enamel is moderately rugose, except near the back border of the crown, where it exhibits a more folded or ridged appearance. The gibbous fang has been over 4 inches in length, with the fore and aft diameter about 2 inches and the transverse diameter 20 lines.

Another mutilated specimen, from Bridge Creek, supposed to be an upper incisor, is represented in Fig. 27, Plate VII. When complete, the tooth has measured over 4 inches in length. The fang is long, conical, and nearly straight. The crown forms with its fang an obtuse bend or angle. It is conical, compressed from without inwardly, and has the lateral borders subacute and somewhat expanded toward the base.

A third specimen, from John Day's River, represented in Fig. 4, Plate II, consists of the greater portion of the crown of an anterior premolar. It is blunted at the apex as the result of wear. When perfect and unworn it has measured about 1¼ inches in length, about 16 lines antero-posteriorly, and about 9½ lines in thickness.

It is not improbable that part or the whole of the specimens pertain to the species named *Elotherium superbum*, from an isolated incisor tooth found in Calaveras County, California, in the same formation in which was discovered the specimen of a lower jaw referred to *Rhinoceros hesperius*.

Solidungula.

ANCHITHERIUM.

ANCHITHERIUM BAIRDI.

The extinct genus of solidungulate animals, Anchitherium, was originally described from remains found in the middle Tertiary formation of France. Abundant remains of a species have also been found in the Mauvaises Terres of White River, Dakota, which have been described under the name of *Anchitherium Bairdi*. The Condon collection contains several specimens, consisting of detached molars and fragments of others, apparently of the same species. One of the best preserved of these, the crown of an upper molar, is represented in Fig. 15, Plate VII. In every respect it agrees with the upper molars of the *Anchitherium Bairdi* of White River.

ANCHITHERIUM CONDONI.

A specimen in the Oregon collection of fossils, consisting of a small jaw-fragment with a mutilated molar, represented in Fig. 5, Plate II, I have referred to a species of Anchitherium, though several points lead me to suspect that it may belong to a different though closely allied genus. The general form and construction of the teeth are the same as in *A. aurelianense* and *A. Bairdi*. The intermediate lobes of the crown are proportionately larger, more distinct from the others, and more prominent than in the species just mentioned. A tubercle springing from the basal ridge between the antero-internal and antero-median lobes is obsolete in the true Anchitherium.

The diameter of the crown in both directions is about three-fourths of an inch. The species was named in compliment to the Rev. Thomas Condon, through whose interest in natural history most of the Oregon fossils have been brought to the notice of the world.

Perissodactyla.

LOPHIODON?

Among the fossils from Bridge Creek, in the Condon collection, there is a small fragment of an upper jaw containing two molar teeth, represented in Fig. 1, Plate II, which probably indicates a tapiroid animal, allied if not actually pertaining to the extinct genus Lophiodon. The teeth, which appear to be the upper back premolars, are much worn, and the last one is mutilated. They belonged to an animal about the size of the common American tapir, (*Tapirus terrestris.*)

The teeth nearly resemble the corresponding ones, as we might suppose them to be in the same state of wear, of *Lophiodon isselense*, of the Eocene formation of France, as represented in Gervais's Plate XVIII, of the Zoologie et Paléontologie francaises; or they would appear to bear a nearer resemblance to those of *Palæosyops paludosus* of the Bridger Tertiary formation of Wyoming.

The teeth are inserted with three fangs, two externally and a broader one internally. The crowns are widest transversely, square without, semicircular within. They are composed of a pair of pyramidal lobes externally and an internal median conical lobe embraced by a thick basal ridge. The antero-external lobe extends in a ridge to the fore part of the base of the inner lobe, and the postero-external lobe appears to have been continuous by a ridge with the base of the inner lobe. A thin basal ridge festoons the outer part of the crown.

In the worn condition of the teeth they present a wide tract of dentine continuously on the outer lobes. In the penultimate premolar the tract extends inwardly from the postero-external lobe on the inner lobe, and from the antero-external lobe to the base of the latter in front. In the last premolar the dentinal surface of the outer lobes extends continuously on the inner lobe.

The penultimate premolar measures 8½ lines antero-posteriorly, and 11¼ lines transversely; the last premolar measured about 10 lines antero-posteriorly, 11 lines transversely.

The size of the specimen, and its apparent relationship with Lophiodon, led me to suspect that it might pertain to the same animal as an isolated molar tooth, from the Mauvaises Terres of White River, Dakota, described under the name of *Lophiodon occidentalis*.

RHINOCEROS.

A number of fossils in the Oregon collection appear to indicate two different species of Rhinoceros, or perhaps the hornless form Aceratherium. One of them was about the size of the *Rhinoceros occidentalis* of the Tertiary of the Mauvaises Terres of White River, Dakota, and was first supposed to belong to that species. A more attentive examination of its remains has led to the detection of several peculiarities which render it probable it may be a distinct species. As the specimens co-ordinate in size with the lower jaw from the California Tertiary, on which was founded the *R. hesperius*, they may perhaps pertain to this species; and in this view I will so consider them. Of course, more ample material may confirm or refute our position, and may determine the fossils to indicate an animal different from *R. occidentalis* and *R. hesperius*.

The second species, a larger animal, intermediate in size to the latter ones, and the *R. crassus* of the Niobrara Pliocene Tertiary, has been distinguished with the name of *R. pacificus*.

RHINOCEROS HESPERIUS?

The fossils of the Condon collection, attributed with some probability to this species, consist of a mutilated portion of an upper jaw an isolated upper molar, and a lower-jaw fragment containing one entire molar.

The upper-jaw specimen contains portions of the fangs of the molars, of which there were seven, occupying a space of about $7\frac{1}{4}$ inches, or about equal to that in *Rhinoceros occidentalis*.

The anterior extremity of the space included by the zygoma extends to a line with the interval of the second and third molars; in *R. occidentalis* it extends only to a line with the back part of the last molar.

The infra-orbital foramen is large, and occupies a position above the second premolar; in *R. occidentalis* it is over the third premolar.

The upper molar, the last of the series, represented in Fig. 8, Plate II, has nearly the size and form of that of *R. occidentalis*. As in this, the crown consists of a pair of lobes diverging inward from the antero-external corner.

A strong bulge projects from the middle of the anterior lobe into the valley of the crown, which is not so well developed in *R. occidentalis*, and a second bulge at the bottom of the valley is absent in the latter. The basal ridge is stronger in front, and internally at the entrance of the valley of the crown it forms two conspicuous, rounded tubercles not seen in a corresponding position in *R. occidentalis*. The presence of these tubercles, however, is, perhaps, merely an individual peculiarity. The tooth measures 15 lines antero-posteriorly and internally, and is estimated to have been 19 lines transversely.

The lower-jaw fragment, containing a molar, represented in Fig. 9, Plate II, exhibits nothing peculiar distinguishing it from the corresponding part either of *R. occidentalis* or *R. hesperius*.

Rhinoceros pacificus.

The fossil specimens indicative of the second species of rhinoceros from the Oregon Tertiary consist of a mutilated right side of the upper jaw with portions of fangs of the molars, except of the first premolar, and several isolated molar teeth.

The specimens indicate a species larger than the preceding, but not reaching *Rhinoceros crassus* of the Niobrara Tertiary, which was about the size of the existing India rhinoceros.

The upper-jaw specimen retains portions of the fangs of six molar teeth, counting from behind. The space occupied by the back two premolars and the molars is estimated at nearly $7\frac{1}{2}$ inches; that occupied by the true molars at rather more than 5 inches.

The fore part of the zygomatic space is on a line with the fore part of the last molar tooth.

Fig. 6, Plate II, represents an upper molar which is supposed to belong to this species. The specimen is broken at its back part, and is labeled "Alkali Flat." The crown at the fore part measures 21 lines in diameter, and is estimated to have measured $1\frac{1}{2}$ inches antero-posteriorly. The bottom of the crown is embraced with a strong basal ridge, which is strongest anteriorly and internally. The inner lobes expand inwardly, but do not bulge in the abrupt manner posteriorly to the same degree that they do in *R. occidentalis*. The bottom of the oblique valley of the crown is expanded, and is complicated by the projection into it of four folds.

Another tooth, represented in Fig. 7, Plate II, likewise labeled "Alkali

Flat," has the appearance in condition of preservation, color, and wear, as if it might have pertained to the same individual as the former specimen. If so, it is related to it apparently as the last premolar to the first molar. The antero-posterior diameter of the crown is nearly 16 lines; the transverse diameter is 19 lines. The basal ridge and inner lobes are as in the former tooth. Traces at the bottom of the oblique valley appear to indicate a disposition to the formation of two folds like those existing in the same position in the larger tooth.

It is not unlikely that this second molar tooth may be a true molar of the preceding species.

The crown of a lower molar tooth, represented in Figs. 24, 25, Plate VII, from Bridge Creek, is supposed to belong to *R. pacificus*. It measures 20 lines fore and aft, and 1 inch transversely, at base.

HADROHYUS.

HADROHYUS SUPREMUS.

Among the Condon collection of Oregon fossils there are several, apparently of a large pachyderm, differing from those previously indicated, and likewise different in character from such as have been heretofore described.

Fig. 26, Plate XVII, represents a fragment of a tooth which I have supposed to be a last upper premolar. The crown of the tooth would appear in its entire condition to have nearly the form and construction of the corresponding tooth of the Oreodonts, but differs especially in the proportionately less degree of development of the inner lobe and the greater degree of production of the inner basal ridge. The remains of the inner lobe have the appearance of being composed of a nearly connate pair, which no doubt would be found better developed and more distinct in the succeeding teeth. In the specimen the inner lobe appears notched, and the dentine is exposed on the outer lobe and the anterior division of the inner lobe.

The transverse diameter of the specimen is $1\frac{1}{4}$ inches. The tooth is labeled "Alkali Flats," and may be regarded as representing the animal to which the above name is given.

Another specimen pertaining to an animal as large as that to which the tooth just described belonged, and perhaps actually belonging to the same, consists of a brain-cast, or rather the cast of the interior of a cranium. The cast has nearly the size and shape of the brain of a horse. The cerebral hemispheres are nearly as much convoluted as in the latter, and measure about 4½ inches in length and breadth.

A third specimen, which may likewise be suspected as belonging to Hadrohyus, is a large atlas, which measures 5 inches in width between the outer acute borders of the articular cups for the occipital condyles, and about 4½ inches from the neural tubercle to the hypapophysis. The vertebra differs in several important points from the atlas of the rhinoceros, horse, ox, &c., but the want of the requisite means of comparison prevents me from determining its nearer relationship.

AN UNDETERMINED CARNIVORE.

A supposed carnivorous animal of large size is indicated by the portion of a large canine tooth, represented in Fig. 26, Plate VII. The specimen pertains to the Condon collection of Oregon fossils.

CHELONIA.

Testudinidæ.

STYLEMYS.

The extinct genus of turtles above named, and originally described from remains found in the Miocene Tertiary formation of the Mauvaises Terres of White River, Dakota, was most nearly related with the existing land-tortoises. The shell is of the simplest form, and is about as prominent as in the less vaulted forms of the living species of Testudo, or the more vaulted ones of the terrapenes. The proportions nearly accord with those of our southern gopher, but the carapace is more uniformly convex.

The carapace is most prominent just back of the middle, and is abruptly rounded posteriorly as usual in the tortoises. The margin is entire, feebly emarginate in front, somewhat expanded and everted over the axillary spaces, and in a less degree everted over the inguinal spaces.

The plastron holds the ordinary proportions to the carapace as in Testudo and Emys. It is for the most part flat, and only moderately turned up in front. The extremities are nearly equal and rounded. The anterior is slightly narrowed; the posterior is moderately notched.

The number, shape, and relations of the bones of the shell are nearly the same as in Testudo and Emys. The number of the vertebral plates is ten, occasionally eleven, from subdivision of the usual eighth plate.

The ninth plate appears like a corresponding pair of costal plates connate in the median line. The tenth plate is lozenge-shaped, and occupies a similar shaped interval of the ninth vertebral and the pygal plates.

The eight pairs of costal plates in their alternate narrowing and widening toward the extremities resemble those of the living tortoises, though the variation is not so great as usual in these.

The interior of the vertebral plates of Stylemys exhibits a deep, narrow, keel-like ridge, as represented in Fig. 6, Plate III, and Fig. 9, Plate XIX, intended for union with the neural arches of the vertebræ. A similar condition exists in the Gallipagos and other living tortoises.

The costal capituli, as seen in Fig. 6, Plate III, and Fig. 9, Plate XIX, are feebly developed as in most species of Testudo, but are not reduced to the rudimental condition observed in our gopher.

The first pair are as well developed as usual. The sixth and seventh pairs unite with processes of the corresponding vertebral plates. The eighth and ninth pairs, better developed than those in advance, unite in the root of the process of the eighth costal plate for the attachment of the pelvis.

The scutes of both the carapace and plastron of Stylemys correspond with those of Testudo. The pygal scute is single as in all living tortoises, except Manouria. The pectoral scutes are very narrow, as usual in Testudo.

The thickness and strength of the shell of Stylemys is greater than ordinarily in the latter, but proportionately not more so than in several living species.

The bones of the limbs, so far as we are acquainted with them, approach in character those of the tortoises. The concavity above the articular surface of the distal extremity of the humerus, but especially of the femur, is deeper than in the living forms.

The remains of Stylemys are apparently referable to three species, all geographically and perhaps geologically separated.

STYLEMYS NEBRASCENSIS.

The remains of this species form one of the most abundant fossils of the Miocene Tertiary deposit of the Mauvaises Terres of White River, Dakota. A multitude of specimens of nearly entire shells have been collected by all explorers of the locality in which they are found. They present a great variety of age, size, and condition of preservation. Many exhibit in their distortion evidences of considerable pressure, while others are so well preserved as to appear entirely unbroken. Their varied conditions, added to slight anatomical variation, led me at first to attribute them to five different species, which I now view as one.

Mature specimens are comparatively rare, at least in an entire condition. One, broken into two pieces, is represented in Plate XXIII of "The Ancient Fauna of Nebraska." A second, more complete, was obtained by Professor Hayden for the Academy of Natural Sciences of Philadelphia.

Very few other bones of *Stylemys nebrascensis*, other than those of the shell, have come under my notice. Among hundreds of shells and fragments of others, I never met with any portions of the skull or jaws.

Fig. 10, Plate XIX, represents a fragment of the scapula with part of its precoracoid. It agrees with the corresponding portion of the bone in Testudo.

Fig. 7, of the same plate, represents the distal extremity of a humerus of a young individual. The hollow above the articular surface is rather deeper than in Testudo.

STYLEMYS NIOBRARENSIS.

Numerous fragments of shells and a few portions of other bones of a second species of Stylemys were discovered by Professor Hayden in the Pliocene sands of the Niobrara River in the year 1857. All the anatomical characters of the specimens indicate the same genus as the former, but several of them point to a different species. It was about the same size as the *S. nebrascensis*.

Fig. 4, Plate III, represents the anterior portion of a plastron of the natural size, and therefore supposed to have belonged to a young animal. The episternals are more prominent forward than in *S. nebrascensis*, and they are deeply excavated beneath the broad scute-covered margin, which is not the case in the species just named.

Fig. 5 represents the last vertebral and the pygal plates of an older animal. It shows that the investing scute is single, as in Testudo. The lower margin of the pygal bone is slightly but decidedly everted, which is not the case in *S. nebrascensis*.

Fig. 6 represents an inner view of a portion of a carapace one-half the natural size. It belonged to a mature animal, and is the most complete portion of the shell of the species which has been submitted to me. It comprises the vertebral plates from the sixth to the ninth inclusive, and portions of the corresponding costal plates on each side. The narrow character of the costal capitula is observable in the sixth and seventh pairs; and the two succeeding pairs are observable as they spring from the strong process for the attachment of the pelvis.

Fig. 8, Plate XIX, represents the distal extremity of a right humerus, and Fig. 6 the same part of a left femur, both half the natural size. The femur would appear to have belonged to a larger animal than the humerus. The concavity above the articular surface is much deeper than in other known turtles. The breadth of the femur, at the condyloid eminences, is 32 lines; that of the humerus, in a corresponding position, has been nearly the same.

STYLEMYS OREGONENSIS.

An isolated vetebral plate, in the Condon collection of Oregon fossils, is supposed to indicate a third species of Stylemys. The specimen was found on Crooked River, and is represented, one-half the natural size, in Fig. 10, Plate XV. It exhibits a transverse groove defining two vertebral scute areas, and on the interior a narrow crest for union with the corresponding neural arch. The plate appears to be the third of the series, and is thicker in proportion with its length and breadth than would appear to be the case in the preceding species of Stylemys. The specimen is 2 inches wide, 1½ inches long, and 7 lines thick.

DESCRIPTIONS OF REMAINS OF VERTEBRATA FROM TERTIARY FORMATIONS OF DIFFERENT STATES AND TERRITORIES WEST OF THE MISSISSIPPI RIVER.

The fossil remains described in the succeeding pages consist mainly of isolated specimens obtained from Tertiary formations in various parts of the country west of the Mississippi River. They are nearly all remains of mammals. Included in the series there are descriptions of a few other Tertiary mammalian fossils, from the country east of the Mississippi, described on account of their relation with the former, and for the most part for the first time.

MAMMALIA.

Order *Carnivora*.

FELIS.

FELIS AUGUSTUS.

Several teeth in fragments of jaws, and portions of other teeth, indicate a species of tiger apparently different from any previously described. The specimens were discovered by Professor Hayden, during Warren's expedition of 1857, on the Loup Fork of the Niobrara River, Nebraska. They belong to the Pliocene Tertiary formation, and were found in association with remains of *Mastodon mirificus*, *Merychyus elegans*, *Procamelus occidentalis*, &c.

The most characteristic of the specimens, represented in Fig. 19, Plate VII, is an upper sectorial molar contained in a small jaw-fragment. The tooth is about the size of that of the Bengal tiger, and is therefore too large to have belonged either to the panther or the jaguar. It is as much too small to have belonged to the extinct American lion, or *Felis atrox*, as its breadth is but little greater than the sectorial molar contained in the lower jaw from which the latter was described. The form of the tooth is the same as in the American panther and Bengal tiger. The breadth of the crown is slightly less, and its thickness proportionately greater than in the corresponding teeth of a skull of the latter with which the fossil was compared.

If the upper sectorial molar of *Felis atrox* had the same proportionate size to the lower one as in the Bengal tiger and other feline animals, it measured nearly an inch and three-fourths in breadth. That of the Loup Fork fossil is a little over an inch and a quarter in breadth. From the difference in size thus indicated between the sectorial molar of the Loup Fork fossil and that of the previously described largest American cats, recent and extinct, we may fairly regard the specimen as characteristic of another species, for which the name heading this chapter has been proposed.

Comparative measurements of the upper sectorial molar are as follows— those from *Felis atrox* being estimated, and that from the jaguar being taken from Plate XIV of De Blainville's Osteographie:

Upper sectorial molar.	F. concolor.	F. onça.	F. augustus.	F. tigris.	F. atrox.
	Lines.	Lines.	Lines.	Lines.	Lines.
Breadth of crown	11	12	15½	16	20
Thickness in front	5½	......	8	7½	10

Another specimen, represented in Fig. 18, Plate VII, consisting of a fragment of a premaxillary retaining the second incisor, the first alveolus, and part of the last one, agrees in size and other characters with the corresponding part in the Bengal tiger.

The remaining specimens are fragments of an upper last premolar and of a canine from the same individual.

A specimen, represented in Fig. 24, Plate XX, found by Professor Hayden on the Niobrara River, but not in proximity with the preceding, consists of the distal extremity of a humerus, probably of the same animal. It has about the same size, proportions, and form as in the corresponding part of the arm-bone of the Bengal tiger. Its diameter at the supracondyles is 3¾ inches; the breadth of the articular surface in front is 2¾ inches. The hole for the brachial blood-vessels and accompanying nerve is quite evident, though the bony bridge defining it is broken.

Felis imperialis.

Among a collection of fossils belonging to the cabinet of Wabash College, Crawfordsville, Indiana, purchased from Dr. Lorenzo Y. Yates, of Centreville, Alameda County, California, there are several which were kindly loaned to me for investigation. The specimens consist of jaw-fragments of a large wolf and tiger.

Professor E. O. Hovey writes me that they are part of a collection of fossil-bones which were obtained from a wash in the side of a hill about twenty-five miles inland from San Leandro, California.

The fossils are not petrified, and indeed have undergone almost no alteration, and are probably quaternary.

The fossil pertaining to a tiger consists of an upper-jaw fragment, represented in Fig. 3, Plate XXXI, one half size. It contains the second premolar, and retains the alveolus of the one in advance and that of the canine.

The specimen indicates a species as large as the largest living Bengal tiger, and, indeed, is slightly larger than the corresponding part of the largest specimen of a skull among many in the Academy Museum of Philadelphia.

The proportions of the specimen indicate a larger animal than the extinct *Felis augustus*, as represented by the fossil-fragments from the Niobrara River of Nebraska. They also indicate an animal as much smaller than the extinct *F. atrox*, as represented by the ramus of a lower jaw found in association with remains of the *Mastodon americanus* and *Megalonyx Jeffersoni*, near Natchez, Mississippi, as the Bengal tiger is compared with the latter.

Taking into consideration the extent of variation in size of the same species, there can be no question that the California fossil might pertain to either the *Felis augustus* or the *F. atrox*. Its associations might aid in the determination whether it was either of these, or whether it is distinct. If found in association with remains of *Mastodon americanus*, it might reasonably be supposed to pertain to a smaller individual of *Felis atrox*; if with any of the peculiar species of the Niobrara fauna, it might be supposed to be a larger individual of *F. augustus*.

Comparative measurements of the fossil with the corresponding portion of the skull of a large Bengal tiger from Hindostan are as follows:

	Fossil.	Bengal tiger.
	Lines.	*Lines.*
Space occupied by the upper premolars and canine	34.0	33.5
From back of last premolar to canine alveolus	21.8	19.0
Antero-posterior diameter of second premolar	12.2	12.0
Diameter of canine alveolus	14.0	13.5

CANIS.

CANIS INDIANENSIS.

The fossil specimen pertaining to a wolf consists of the right ramus of a lower jaw, represented in Fig. 2, Plate XXXI. The specimen indicates an animal larger than any individuals of the recent wolves of North America and Europe, as represented by skulls I have had the opportunity of examining in our Museum of the Academy. It, however, indicates a less robust animal than that formerly described by me under the name of *Canis primævus*, and subsequently as *C. indianensis*, from an upper-jaw fragment, found in association with remains of Megalonyx, &c., on the banks of the Ohio River, Indiana.

The specimen likewise indicates a less robust species than the *C. Haydeni*, of the Pliocene formation of the Niobrara River, but a larger one than *C. sævus*, of the same formation.

I am disposed to view the specimen as pertaining to the *C. indianensis*, and perhaps it was not different from the existing *C. occidentalis*.

Measurements of the fossil, in comparison with those of the skull of a large wolf from the Columbia River, Oregon, and of another from Germany, are as follows:

	Fossil jaw.	Oregon wolf.	European wolf.
	Lines.	*Lines.*	*Lines.*
Length of jaw from condyle to fore part of canine	96.0	90.0	86.0
Depth of jaw at condyle	21.2	20.5	20.5
Depth of jaw at coronoid process	40.0	36.4	37.5
Depth of jaw at sectorial molar	18.0	17.0	15.0
Depth of jaw at second premolar	16.6	14.0	12.5
Length of molar series with canine	66.0	61.0	55.4
Length of molar series	54.0	50.0	45.0
Antero-posterior diameter sectorial molar	16.4	14.6	14.3
Antero-posterior diameter canine	8.4	6.8	6.2

LUTRA?

A specimen of a tibia, submitted to my inspection by the Smithsonian Institution, is represented in Fig. 4, Plate XXXI. It was presented by Clarence King, and was obtained by him on Sinker Creek, Idaho, in association with remains of *Equus excelsus* and *Mastodon mirificus*.

The tibia pertains to a carnivore, and resembles that of an otter more than that of any other animal with which I have an opportunity of comparing it. Its differences, excepting size, are trifling. The tubercle for insertion of the quadriceps extensor is less prominent, so as to give the head of the bone proportionately less thickness in relation with its breadth. The ridge for the attachment of the interosseous membrane at the lower part of the bone is more prominent and sharper. The distal end in front just above the articulation is flatter, and the groove for the flexor tendons behind is deeper.

	Lines.
Length of the bone internally	59
Width of the head	15
Thickness at the inner condyle	10½
Width of the distal end between the most prominent points	11
Thickness at the inner malleole	8

Order *Proboscidea*.

MASTODON.

MASTODON OBSCURUS.

Besides the well-known American mastodon, *M. americanus*, of the post-Tertiary period, there appear to have been at least three others which inhabited this continent. Characteristic remains of a species, to which the name of *M. mirificus* was given, were discovered by Professor Hayden, in association with an abundance of remains of many other extinct animals in the Pliocene formation of the Loup Fork of Platte River, Nebraska. Remains also, apparently of the same species as the South American *M. andium*, have been found in Central America. For an account of the remains of the two species last named, the reader is referred to the "Extinct Mammalian Fauna of Dakota and Nebraska."

In the Museum of the Academy of Natural Sciences of Philadelphia, there is a cast in plaster of a mastodon tooth, the original of which is reputed to have been found in the Miocene formation of Maryland. The original specimen having been lost, the cast is represented in Fig. 13, Plate XXVII, of the work just named. This, together with the fragment of a similar tooth, represented in Fig. 16, of the same plate, has been taken in evidence of the existence of a fourth species, to which the name of *M. obscurus* has been given.

Dr. Lorenzo G. Yates, of Centreville, Alameda County, California, has communicated to the writer a list of localities in which he has discovered remains of mastodons in that State. Specimens collected by him were sent to Professor C. U. Shepard, of Amherst, Massachusetts, who has submitted them to the examination of the author.

One of the specimens, a last inferior molar tooth, represented in Figs. 1, 2, Plate XXI, was found together with the mutilated lower jaw and upper molars, at Oak Springs, in Contra Costa County. The remains were obtained from the rock at the base of one of the rounded hills, of Tertiary age, mentioned in Professor Whitney's Geological Survey of California, p. 32, stretching along near the edge of the San Joaquin plain. According to Mr. William M. Gabb, the formation belongs to the Pliocene Tertiary period.

A small photograph, sent to me by Dr. Yates, exhibits the lower jaw without the ascending portions behind, and with straight tusks projecting with an upward direction. The tusks appear to be as long as the jaw was in its complete condition.

The molar tooth has the same general form and constitution as the corresponding one of the American mastodon, but is smaller than is usual in this species. It resembles the plaster-cast above mentioned sufficiently to render it probable that it belonged to the same animal.

The crown of the tooth is composed of four transverse pyramidal ridges, each consisting of a pair of lobes, and conjoined in a common, broad, low base, without a conspicuous offset or heel. As in the cast of the Maryland tooth, the inner lobes are more mammillary or less angular than in *M. americanus*. In this respect they approach the condition, even more marked, however, in the *M. angustidens* of Europe, and they are well separated to their base as in *M. americanus*. The outer lobes of the crown have the same form as in the latter, but are provided with distinct offsets projecting from their inner part fore and aft. The contiguous offsets come into contact, and thus obstruct the transverse valleys of the crown. This arrangement accords with that of the cast of the Maryland tooth. In *M. americanus* similar offsets from the outer lobes are usually but feebly developed, and scarcely obstruct the bottoms of the transverse valleys.

The enamel worn from the summits of the anterior of the inner lobes leaves a transverse ellipsoidal cup of exposed dentine, as usual in the same position in the American mastodon. A greater degree of wearing on the corresponding outer lobes has produced quadrilobate excavations of dentine, in which the specimen agrees with the plaster-cast. In the same stage of wear in *M. americanus*, the excavations have a more lozenge-like outline.

The anterior three divisions of the crown are nearly alike in size and construction. The fourth division is less well developed, and consists of a pair of conical lobes, but the inner is much smaller than the other, and is connate, with a supplemental lobe in advance. Back of these there is a small conical tubercle, corresponding with the heel or rudimental fifth division of the crown in *M. americanus*.

In the plaster-cast of the Maryland tooth, the fourth division of the crown consists of a pair of nearly equal conical lobes, embracing a smaller pair at their fore part. Behind these there is a pair of conical tubercles corresponding with the single one in the California tooth. The differences indicated between the posterior extremity of the crown of the latter and the cast of the Maryland tooth are not greater than those observed between the same teeth of different individuals of *M. americanus*, and are therefore unimportant as distinctive characters.

Well-developed elements of a basal ridge, in the California tooth, occupy the outer fore part of the crown and the intervals of the outer lobes. Between the posterior three divisions of the crown they are better marked than in the cast of the Maryland tooth, or even than is usually the condition in the American mastodon. The ridge is also distinctly produced around the outer part of the third and fourth external lobes and the back of the crown, which is not the case in the cast, nor usually in *M. americanus*.

The jaw-fragment containing the tooth is too much mutilated to ascertain anything of importance in regard to it, other than that it measured about $6\frac{1}{2}$ inches in depth at the fore part.

Comparative measurements of the California tooth, the cast of the Maryland tooth, and two teeth of *M. americanus* are as follows:

Last lower molar.	California.	Maryland.	M. americanus.	
			Female.	Male.*
	Lines.	*Lines.*	*Lines.*	*Lines.*
Fore and aft diameter of crown	75	75	75	90
Breadth at anterior three ridges	33	43	38–42	41–45
Breadth at fourth ridge	25	28	33	39
Height of third inner lobe, unworn	25	26	28	34
Height of fourth inner lobe	16	20	20	30
Height of fourth outer lobe	24	22	22	30

* The specimen of the tooth of a male has five transverse divisions to the crown in addition to a small heel.

The second specimen received from Professor Shepard consists of the fragment of a tusk, from Dry Creek, Stanislaus County, California. It was discovered by Dr. Yates imbedded in the bluff of a hill, about ten feet above the bed of the creek. The hill, upward of a hundred feet in height, is one of those mentioned in Professor Whitney's Geological Survey as being scattered over the San Joaquin plain, at the base of the foot-hills of the Sierra Nevada.

The specimen is represented in Figs. 3, 4, Plate XXI, and is remarkable from its exhibiting characters which indicate the species to have been nearly related with the *Mastodon augustidens* of Europe. The molar tooth from Contra Costa County, likewise presents a form which approximates it to the same animal, so that it is probable both specimens may belong to the same species.

The fragment is six inches long, slightly curved in two directions, and in transverse section (Fig. 3) is ovate, with the anterior pole acute. The pulp-cavity, opening half the diameter of the tusk at its larger broken end, extends half the length of the specimen. On one side of the tusk, as in *Mastodon augustidens*, there is a broad layer of enamel, which extends from the acute border two-thirds the width of the specimen. The enamel is somewhat rugose, and is two-thirds of a line thick. In one position, near the smaller end of the fragment, it has been worn through irregularly for the extent of about 1½ inches. The convex or thicker border of the tusk has also been worn off to an extent of two-fifths of the surface. The broken ends of the fragment exhibit the usual decussating lines of structure of the dentine so characteristic of the ivory of the great proboscidians.

The entire length of the tusk appears to have approximated two feet. The other dimensions are as follows:

	Lines.
Long diameter of the larger extremity	28
Short diameter of the larger extremity	19
Long diameter of the smaller extremity	22
Short diameter of the smaller extremity	16
Breadth of enamel layer at larger extremity	22
Breadth of enamel layer at smaller extremity	19

In the form of the tusk and the possession of an enamel band it resembles the same organ in the *Mastodon augustidens*. The specimen when first described was viewed as probably representing a species distinct from that to which the Contra Costa specimens pertained, and was therefore referred to an animal with the name of *Mastodon Shepardi*, in honor of Professor C. U. Shepard.

Since writing the above, I have received another specimen from Professor Shepard, consisting of a last inferior molar tooth, obtained by Dr. Yates in Contra Costa County, California. It is almost identical in form and size with the one previously described from the same locality, but appears to have belonged to an older individual, as indicated by the more worn condition.

From the Smithsonian Institution I have recently received for examination some remains of a mastodon and an elephant, which were found near Santa Fé, New Mexico, and were presented to the institution by the Hon. W. F. M. Arny. The mastodon remains consist of three fragments of a lower jaw, a vertebral body, and a rib-fragment. They are white, and from adherent portions of matrix appear to have been imbedded in an indurated clay. The cancellated structure of the bones is filled with the same matter together with crystalline calcite.

The lower-jaw fragments appear all to have pertained to the same specimen. One of them, represented in Fig. 1, Plate XXII, consists of a portion of the right ramus containing the last molar tooth nearly of the size of the corresponding part in the American mastodon. The molar tooth, represented in Fig. 4 of the same plate, has lost the portion back of the third ridge of the crown. The portion preserved sufficiently resembles in its construction the corresponding portion of the California tooth above described to belong to the same species, which I suspect actually to be the case. It also resembles more nearly the corresponding portion of the same tooth of *M. augustidens* of Europe than it does that of the *M. americanus*.

The other jaw-fragments, represented in Figs. 2, 3, form together the anterior extremity of an enormously prolonged symphysis, like that of *M. augustidens*. The specimen is rather more than a foot in length, and contains portions of tusks extending through the pieces and broken off on a level with the extremities of the symphysis. This has been somewhat crushed laterally, so as to disarrange the proper relative position of the two tusks. It is of nearly uniform width, but widens at the posterior extremity. Below, it is slightly convex or nearly straight longitudinally, and is depressed along the median line. The sides are convex, and extend upward in ridges which form the boundaries of a deep groove at the upper part of the symphysis. The groove is narrower behind, and becomes shallow in front. The tusks are slightly compressed cylindrical, and curved in their course. They are oval in transverse section, with the long diameter directed from within upward and outward. They are unprovided with enamel, and at the broken ends exhibit the decussating curved lines of structure of the ivory enveloped in a thick layer of dense cementum. At the posterior extremity the broken ends exhibit the pulp-cavity occupied with matrix and surrounded with a margin of about a line in thickness, so that the symphysis is broken off near the bottom of the incisive alveoli.

From the thinning of the anterior alveolar borders of the symphysis it would appear as if the latter was nearly complete, so that if we suppose the lower tusks projected about 6 inches from the jaw, it would give them an entire length of about 20 inches.

The breadth of the fore part of the symphysis, in its complete condition, is rather more than 5 inches. At the back part, corresponding with the position of the bottom of the incisive alveoli, it has been about an inch wider.

The long diameter of the tusks, at their anterior broken ends, is about 20 lines; the short diameter 17 lines. These diameters are nearly uniform throughout as existing in the specimen.

The fore and aft diameter of the last molar tooth, when complete, is estimated to have been full 6½ inches. The width of the crown at the base of the second and third ridges is 35 lines. The measurements indicate the proportions of the tooth to be slightly greater than in the corresponding California tooth or the cast of the Maryland tooth.

The depth of the lower jaw below the second ridge of the last molar is 6½ inches; and the thickness is 5 inches.

I think it probable, without being positive in the matter, that the Mastodon remains above described, which have been referred to species under the names of *Mastodon obscurus* and *M. Shepardi*, including those from New Mexico, belong to one and the same species. This, from the form of the molar teeth, the constitution of the upper tusks, and the prolonged symphysis of the lower jaw, was clearly a near relation of the *Mastodon augustidens* of Europe.

In a note, on page 74, of volume II, of the Palaeontological Memoirs of the late Dr. Falconer, it is stated that at Genoa he had seen a cast of a lower jaw of a mastodon from Mexico, with an enormous *bec* abruptly deflected downward, and containing one very large incisor. The beak is much thicker than in

M. angustidens and larger than in *M. longirostris*. "The outline of the jaw resembles very much the figure in D'Orbigny's voyage, described by Laurillard as *M. andium*. The Genoese paleontologists had named it Rhynchotherium, from the enormous development of the beak, approaching Dinotherium."

Perhaps this Mexican specimen of a lower jaw may pertain to the same species as the specimens above described, though the beak of the New Mexican specimen is unlike that of the figure above alluded to in the work of D'Orbigny.

The vertebral body and rib-fragment accompanying the jaw-fragments from New Mexico present nothing remarkable. The former is of a lumbar vertebra, and would indicate an animal about as large as the living Asiatic elephant. Its length is 34 lines; its breadth is about 5 inches at the posterior border; and its height is $3\frac{1}{2}$ inches.

MASTODON MIRIFICUS.

Some small fragments of jaws and teeth, apparently referable to this species, in the museum of the Smithsonian Institution, were obtained by Mr. Clarence King, from Sinker Creek, Idaho.

MASTODON AMERICANUS.

Among a collection of remains of the American mastodon, from Benton County, Missouri, deposited in the museum of the Academy of Natural Sciences of Philadelphia by the American Philosophical Society, there is a singular tooth, which I suppose to be of abnormal character and to pertain to the *Mastodon americanus*. The specimen is in the same state of preservation as the associated remains, and is represented in Figs. 5 and 6, Plate XXII. It consists of the complete crown of a molar tooth without the fangs. Its shape is so peculiar that I can form no clear idea as to the relative position it occupied in the jaws, or as to its homologous character in comparison with normal teeth.

The crown in transverse outline is irregularly oblong oval, more bulging on one side than the other, and somewhat prolonged at the extremities. From a thick expanded base there project four conical lobes, of which the intermediate two are nearly equal and nearly twice the size of the others, also nearly equal in size. The basal ridge on the more prominent side of the crown is mammillated, and twice the depth that it is upon the other side, in which position it is comparatively smooth.

The long diameter of the crown is 55 lines; the short diameter, 29 lines.

A small collection of fossil teeth, from near Pittstown, on the Susquehanna River, in Luzerne County, Pennsylvania, now preserved in the Museum of the Academy of Natural Sciences, is of interest on account of the association. The specimens consist of two molars of *Equus major*, hereafter described, a molar of *Bison latifrons*, also to be described, and three first premolars, apparently from as many different individuals of *Mastodon americanus*. Of these one is represented, of the natural size, in Fig. 9, Plate XXVIII.

ELEPHAS.

ELEPHAS AMERICANUS.

In the preceding account of the remains of mastodon from near Santa Fé, New Mexico, those of an elephant are referred to which were found in association with them. There is but one specimen, consisting of the back part of a molar tooth, apparently the last upper one. It is composed of eight unworn lobes, decreasing successively in length. They present the ordinary thin, elongated, palmate appearance, with the digitate extremities curving forward and ending in mammillary points. The eight plates occupy a space of $4\frac{5}{8}$ inches. The second of the plates is $3\frac{3}{4}$ inches broad near the middle, and when entire was upward of 7 inches in length.

The specimen is insufficient to determine whether it pertained to a species different from the ordinary *Elephas americanus*, and it presents nothing peculiar. The thickness of the lobes, or double plates, indicates the coarse-plated variety of teeth of the American elephant, named by Dr. Falconer *Elephas columbi*.

Since the above was written I have received, from the Smithsonian Institution, for examination some remains of an elephant from Chihuahua. Professor Baird reports that the remains came from an ancient lagoon-bed at Potos Spring, seventy-five miles south of El Paso, Texas, and were presented to the institution by General Carleton, United States Army.

The specimens consist of fragments of molar teeth with adherent gravel, and with the exterior cementum much worn away by water action. They indicate the coarse-plated variety of teeth of the American elephant. One of the better preserved specimens consists of the fore part of a last lower molar about one-third worn down. It comprises about eight lobes, or double plates, included in a space of $5\frac{3}{8}$ inches. The width of the sixth lobe is $3\frac{3}{4}$

inches. The first lobe is nearly obliterated, and its back plate conjoins the contiguous one of the second lobe.

Another specimen consists of the back part of a molar with six lobes, occupying a space of nearly $4\frac{1}{2}$ inches. The lobes exhibit the same narrow, elongated, palmated form, with curved digitate extremities, as in the molar fragment from New Mexico. The first of the six lobes is worn off at the summits of the digitate ends. The others are unworn, and the second plate is $3\frac{3}{4}$ inches wide near its middle.

MEGACEROPS.

MEGACEROPS COLORADENSIS.

An imperfectly known extinct animal, which was supposed to be related with the great ruminant, the Sivatherium of the Tertiary formation of the Sewalik Hills of India, is indicated by a singular looking fossil discovered in Colorado. The specimen belonged to Dr. Gehrung, of Colorado City, by whom it was presented to Professor Hayden. It is represented one-half the natural size in Figs. 2, 3, Plate I, and Fig. 2, Plate II, and was originally described in the Proceedings of the Academy of Natural Sciences of Philadelphia for January, 1870, under the name heading this article.

The fossil is singularly puzzling in its character, and possesses so little in common with the homologous portion in ordinary animals that its relationship would have remained unknown, or entirely conjectural, had we not been previously acquainted with the Sivatherium. The specimen appears to correspond with that portion of the face of the latter which comprises the upper part of the nose, together with the forehead and the anterior horn-cores. As is described to be the condition in the corresponding portion of the skull of Sivatherium, all the bones entering into the constitution of the fossil are completely co-ossified, so as to leave no traces of the original course of the sutures. The nasal and contiguous bones are of great thickness, and as solid as those generally in the living Sirenians.

The horn-cores of the Colorado fossil resemble the anterior ones of Sivatherium both in form and relative position. They are large, dense, conical knobs, somewhat trilateral, and with a rounded, dome-like summit, which is more porous on the surface than any other part of the fossil. They are nearly straight, and divergent from each other, and their summits project more over their base externally than in Sivatherium.

The space between the horn-cores extending across the forehead forms a deep concavity divergent outwardly. The surface of the forehead from the broken border of the specimen behind to the end of the nasals forms a moderate uninterrupted convexity. In Sivatherium, the rhinoceros, and the tapir, the corresponding surface is interrupted by a concavity at the root of the nose.

The face, as formed by the nasals and their apparent conjunction with the maxillaries in advance of the position of the horn-cores, is exceedingly short in comparison with the corresponding part in Sivatherium and the rhinoceros, and is more like that in the tapir.

The nasals together form a strong, thick, tongue-like process, projecting free from their conjunction with the frontals in advance of the horn-cores. The overhanging process of the nose is proportionately wider, thicker, and longer than that of Sivatherium. Its upper surface is not vaulted as in the latter and the rhinoceros, but simply continues the convexity of the forehead. The lateral margins are somewhat expanded, (not sufficiently expressed in Fig. 2, Plate I,) and are thinner than elsewhere. The end is thicker than at the sides, is more obtuse than in Sivatherium or the tapir, and is roughened and porous, probably to have given firmer attachment to a proboscis. A notch occupies the extremity of the obliterated internasal suture.

One of the most remarkable characters of the Colorado fossil is the great comparative extent of the lateral nasal notch. It not only exceeds that of Sivatherium, but also that of the rhinoceros and tapir. In the former its bottom is far in advance of the position of the horn-cores, and in the rhinoceros it holds nearly the same relative position. In the tapir the notch extends back over the position of the orbits. In the Colorado fossil it extends far back beneath the position of the horn-cores, where the nasals apparently conjoin the maxillaries. The relative position of the orbits cannot be ascertained in our fossil, as all the contiguous parts are broken away. They appear as if they had been situated farther posteriorly in relation with the position of the horn-cores than in Sivatherium. The horn-cores, projecting forward and outward, overhang a large recess, which would appear to have been just in advance of the orbit, and is situated externally above and behind the lateral nasal notch.

The broad and stout projecting nasals were probably intended as a point of attachment for a movable snout or proboscis, intermediate in degree of development to that of the tapir and elephant or mastodon. The similar constitution of the nose of Sivatherium led its discoverer and describer, Dr. Falconer, to attribute a like prehensile organ to that animal. The strength and co-ossification of the nasals, together and with the frontals and maxillaries, are also no doubt related with the unusual position of the horn-cores, just as a similar condition of things in the rhinoceros is related with the support of a horn on the nose.

Megacerops coloradensis is estimated to have approximated two-thirds the size of the *Sivatherium giganteum*.

Measurements from the fossil referred to *Megacerops coloradensis* are as follows:

	Inches.	Lines.
Distance from the summit of one horn-core to the other	10	6
Length of curve between the same two points	13	10
Length of lateral nasal notch from end of nasals	1	6
Distance from end of nasals to center of space between horn-cores	6	0
Breadth of nasals 2½ inches behind the end	1	3
Thickness of nasals where co-ossified	1	3
Diameter of horn-cores 2½ inches from summit fore and aft	2	10
Diameter of horn-cores 2½ inches from summit transversely	2	5
Breadth of face below horn-cores	8	8
Breadth at bottom of lateral nasal notches	5	6

Since writing the above, I have recalled to mind a specimen of a horn-core which was obtained by Dr. John Evans from the Mauvaises Terres of White River, Dakota, and which is noticed in the account of Titanotherium, on page 216 of the "Extinct Mammalian Fauna of Dakota and Nebraska." The reference of the specimen to any particular animal was considered very uncertain, though it was suspected that it might pertain to Titanotherium. It is now represented in Fig. 3, Plate XXVIII, and is seen by comparison to bear a near resemblance to the horn-cores of Megacerops. It is rather larger and slightly more tapering and curved than in the latter. The specimen may, perhaps, belong to another species of Megacerops.

Since the foregoing was written, Professors Marsh and Cope have reported the discovery of remains of several huge mammals in the Bridger Tertiary beds, which they have described under the names of Tinoceras, Dinoceras, Eobasileus and Loxolophodon. The ordinal relations of these is a matter of dispute, and it is a question especially whether they are proboscideans, or are representatives of a previously unknown order. One of their most remarkable peculiarities is the possession of several pairs of bony protuberances to the skull, which are viewed as horn-cores.

In a recent paper entitled "On the Gigantic Fossil Mammals of the Order Dinocerata," by Professor Marsh, published in the American Journal of Science for February, 1873, there is a representation of an almost complete skull, described under the name of *Dinoceras mirabilis*. This skull, which appears to agree with the corresponding parts, including the teeth, described in the preceding pages under the name of *Uintatherium robustum*, is represented with three pairs of bony protuberances, or horn-cores. In comparing the Colorado fossil, it would appear that the horn-cores accord with the second pair of the Wyoming fossil, in which they are seen to spring from the upper part of the maxillaries, where these join the nasals.

The resemblance between the specimen belonging to Megacerops and the skull described by Professor Marsh renders it probable that the former belongs to the same order, instead of to the ruminants, as previously supposed.

Order *Solidungula*.

EQUUS.

Equus occidentalis.

The remains of equine animals which of late years have been discovered both in North and South America indicate a number of species and genera really wonderful, when we take into consideration that neither continent possesses a single living indigenous species. The remains from many parts of North America, mainly consisting of isolated molar teeth, which have come under my observation, exhibit so much difference in size and variation of the enamel-folding, as displayed on the worn triturating surface, that in many cases I have failed to refer them to species with any degree of certainty.

In the Proceedings of the Academy of Natural Sciences of Philadelphia for 1865, page 94, I have given a notice of two specimens of upper molars from California, submitted to my examination by Professor J. D. Whitney, which were referred to a species with the name of *Equus occidentalis*. One of these specimens is represented in Fig. 2, Plate XXXIII, and was obtained from auriferous clay, at a depth of thirty feet from the surface, in Tuolumne County, California.

Subsequently, in the same Proceedings for 1868, page 26, and in the "Extinct Mammalian Fauna of Dakota," &c., I described a number of remains obtained by Professor Hayden on Pawnee Loup Fork of the Platte

River, and on the Niobrara River, Nebraska, which I referred to a species with the name of *E. excelsus*. A characteristic specimen referred to the latter consists of a portion of the upper jaw containing the back four molars, represented in Fig. 31, Plate XXI, of the work last named. The teeth in this specimen are so nearly identical in character with those from California, referred to *E. occidentalis*, as may be seen by comparing the figure with Figs. 1, 2, Plate XXXIII, of the present work, that there can be little doubt of the two named species being the same.

Since the original description of the two specimens referred to *E. occidentalis*, I have seen others of half a dozen different individuals from California. All these present sufficient correspondence in peculiarity of character as to render them fairly representative of an extinct species, for which the name of *E. occidentalis* is appropriate. Fig. 1, Plate XXXIII, represents a series of the anterior four upper molars contained in a jaw-fragment. The specimen, together with another similar one from a second individual, and containing all the molars except the last one, were obtained by Dr. George H. Horn from an asphaltum deposit near Buena Vista Lake, California, and presented to the Academy. Similar specimens have also been submitted to my examination, obtained at the same locality by Professor Whitney.

The upper molar teeth of *E. occidentalis* are about the size of those of the larger varieties of the domestic horse. From them they are in general readily recognizable by the greater simplicity in the course of the enamel lines, as displayed on the worn triturating surface, and in the absence of the small enamel-fold, directed inwardly, at the bottom of the oblique valley between the inner principal folds of the crown, in which point these teeth accord with those of the existing ass.

The measurements of the specimens referred to *E. occidentalis* and represented in the figures are as follows:

Specimen of Fig. 1.	Antero-posterior.	Transverse.
	Lines.	*Lines.*
Diameter of first molar	18	13
Diameter of second molar	11½	14½
Diameter of third molar	14½	14½
Diameter of fourth molar	12½	13½

Specimen of Fig. 2, representing a second or third molar:

	Lines
Length externally	27
Antero-posterior diameter of triturating surface	15½
Transverse diameter	13½

A tooth in the collection of the Smithsonian Institution, apparently referable to the same species, was discovered by Mr. Clarence King on Sinker Creek, Idaho.

EQUUS MAJOR.

Figs. 3 to 17, Plate XXXIII, represent specimens, from different localities of the United States, which are viewed as pertaining to an extinct horse, originally referred by the author to a species under the name of *Equus complicatus*, and which is suspected to be the same as that which was first designated by Dr. Dekay under the name of *Equus major*.

Figs. 3, 4, 7 to 10, 12, 13, represent specimens of teeth submitted to my examination by Messrs. D. G. Elliot and George N. Lawrence, of New York. They were obtained from an asphaltum-deposit and from a stratum of clay beneath, in Hardin County, Texas, and were found in association with remains of mastodon and other extinct animals.

Figs. 3, 4 represent a first upper molar of the right side. It differs in no important degree from the corresponding tooth of the domestic horse, but is somewhat larger than usual, and is less simple in the course of the enamel lines on its triturating surface.

Figs 5, 6 represent a similar tooth, from Illinois Bluffs, Missouri, six miles west of Saint Louis. According to the late Dr. B. F. Shumard, it was derived from the quaternary formation of Missouri.

Figs. 7, 8 represent a last superior molar of the right side, accompanying the first molar, from Hardin County, Texas. It is remarkable for its great extent of curvature compared with the corresponding tooth in the recent horse. The arrangement of the enamel is similar to that in the latter, and is but little more complex than usual.

Fig. 9 represents a last lower molar, and Fig. 10 a fifth lower molar. These present nothing peculiar distinguishing them from the corresponding teeth of the recent horse.

Fig. 11 represents a second or third upper molar of the right side. The specimen was found by Dr. Thomas H. Streets, in a gully of Galveston Bay,

Texas, and presented by him to the Academy of Philadelphia. In the complexity of folding of the enamel, as seen on the triturating surface, this tooth is quite characteristic of *Equus complicatus*.

Fig. 12 represents a first lower temporary molar, one of the specimens from the asphaltum-bed of Hardin County, Texas.

Fig. 13 represents an upper last temporary molar, another of the specimens from the locality just indicated.

Fig. 14 represents an upper molar of *Equus complicatus* from the "phosphate-beds" of Ashley River, South Carolina.

Fig. 15 represents an inferior molar from the same locality. The upper molar, in the complex condition of its enamel-folding, is characteristic of the species. The lower molar presents nothing distinctive from those of the recent horse.

Teeth of horses are frequently found in the Ashley phosphate-beds, mingled with abundance of fossil shark-teeth, remains of mastodon, elephant, &c. Many of them are undistinguishable from those of the recent horse, but others in size and complexity of the enamel-folding in the superior molars are sufficiently characteristic of *Equus complicatus*.

Figs. 16 and 17 represent an upper and a lower molar, which were found associated with remains of mastodon at Pittstown, on the banks of the Susquehanna River, Luzerne County, Pennsylvania. The teeth are more than half worn away. Their size, and a rather greater degree of complexity than usual in the enamel lines of the triturating surface of the upper molar, would probably indicate that they belong to *Equus complicatus*.

Measurements of the specimens, represented in Figs. 3 to 17, and referred to *E. complicatus*, are as follows:

First upper molar, Figs. 3, 4.—Length of crown externally, 35 lines; antero-posterior diameter, 21 lines; transverse diameter, 15 lines.

First upper molar, Figs. 5, 6.—Length of crown, 33 lines; antero-posterior diameter, 21 lines; transverse diameter, 15½ lines.

Last upper molar, Figs. 7, 8.—Length at antero-external border from end of fang, 36 lines; length posteriorly, 19 lines; breadth of triturating surface, 19 lines; width, 21 lines.

Last lower molar, Fig. 9.—Breadth, 16½ lines; width, 7 lines.

Fourth or fifth lower molar, Fig. 10.—Length of crown, 31 lines; breadth, 13½ lines; width, 7¾ lines.

Upper second or third molar, Fig. 11.—Length of crown, 32 lines; breadth, 14 lines; width, 15 lines.

First lower temporary molar, Fig. 12.—Breadth, 17½ lines; width, 8 lines.

Last upper temporary molar, Fig. 13.—Breadth, 17 lines; width, 10 lines.
Upper second or third molar, Fig. 14.—Length of crown, 25 lines; breadth, 14 lines; width, 13 lines.
An intermediate lower molar, Fig. 15.—Length of crown, 24 lines; breadth, 13¼ lines; width, 9¾ lines.
Upper molar, Fig. 16.—Length of crown, 22 lines; breadth, 13½ lines; width, 11½ lines.
Lower molar, Fig. 17.—Length of crown 22⅜ lines; breadth, 14⅖ lines; width, 11½ lines.

Among a small collection of fossils from Texas, submitted to my examination by Professor S. B. Buckley, there is a specimen of an upper molar tooth of a horse of peculiar character, represented in Fig. 18, Plate XXXIII. The exact locality from whence the specimen was obtained is unknown. The tooth is apparently a fourth or fifth of the series, and is only sufficiently worn to exhibit the course of the enamel layers on the triturating surface. The tooth is longer than in the domestic horse, and is rather narrower than usual in relation with its fore and aft diameter. The folding of the enamel defining the median lakes of the triturating surface is as complex as in *Equus complicatus*, but in a different position. In the latter the folding is greatest on the contiguous sides of the lakes, as seen in Figs. 11 and 14, but in the tooth under consideration the contiguous sides of the lakes are less folded than usual even in the domestic horse, while the enamel border at the inner sides of the lakes is folded in an unusual degree. Further, the broad inner peninsular fold of the triturating surface, which in the domestic horse and other known species has a simple oval, elliptical, or reniform outline, in this specimen is of extreme width, narrow, and folded at the extremities. The width of this inner fold or column is uniform throughout the length of the crown.

The length of the crown of this tooth, without the fangs, in the entire condition has been upward of 4 inches. Its fore and aft diameter at the triturating surface is 16 lines; its transverse diameter at the middle of the same is 1 inch.

Fig. 19, Plate XXXIII, represents a fragment of an upper molar submitted to my examination by Professor J. S. Newberry. It was obtained from the lignite-beds of Shoalwater Bay, Washington Territory. It presents nothing which distinguishes it from the corresponding part of the molars of the domestic horse.

The length of the crown externally is 2¾ inches, and the fore and aft diameter of the triturating surface is 14 lines.

HIPPARION.

A small collection of fossils, submitted to my examination by Professor S. B Buckley, mainly consist of equine remains, of which the determination is uncertain and the near relations obscure. Most of them were obtained in Washington County, Texas, a few in the contiguous county of Bastrop, and several others in Navarro County. They were usually found in digging wells, at the depth of from 25 to 30 feet, imbedded in a rocky stratum. Most of the specimens are free from matrix, but several have attached portions of a hard arenaceous limestone. From the character of the fossils, I suppose the formation to be of contemporaneous age with that which has been called Pliocene Tertiary of the Niobrara River, Nebraska, Little White River, Dakota, and that noticed in the preceding pages of the Sweetwater River, Wyoming.

Fig. 14, Plate XX, represents a specimen labeled as having been obtained from an ossiferous rock at a depth of 25 feet, in Washington County, Texas. It is a last upper molar of a small equine animal, and is moderately worn away at the triturating surface. It is strongly curved, and is nearly twice the length antero-externally that it is postero-internally. In the isolation of the antero-internal column from the antero-median column, as seen on the triturating surface, it accords with the character of Hipparion. It sufficiently resembles in its relative proportions, and the complexity of arrangement of its enamel-folds, the fragment of a tooth, represented in Fig. 17, Plate XVIII, of the Extinct Mammalian Fauna of Dakota, &c., to belong to the same species. The latter specimen was also obtained in Washington County, Texas, and has been referred to *Hipparion speciosum*, a species originally proposed from specimens discovered at Bijou Hill, Dakota, and represented in Figs. 16, 18 and 19 of the work just indicated.

The measurements of the specimen are as follows:

	Lines.
Length of crown antero-externally	11½
Length of crown postero-internally	6½
Breadth of crown antero-posteriorly	9¼
Breadth of crown transversely	7¼

Another specimen, consisting of the middle portion of an upper molar, from its proportions and the folding of the enamel lakes of the triturating surface, is supposed to belong to the same species. It was obtained in Navarro County, Texas.

Fig. 15 represents a specimen found in association with that of the previous figure of the same plate. It appears to be a third or fourth upper molar, and, from the size and arrangement of enamel on the triturating surface, might be supposed to belong to the same animal as the former specimens. In the proportions of the tooth it resembles those of Merychippus more than it does those of Hipparion. The crown is quite short, and exhibits a considerable degree of curvature. It is about 2 lines long on the inner side, and three times that length on the outer side. On the triturating surface the antero-internal column appears as an elliptical ring, as in Hipparion, but it exhibits a pointed process indicative of continuity at a later period with the antero-median column, as in Protohippus and Merychippus. The tortuous enamel-line on the inner part of the triturating surface presents no median fold directed toward the elliptical ring, as is the case also in the fourth molar of Protohippus, as seen in Fig. 2, Plate XVII, of the Extinct Mammalian Fauna of Dakota, &c.

The antero-posterior diameter of the tooth is $7\frac{3}{4}$ lines, and its transverse diameter $8\frac{1}{2}$ lines.

Another specimen, consisting of a mutilated, unworn molar, from its proportions, is supposed to belong to the same species as the former. It was obtained at a depth of 30 feet from the surface in Washington County, Texas. The crown internally is $5\frac{1}{2}$ lines long, and has measured externally about 10 lines. Its breadth is 9 lines, and its transverse diameter has been but little less.

PROTOHIPPUS (?) s. MERYCHIPPUS?

Among the Texan collection of fossils there are several which are suspected to belong to one or other of the equine genera above named.

Fig. 16, Plate XX, represents a specimen obtained from a well, at a depth of 32 feet, at Independence, Washington County, Texas. It is an upper molar, apparently the second or third of the series of the usual complement of six large teeth in equine animals. In its proportions it would appear to belong to the genus Merychippus rather than Protohippus. The crown is from $3\frac{1}{2}$ to 4 lines in length on the inner side, and from 7 to 8 lines on the outer side. The median enamel lakes of the triturating surface are of simple character, and widely gaping, apparently indicating but comparatively little wear, notwithstanding the shortness of the crown. In the appearance of the triturating surface it resembles more the teeth of *Protohippus perditus*, as represented in Fig. 2, Plate XVII, of the Extinct Mammalian Fauna of Dakota, &c., than it does those of Merychippus, represented in Figs. 5 and 9 of the same plate. On the other hand, it bears a near resemblance to the teeth from Little White River, Dakota, represented in Fig. 1, Plate XXVII, of the work just quoted, which were supposed to pertain to *Merychippus mirabilis*.

Another specimen, from Bastrop County, Texas, consists of an upper molar with the portion internal to the median enamel lakes broken away. It is rather smaller than the preceding, and would appear to hold the relation with it in the series of a fourth or fifth molar.

A third specimen, accompanied by a label in the handwriting of Dr. Shumard, is marked ("Eocene,) Trinity River, Navarro County, Texas." It is a lower molar, represented in Fig. 20, Plate XX, and may perhaps belong to the same species as the preceding.

The measurements of the specimens are as follows:

	Lines.
Breadth of the second upper molar	9¼
Width of the second upper molar	10¼
Breadth of the fourth upper molar	8
Breadth of the lower molar	9
Width of the lower molar	5¾

Fig. 17 represents a specimen found in association with those of Figs. 14, 15, at a depth of 25 feet, in Washington County, Texas. It is an upper molar of the right side, probably the fourth of the series. It is but moderately worn, and is imperfect at the back inner corner. In its proportions and degree of curvature it agrees with the teeth of Protohippus. In size and arrangement of the enamel it approaches in character some of those referred to *Protohippus placidus*, represented in Figs. 39 to 48, Plate XVIII, of the Extinct Mammalian Fauna of Dakota, &c.

The measurements of the specimen are as follows:

	Lines.
Length on inner side	8
Length at antero-external corner	12
Breadth of triturating surface	7½
Transverse diameter of surface	7

Fig. 18 of the same plate represents a specimen from "Little's Well," 30 feet in depth from the surface, in Bastrop County, Texas. It is a first upper molar, and is sufficiently like the former to belong to the same species. It also resembles the corresponding tooth of Fig. 6, Plate XXVII, of the Extinct Mammalian Fauna of Dakota, &c., sufficiently to pertain to the same species. This specimen was obtained on Little White River, Dakota, and was referred to *Protohippus placidus*. The proportions of the former are the same, but, being more worn, it is shorter, and appears larger at the triturating surface. At the same stage of abrasion they would even bear a greater resemblance to each other, as the open fold on the posterior part of the Little White River specimen, in a more worn condition, would then form an islet on the triturating surface, as in the Texas specimen.

The measurements of the specimen, in comparison with those of the Little White River specimen, are as follows:

	Texas specimen.	Dakota specimen.
	Lines.	*Lines.*
Length of crown at the antero-internal column	6	8
Length of crown at the postero-external column	10	12½
Breadth of crown at the triturating surface	8½	9
Width of crown at the triturating surface	7	8

The Museum of the Smithsonian Institution contains several specimens of teeth apparently of *Protohippus perditus* and *Merychippus mirabilis*, obtained by Mr. Clarence King in Utah.

ANCHITHERIUM.

ANCHITHERIUM (?) AUSTRALE.

Among the Texan collection of fossils there is a specimen of peculiar character represented in Fig. 19, Plate XX. It was found in association with that of Fig. 16, of the same plate, in Washington County, Texas. It is the first of the series of six large upper molars as existing in equine animals, but exhibits in front the impress of a premolar larger than usual in members of the order. The specimen is broken at its outer part, but the remainder is nearly as characteristic as if the whole were complete. The crown is so worn away that the dentine is continuous upon all the constituent lobes. An oblique valley extends from the inner side and ends in a foot-like expansion near the center of the triturating surface, and back of the center there remains a crescentic enamel lake.

The tooth is devoid of cementum, and resembles in its constitution the corresponding one of Anchitherium nearer than it does that of other known equine animals. The inner and intermediate lobes appear somewhat fuller than in Anchitherium, and the intermediate spaces narrower and less convergent at bottom.

It may perhaps belong to Anchippus, founded on an imperfect tooth from the same locality, and represented in Fig. 13, Plate XXI, of the Extinct Mammalian Fauna of Dakota, &c. It presents important peculiarities, but these may depend on the difference of position of the tooth in the series. There is, however, one feature in the tooth of Anchippus which is absent in the specimen under consideration, rendering it probable that the teeth pertain to different genera. The feature to which I allude consists of a conspicuous fold or offset from the postero-median lobe projecting into the oblique valley of the crown toward the antero-median fold. In Parahippus the same fold exists in a more complex condition.

The tooth in question likewise resembles that represented in Fig. 11, Plate XXI, of the work above quoted, as characteristic of the genus Hypohippus, nearly as much as it does those of Anchitherium, and may, perhaps, belong to a smaller species of the former.

In the uncertainty as to the nearer generic relationship of the specimen it may be regarded as indicative of a species of Anchitherium with the name given at the head of the chapter. The species was about as large as the *Anchitherium aurelianense* of the Eocene Tertiary deposits of France.

The estimated size of the tooth is 11 lines in diameter antero-posteriorly and nearly the same measurement transversely.

ANCHITHERIUM AGRESTE.

During Professor Hayden's exploration in Montana, he discovered several fossil jaw-fragments of a species of Anchitherium. They were found in association with a Helix, partially imbedded in an indurated, gray, arenaceous marl, and were derived from a lacustrine Tertiary deposit on Red Rock Creek, one of the head branches of the Jefferson Fork of the Missouri River.

The jaw-specimens belonged to a species considerably larger than the *Anchitherium Bairdi* of the Miocene Tertiary of White River, Dakota, and approached in size the *A. aurelianense* of the Miocene Tertiary of France.

The teeth in the specimens, as represented in Figs. 16, 17, Plate VII, are considerably worn, but retain their anatomical characters sufficiently to show that they are identical in form with those of the two species just named. They nearly accord in size with the mutilated upper molar, represented in Fig. 5, Plate II, from Oregon, referred to *Anchitherium Condoni*. In the doubt whether the latter is really a true species of the genus in which it has been placed, the lower-jaw fragments in question are regarded as representing a species with the name heading the chapter.

Measurements from the specimens are as follows:

	Lines.
Space occupied by the back four molars	34
Space occupied by the back three molars	27
Fore and aft diameter of last premolar	8
Transverse diameter of last premolar	6¼
Fore and aft diameter of first molar	7¼
Transverse diameter of first molar	7
Fore and aft diameter of last molar	11
Transverse diameter of last molar	5¼

ANCHITHERIUM (?) ———.

In digging a well at Antelope, Nebraska, in the summer of 1868, at the depth of 60 feet a stratum was found which was stated to be remarkable for the number of fossil-bones it contained. The relative age of the stratum has not yet been ascertained, but from the character of the fossils it is suspected to be contemporary with the Mauvaises Terres formation of White River, Dakota, or perhaps with the later formation of the Niobrara River, Nebraska. From among the specimens collected at the time, my friend Dr. John L. Le Conte obtained a coronary bone of a small equine animal, which he sent to me for examination.

The specimen was exhibited to the Academy of Natural Sciences, and is noticed in its Proceedings for August, 1868. Subsequently, some remains, apparently of the same animal, from the same locality were described by Professor Marsh in the American Journal of Sciences for October, 1868, and referred by him to a diminutive horse with the name of *Equus parvulus*.

The specimen of the coronary bone is represented in Fig. 23, Plate XX. It is only a little over half the length and is considerably less than half the breadth of the corresponding bone of the horse, so that it indicates an animal of little more than half its height and of more slender proportions. Its size would about accord with *Anchitherium Bairdi* of the White River Tertiary formation of Dakota.

The measurements of the specimen are as follows:

	Lines.
Length in the axis	9
Breadth at the upper extremity	9
Thickness at the upper extremity	6
Breadth at the lower extremity	8

Order *Ruminantia*.

BISON.

BISON LATIFRONS.

Remains of large oxen which were contemporaneous with the American mastodon have been discovered in several parts of North America. They have been referred to several extinct species, but the materials have been too incomplete to determine the question with any degree of satisfaction whether they pertain to more than one. The fossils indicate individuals very greatly differing in size, but the difference is perhaps sexual rather than specific. The more robust specimens probably belonged to males, and the smaller ones to females.

The most complete specimen which the author has had the opportunity of examining is the cranium, retaining the horn-cores, represented in Figs. 4, 5, Plate XXVIII, one-fifth the natural size. It was discovered by Mr. Calvin Brown and his son Wilfred, of San Francisco, California, while engaged as engineers in the construction of the Spring Valley water-works of that city, and by these gentlemen was presented to the Academy of Natural Sciences of Philadelphia. Mr. Calvin Brown informs me that the cranium was found in a bed of blue clay, 21 feet below the surface, in Pilarcitos Valley.

The specimen resembles the corresponding part of the skull of the living buffalo (*Bison americanus*) so closely that it will be unnecessary to describe it in detail. Besides being larger, the horn-cores are especially disproportionately larger, and are more transverse in their direction, or are less inclined backward. The occiput appears proportionately wider and lower from the less degree of prominence of its summit. The latter is, however, wider, and is more distinctly defined from the posterior occipital surface by the rougher and more prominent protuberance of attachment for the nuchal ligament. The occipital foramen is no larger than in the buffalo, and the notch below, between the condyles, is more contracted. The forehead, near its middle, is rather more protuberant than in the buffalo.

Comparative measurements of the fossil with the corresponding part of the skull of a large buffalo are as follows:

	Bison latifrons.	Bison americanus.
	Inches.	Inches.
Distance between tips of horn-cores	36	26
Distance between bases of horn-cores	15½	12
Circumference at bases of horn-cores	14	11
Length of horn-core along lower curvature	14½	12
Breadth of forehead where narrowest	13½	11
Breadth of forehead at back of orbits	16	13½
Length of forehead from occiput to fronto-nasal suture	13½	10
Breadth of occiput	12½	10
Depth of occiput	7	6½
Breadth of condyles together	6	5
Transverse diameter of occipital foramen	2	2
Vertical diameter of occipital foramen	1¾	1¾
Distance between ends of paramastoid processes	5½	4¾
Length of temporal fossa	5¾	4¾

The California collection of fossils, belonging to the cabinet of Wabash College, Indiana, contains several specimens of teeth which I suppose to belong to *Bison latifrons*. They were loaned to me for examination through the kindness of Professor E. O. Hovey, and are represented in Figs. 6, 7, Plate XXVIII. They consist of the second and third upper molars, and agree in constitution with the corresponding teeth of the recent buffalo, and in size correlate with the skull above described and referred to *B. latifrons*.

The measurements of the specimens, in comparison with those of the buffalo, are as follows:

	Bison latifrons.	Bison americanus.
	Lines.	Lines.
Second upper molar:		
Antero-posterior diameter of triturating surface	16½	15
Transverse diameter of triturating surface	11	10¾
Transverse diameter of crown near base	14	12¾
Third upper molar:		
Antero-posterior diameter of triturating surface	18	15
Transverse diameter of triturating surface	10¼	9¾
Transverse diameter of crown near base	13½	11½

An isolated upper second molar of *Bison latifrons*, found in association with remains of *Mastodon americanus* and *Equus major* at Pittstown, on the Susquehanna River, Luzerne County, Pennsylvania, is represented in Fig. 8, Plate XXVIII. It is considerably worn, the usual median internal fold of the tooth, in a less worn condition, being seen in the specimen as an oval islet.

The fore and aft diameter of the specimen is $16\frac{1}{2}$ lines, and its transverse diameter at the triturating surface 12 lines.

A specimen of a last inferior molar of a Bison, represented in Fig. 4, Plate XXXVII, and a metacarpal bone of the *Megalonyx Jeffersoni*, presented to the Academy by Dr. Edward D. Kittoe, of Galena, were obtained, together with some additional bones, from a crevice of the lead-bearing rocks, at a depth of 130 feet from the surface, near Elizabeth, Jo Daviess County, Illinois. The tooth is about the size of that of the recent buffalo, and may pertain to that species, though it is not improbable it may have belonged to a small individual of *Bison latifrons*.

The specimen is but little worn. The length of the crown at its fore part is $2\frac{1}{2}$ inches; its breadth 23 lines; its thickness at the base anteriorly 10 lines; and near the triturating surface 7 lines.

AUCHENIA.

AUCHENIA HESTERNA.

In the Proceedings of the Academy of Natural Sciences for 1870, page 125, the writer described some fossil remains from California, submitted to his inspection by Professor J. D. Whitney. Among the fossils were several which were attributed to a large extinct llama, with the name of *Auchenia californica*. The specimens upon which the species was founded consisted of a metacarpal bone, the fragment of another, the proximal end of a femur, an acetabulum, and portions of a tibia. The species indicated was much larger than the camel, as the head of the femur is 3 inches in diameter, and the metacarpal is 19 inches long, whereas the latter in the camel is but 13 inches long.

In the Philosophical Transactions of London for 1870, Professor Owen has described some remains of a large extinct llama from Mexico, under the name of *Palauchenia magna*. This animal approximated in size the camel, whereas the remains attributed to *Auchenia californica* much exceeded it.

Of the remains referred by Professor Owen to Palauchenia, there is a series of molar teeth described and figured from casts and photographs. The teeth are considered as pertaining to the lower jaw, but from a view of the figures I cannot avoid the suspicion that they really belong to the upper jaw. In the form and proportions of the molars, but especially in the form, constitution, and number of the premolars, the series appears to me to resemble more the upper one of the camel and llama than it does the lower one. In one respect one of the molars, the last of the series, approaches in character the last lower molar of the camel and llama. This is in the possession of a fifth lobe, which is, however, much less well developed than in the latter animals. If the view I have taken is not erroneous, Palauchenia, so far as we know it from its remains, would not present sufficient distinctive character to be regarded as of a different genus from Auchenia.

Among the collection of fossils from California, belonging to the cabinet of Wabash College, Indiana, there is a well-preserved series of lower molar teeth, represented in Figs. 1, 2, Plate XXXVII. These, from their size and constitution, would appear to belong to a species of llama exceeding in size not only the existing llama, but also the camel and the Palauchenia.

The question at once arises whether these teeth belong to *Auchenia californica*, *Palauchenia magna*, or to a third species.

The proportions of the bones upon which the former was founded indicate an animal one-third larger than the camel, but the teeth above noticed might belong to an animal but little exceeding a large camel or the *P. magna*. If the characters assigned to the latter as a genus are correct, it is clear that the series of teeth from California do not belong to the same animal, and they then could only pertain to a small individual of *Auchenia californica*, or to another species rather larger than the existing camel. Under the circumstances, until further light is thrown on the subject by the discovery of additional material, we may suppose that two large species of llama, perhaps exclusive of *Palauchenia magna*, were once inhabitants of the western portion of the North American continent, contemporaneously with the *Mastodon americanus*. One of these species, a third larger than the existing camel, is the *Auchenia californica*; the second, intermediate in size to the two latter, may be named *A. hesterna*.

The teeth in question indicate an animal which had arrived at maturity. While the first molar, which earliest acquired its functional position, is much worn, the last molar has its fifth lobe unabraded, and the premolar has but partially lost its summit.

The molars show no characteristic differences from those of the llama and camel. The narrow fold seen projecting outwardly in advance of the antero-external lobe of the last molar, and in a less degree in the second molar, in the llama, is nearly obsolete in the fossil.

The premolar presents some difference from the corresponding tooth in the llama. The crown is thickest, and is rounded behind, and it narrows forward to the anterior subacute border, which is convex longitudinally, and is thickened toward the bottom. The outer side is not impressed at the back part, as in the llama, and is feebly impressed at the fore and upper part. The inner side also is but moderately impressed along the middle, compared with its condition in the llama. A deep enameled pit occupies the inner back part of the crown, penetrating from the triturating surface, as in the latter. The pit opens backward for a considerable portion of its depth, and is closed in this position by apposition with the succeeding tooth.

The measurements of the teeth, in comparison with those of the camel and llama, are as follows:

	Auchenia besterna.	Auchenia lama.	Camel.
	Lines.	*Lines.*	*Lines.*
Fourth premolar:			
* Breadth of crown where greatest	13	5½	12
* Width of crown where greatest	6	3	7
Length of crown to origin of fangs	20	6	14
First molar:			
Breadth of triturating surface	20	7	18
Width of triturating surface	10½	5	9
Length of crown	20	5	5
Second molar:			
Breadth of triturating surface	26	9	23
Width of triturating surface	8¾	5¾	10
Width of crown where greatest	10	5¾	10
Length of crown	36	6	16
Third molar:			
Breadth of crown where greatest	31	13	28
Width of crown where greatest	10	5½	10
Length of crown	41	7	17

* For brevity I have used breadth for the antero-posterior diameter, and width for the transverse diameter.

The length of the series of lower molars and premolars together, in the different species, is as follows:

	Lines.
Length of the series in the llama	32
Length of the series in the camel	66
Length of the series in the Auchenia hesterna	84

Accompanying the inferior molar specimens from California there is a specimen of an upper molar represented in Fig. 3, Plate XXXVII, which, from its constitution and size, is supposed to belong to the same species, if not the same individual.

It is a first or second true molar of the left side, and closely resembles the corresponding teeth of the llama.

Its comparative measurements are as follows:

Second upper molar.	Llama.	Camel.	Auchenia hesterna.	Palauchenia magna.*
	Lines.	Lines.	Lines.	Lines.
Breadth of the triturating surface	9¾	20	23½	21
Width of the triturating surface	8	11	12	11½
Width of crown near base	8	13	14½	....
Length of crown	6½	10	29	16½

* Professor Owen's measurements given as those of the second lower molar.

PROCAMELUS.

The genus Procamelus, or Protocamelus, was originally named from remains discovered by Professor Hayden, in the Tertiary sands of the Niobrara River, Nebraska. Three species were indicated from the locality under the names of *Procamelus robustus*, *P. occidentalis*, and *P. gracilis*. The specimens show that Procamelus possessed a series of four premolars and three molars to the lower jaw, from which we may infer an equal number to the upper jaw. The molars and last premolar have the same form as those of the camel.

Among the Texan collection of fossils, loaned by Professor Buckley, there is a specimen of a tooth supposed to belong to Procamelus. It is represented in Fig. 21, Plate XX, and was found in association with the equine teeth before described, and represented in Figs. 14, 15, and 17 of the same plate. It is a first or second upper molar, and sufficiently resembles the corresponding tooth of *P. occidentalis*, as we may suppose it would appear in the same stage of wear, as to render it probable that it may belong to the same species.

The tooth is much worn, leaving two narrow crescentic enamel pits in the middle of the triturating surface. No trace of an internal column or tubercle exists in the interval internally of the inner lobes of the crown.

The specimen measures 11 lines antero-posteriorly, and nearly the same extent transversely.

Fig. 22 represents an astragalus found in association with the molar tooth just described, and probably belonging to the same animal. It has nearly the size and form of those of the common deer, but is proportionately a little longer and narrower.

Another specimen in the same collection consists of a cubo-navicular bone of a ruminant a fourth smaller than the common deer. It was found in association with the equine tooth above described, and represented in Fig. 16, Plate XX.

An additional specimen consists of a last lumbar-vertebra, apparently of a ruminant. It was obtained in Washington County, at a depth of 30 feet, from a hard arenaceous limestone. It is white in color, crushed downward, and has a portion of the matrix adherent. The vertebra has nearly the size and form of the corresponding bone of the camel, and may have pertained to the largest species of Procamelus, named *P. robustus*.

Procamelus virginiensis.

I may here indicate the recent discovery of some remains, apparently of a species of Procamelus, in the Miocene Tertiary formation of Virginia, the first which have yet been noticed of the family in any locality east of the Mississippi River.

Mr. C. M. Smith, of Richmond, Virginia, while engaged in excavating a tunnel beneath the city, discovered a number of bones and teeth, which he has loaned to me for investigation. They were found imbedded in blue clay containing numerous infusorial remains, among which the beautiful frustules of a Coscinodiscus are especially conspicuous. The fossil-bones mainly consist of those of cetaceans and fishes, but among them are a few of land-animals, and also a portion of a humerus of a bird. The formation from which the fossils were derived is probably an estuary deposit of Miocene age. Among the fossils there are several teeth, which are supposed to belong to a species of Procamelus. The specimens, consisting of a last premolar, and the first and last molars of the lower jaw, are represented in Figs. 26 to 29, Plate XXVII. The teeth have the same form and constitution as those of the western species of Procamelus above named, and they appear to indicate an additional species, which was about the size of the existing llama, and intermediate in size to *P. occidentalis* and *P. gracilis*.

The measurements of the teeth are as follows:

	Lines.
Antero-posterior diameter of last premolar	7
Transverse diameter of last premolar	4
Antero-posterior diameter of first molar	7¾
Transverse diameter of first molar	6
Antero-posterior diameter of last molar	12½
Transverse diameter of last molar	6

MEGALOMERYX.

MEGALOMERYX NIOBRARENSIS (?)

The genus to which the above name was applied has not been determined by positive characters, and may prove not to be distinct from Procamelus. It was proposed on two specimens of teeth of a large ruminant, apparently of the camel family, discovered by Professor Hayden in the Pliocene Tertiary sands of the Niobrara River, Nebraska. The teeth, both lower molars, are described in the "Extinct Mammalia of Dakota and Nebraska," page 161, and are represented in Figs. 12-14, Plate XIV, of that work.

A similar tooth was submitted to my examination, by Professor J. D. Whitney, from the Pliocene Tertiary of Tuolumne County, California.

Figs. 24, 25, Plate XXVII, represent a mutilated lower molar, apparently of the same species. This was found in L'Eau qui Court County, in Northern Nebraska, and was presented to Swarthmore College by George S. Truman.

CHELONIA.

EMYS.

EMYS PETROLEI.

An extinct species thus named is indicated by a number of fragments of several turtle-shells, which were found in association with remains of Mastodon, Megalonyx, Equus, *Trucifelis fatalis*, &c., in Hardin County, Texas. They were obtained from a stratum of clay beneath a bed of bitumen, and, like most of the other fossils accompanying them, are thoroughly saturated with bitumen.

The most characteristic specimens consist of two isolated episterna, represented in Fig. 7, Plate IX. They indicate an animal about the size of the recent *Emys scabra* of the Southern States, but the bones are proportionately more robust than in that species. They abruptly project in advance of the lateral grooves defining the gular scutes, and are squarely truncated. The upper gular surface is nearly square, and slopes forward to an acute edge. In one specimen it is wider fore and aft than transversely; in the other rather less. Behind the gular surface, the bone is deeply hollowed into a concavity.

The measurements of the specimens are as follows:

	Lines.	Lines.
Width of episternal at the front border	10	11
Length of internal border	12	14
Length of postero-lateral border	12	15
Greatest thickness of the bone	5½	5¼

A hyposternal bone about the middle is 28 lines fore and aft, 26 lines wide behind the inguinal notch, and half an inch where thickest internally.

The fore part of a nuchal plate of the carapace resembles the corresponding portion in *Emys scabra*, but is more deeply indented. Its width in front is an inch; the length of its median ridge is 10½ lines; and its thickness where greatest is half an inch.

FISHES.

The following species of extinct fishes were first described by the writer in the Proceedings of the Academy of Natural Sciences of Philadelphia for June, 1870. The specimens were borrowed for my examination from a gentleman of New York, by my friend Mr. George N. Lawrence, of the same city. The locality of the specimens was not ascertained other than that they came from the Rocky Mountains. They were accompanied with some shells, evidently of the later Tertiary period, and also with a coronary bone, apparently of *Equus excelsus*. The fish-remains consisted of eight detached pharyngeal bones of a cyprinoid, and a single dermal bone of a ray.

Subsequently, while a notice of these fossils was in press, the writer received from Professor Hayden a pharyngeal bone of the same species and appearance as the former, which was labeled "Castle Creek, Idaho."

More recently, Professor J. S. Newberry sent to me a small collection of fossils, among which were seven additional specimens of pharyngeal bones, identical in appearance with the former, which were stated to have been found at Castle Creek, Idaho.

Later, Professor Cope described, in the Proceedings of the American Philosophical Society, a number of species and genera of extinct cyprinoid fishes from Catharine's Creek, Idaho. Among these he indicates the same species as that to which the above-mentioned pharyngeals have been attributed, and which have been referred to a previously undescribed genus, as follows:

Family *Cyprinidæ*.

MYLOCYPRINUS.

MYLOCYPRINUS ROBUSTUS.

The specimens, consisting of detached pharyngeal bones with teeth, from which the genus and species were originally described, were all imperfect. Having attempted the description without a previous comparison with the corresponding bones of a recent cyprinoid, I find I have been so careless as to have described them in an inverted position. The specimens later received are better preserved, and among them are five complete ones. All the specimens together exhibit such a variety in size and detail as to lead one to suspect they may represent several different species, though I view them as belonging to a single one, the differences being, as I suppose, mainly due to a difference of age. Six specimens, from Professor Newberry's collection, are represented, of the natural size, in Figs. 11 to 17, Plate XVII, all of them, excepting Fig. 16, being views beneath with the back part directed upward. Fig. 16 represents an inner view, exhibiting the masticatory surfaces of the teeth.

The principal row of teeth consisted of five, as may be seen by the organs themselves and their remains in Figs. 11 to 14, inclusive. They are all of the true masticatory type, and are directed inwardly, opposed to those of the other side. The first and last of the series are the smallest, and the intermediate ones are comparatively large.

In the smallest specimens, and the youngest, as I suppose them to be, the second tooth is the largest, and from this they successively decrease in size to the last, as seen in Fig. 13. In the largest and oldest specimens, the intermediate three teeth are nearly equal in size, as seen in Figs. 16, 17. In the specimens of intermediate size and age we notice some irregularity, but generally a disposition to increasing uniformity of size in the corresponding teeth.

The first tooth is directed backward toward those behind; the others are parallel in their direction inwardly.

The crown of the terminal teeth is more mammillary than in the intermediate ones, in which it is oval with the longer diameter directed from above downward, and the short diameter fore and aft. The masticating surface of the teeth is broad, oval, moderately convex, sometimes nearly flat, and usually slightly depressed at the middle or at the center. The crowns resemble strikingly those of worn human premolars, and are covered by thick, smooth enameloid substance.

The teeth are supported on strong bony columns as long as the crowns They project from the lower ramus of the pharyngeal below the position of the upper or posterior ramus. The last of the series projects backward and inward from the conjunction of the two branches, as usual in cyprinoids.

In the older specimens, it would appear that the first tooth of the series was after a certain time not replaced.

Most of the specimens present evidences of the existence of two minute teeth forming a second row above the principal one.

The pharyngeal bones, in accordance with the strong crushing teeth they sustain, are stronger than usual in the ordinary living carp-like fishes.

The pharyngeal bone is widest opposite the larger teeth. The oblique surface directed forward and outward exhibits the usual deep hollows extending to the bases of the teeth, or through the bone in some cases when the latter are absent or shed. The posterior and inferior surfaces are flat, and transversely striated, or, in the older ones, more or less strongly ridged. The anterior border is vertically concave. The external border, acute below and obtuse behind, is unusually thick. The inner border, extending backward beyond the conjunction of the two branches of the bone, is that which sustains the teeth.

The upper or posterior ramus is comparatively short, bent forward and inward, and ends in a point by which it was suspended from the occiput. The extremity of the lower or anterior ramus, extending in advance of the teeth, ends in a triangular process with a lozenge-like articular surface for symphysial attachment with the bone of the opposite side.

Measurements derived from seven specimens are as follows:

Measurements.	Spec. 1.	Spec. 2.	Spec. 3.	Spec. 4.	Spec. 5.	Spec. 6.	Spec. 7.
	Lines.	Lines.	Lines.	Lines.	Lines.	Lines.	Lines.
Length of series of five teeth........	....	....	....	....	$7\frac{3}{4}$	7	....
Length of series of four teeth, excluding the first	15*	12	$9\frac{1}{2}$	9	6	6	5
Length of series of intermediate three teeth	12	10	8	7	....	....	4
Length of series of anterior three teeth.	....	....	....	....	$5\frac{1}{2}$	$4\frac{1}{4}$	....
Long diameter of crown of first tooth...	....	....	....	....	$1\frac{1}{4}$	1	....
Long diameter of crown of second tooth.	$4\frac{1}{2}$	$4\frac{1}{4}$	$3\frac{3}{4}$	3	$2\frac{1}{4}$	$2\frac{1}{2}$	$2\frac{1}{4}$
Short diameter of crown of second tooth.	$3\frac{3}{4}$	$3\frac{1}{4}$	$2\frac{1}{2}$	$2\frac{1}{2}$	2	2	$1\frac{1}{2}$
Long diameter of crown of third tooth...	....	$4\frac{1}{2}$	$4\frac{1}{4}$	$3\frac{1}{4}$	$2\frac{1}{2}$	·2	$1\frac{3}{4}$
Short diameter of crown of third tooth..	....	$3\frac{1}{2}$	$2\frac{3}{4}$	$2\frac{1}{2}$	2	$1\frac{1}{2}$	$1\frac{1}{4}$
Long diameter of crown of fourth tooth.	5	$4\frac{1}{2}$	4	3	....	....	$1\frac{1}{2}$
Short diameter of crown of fourth tooth.	$3\frac{1}{2}$	3	$2\frac{1}{4}$	2	....	....	1
Long diameter of crown of fifth tooth...	....	3	....	....	....	....	1
Short diameter of crown of fifth tooth..	....	2	....	....	....	....	$\frac{3}{4}$
Length of lower branch from back of pharyngeal...................	24*	20*	17	16	14	11	$10\frac{1}{2}$
Depth of upper branch to bottom of pharyngeal................	17*	$14\frac{1}{2}$	$11\frac{1}{2}$	11	10	....	....
Width of pharyngeal inferiorly........	14	$12\frac{1}{2}$	10	9	$8\frac{1}{2}$	8	7

* Estimated.

Family *Raiæ*.

ONCOBATIS.

Oncobatis pentagonus.

An extinct ray, before alluded to, is indicated by a single dermal bone, of which two views are given, of the natural size, in Figs. 18, 19, Plate XVII. The bone has a pentagonal outline with curved margins, with the under or inner surface strongly convex and smooth. The upper or free surface presents five sloping planes more or less well defined by prominent borders. Less than half the extent of the external surface at the center is occupied by an areola of thin enameloid substance which is smooth and shining and marked with concentric lines. The summit of the bone, in the center of the areola, projects as a point of harder and more translucent osseous substance.

The measurements of the specimen are as follows:

	Lines.
Greater diameter of the dermal bone	16
Shorter diameter of the dermal bone	15
Thickness from summit	8

The many more fossil-remains of fishes from the Tertiary formation of Idaho, described by Professor Cope, he attributes to two additional species of Mylocyprinus, seven species of four other genera of Cyprinidæ, and a species of Salmonidæ. Fossil-shells described by Mr. Meek from the same formation, as well as the cyprinoid fishes, indicate a fresh-water deposit. The presence of a ray may probably indicate an easy communication with salt water.

DESCRIPTION OF REMAINS OF REPTILES AND FISHES FROM THE CRETACEOUS FORMATIONS OF THE INTERIOR OF THE UNITED STATES.

The Cretaceous formation in the interior of the United States covers an area reaching southerly into Texas, and extending over a large portion of the eastern slope of the Rocky Mountains, northerly along the region of the Upper Missouri River to its sources. Exposed to view over a great extent of this area, a still larger portion underlies the vast Tertiary deposits of the country. Its thickness ranges from 800 to 2,500 feet, and it consists of various colored strata of indurated clays and sandstones, and indurated marls and limestones. So far as known, most of them are of marine origin, and contain an abundance of characteristic fossils. Some of the strata contain remains of terrestrial plants, proving that the country in the vicinity of the great Cretaceous seas was clothed with forests resembling, in the generic characters of the trees, the forests of our own time. Species of sweet-gum, poplar, willow, birch, beech, oak, sassafras, tulip-tree, magnolia, maple, and others have been described from the fossils. With such a vegetation we would expect the contemporaneous existence of some forms of mammalian life, but as yet, in these as well as in other Cretaceous deposits of the world, no remains of mammals have been discovered. We are, however, still on the lookout for some lacustrine or river deposit of the Cretaceous era which perhaps will reveal early forms of mammals—forms which may more nearly relate the mammal with the reptile than any now known to us.

Remains of birds have been found in the Cretaceous formation of Kansas, and have been described by Professor Marsh. Two genera indicated by him under the names of Ichthyornis and Apatornis are the most remarkable of their kind, and may be viewed as the most interesting and important paleontological discovery yet made in the West. They have biconcave vertebræ, and the jaws are furnished with teeth. Like the Archæopteryx of the Solenhofen limestone, they make the relationship of birds to reptiles much nearer than appears among existing forms.

In remains of reptiles and fishes the western Cretaceous formation abounds. Many of these have been described by Professor Cope and Professor Marsh. Among the reptiles are some of the largest and most wonderful of their kind, represented by great turtles allied to Atlantochelys; numerous species of Mosasaurus and closely related genera; the Polycotylus and the long-necked Discosaurus allied to Plesiosaurus; and Pterodactyls, with an enormous expanse of wings.

The following pages contain descriptions of remains of reptiles and fishes which have come under the observation of the author mainly from the western Cretaceous deposits. A few of the remains are doubtful as to the formation from which they have been derived, but are believed to be Cretaceous fossils. As intimately related with the western Cretaceous fossils, descriptions of a few others are included from eastern localities.

Most of the fossils were submitted to the examination of the author by the Smithsonian Institution, and form part of a collection from the Smoky Hill River, Kansas, and from the Indian Territory, presented to the Army Medical Museum of Washington by Dr. George M. Sternberg, United States Army. Others from the Smithsonian Institution were collected in the vicinity of Fort McRae, New Mexico, and were presented to the Army Medical Museum by Dr. W. B. Lyon, United States Army. Many of the fossils were collected during the explorations of Professor Hayden. The remainder form part of the Museum of the Academy of Natural Sciences and Swarthmore College, or have been contributed by Dr. William Spillman, Dr. John L. Leconte, Professor George H. Cooke, William M. Gabb, George H. Truman, and others.

REPTILES.

Order *Dinosauria*.
POICILOPLEURON.

POICILOPLEURON VALENS.

During Professor Hayden's expedition of 1869, a fossil was given to him as a "petrified horse-hoof." The specimen was found in Middle Park, Colorado, and according to Professor Hayden was probably derived from a formation of Cretaceous age. Similar specimens were reported not to be uncommon, and were known as above designated. Indeed the writer has seen a second specimen, which was also called a fossil horse-hoof, but unfortunately his notes in relation to it have been mislaid.

The fossil in question consists of one-half of a vertebral body as represented in Figs. 16 to 18, Plate XV. When resting upon the articular face, it is not surprising that it should have been taken for a "petrified horse-hoof" by those not conversant with anatomy.

The vertebral body in its entire condition would resemble in form those of Megalosaurus, and in shape and other characters resembles those of *Poicilopleuron Bucklandi*. This is an extinct reptile, from the Oolitic formation of Caen, Normandy, described by M. Deslonchamps; and remains apparently of the same animal, from the Wealden formation of Tilgate, England, have been noticed by Professor Owen. It has been viewed as a crocodilian, and is estimated to have been about 25 feet in length.

The Colorado fossil would indicate an animal approximating 40 feet in length.

One of the most remarkable characters of Poicilopleuron is the presence of a large medullary cavity within the bodies of the vertebræ, as well as in the long bones of the limbs. Among living animals I know of a similar condition in the vertebræ of none except in the caudals of the ox. This curious feature is a striking one in the Colorado fossil, as represented in Fig. 18. The lower two-thirds of the body appear occupied by a large cavity, crossed by a few osseous trabeculæ. The cavity is bounded by a thick lateral and inferior wall of compact substance, resembling that of the shaft of the long bones of most mammals. The wall is about 2 lines thick, and thins away at the upper part of the body where this is occupied by the ordinary spongy substance. The latter extends into the abutments of the neural arch, and is here more dense in character. The cavernous structure of the fossil is filled with crystalline calcite.

The estimated length of the vertebral body is about 6 inches. At the sides and beneath it is much constricted or narrowed toward the middle. The transverse section approaching the latter position is vertically ovoid, with the lower and narrower end forming an acute angle.

The articular end of the specimen, Fig. 16, is moderately depressed its greater extent, most so above and becoming more superficial below. Its upper border overhangs the deepest portion of the surface; the lateral borders are obtusely rounded, and widen below in a strongly convex ledge, probably for the accommodation of a chevron bone. The breadth of the articular surface is nearly 4 inches; its vertical extent a little over that measurement.

The abutments of the neural arch are firmly co-ossified with the body, but their sutural connection is plainly visible. Just below the suture, the side of the body presents a concavity. The beginning of a groove or narrow concavity is also seen extending forward beneath the body. The lateral surfaces of the specimen are smooth, excepting near the everted articular border of the body, where they are roughened for the firmer attachment of ligaments.

Poicilopleuron was probably a semi-aquatic Dinosaurian, an animal equally capable of living on land or in water, and perhaps spending most of its time on shores or in marshes. Whether the cavernous structure of its skeleton was related to pneumatic functions, as in birds, flying reptiles, and some others, or whether it was only occupied with ordinary marrow, is a question that appears uncertain while our knowledge of the skeleton itself is so incomplete.

Order *Chelonia*.

Among Dr. Sternberg's collection of fossils from the Smoky Hill River, Kansas, there are several which appear to be the limb-bones of a turtle. Similar bones from the Cretaceous formation of New Jersey and Mississippi I formerly attributed to species of Mosasaurus, but the recent discoveries of characteristic portions of the skeleton of this and allied animals, retaining the limbs, have proved that view to be erroneous.

A huge turtle, represented by the proximal extremity of a humerus found in the green sand of New Jersey, was named by Professor Agassiz *Atlantochelys Mortoni*. Professor Cope has described some remains of a species nearly as large as the former, from Kansas, under the name of *Protostega gigas*; and an arm-bone of a smaller turtle, from the Cretaceous formation of Mississippi, he has referred to a species with the name of *P. tuberosa*. Remains of a turtle, about the size of the Mississippi snapper, from Kansas, he has attributed to another genus with the name of *Cynocercus incisus*. The specimens of limb-bones above mentioned, and represented in Figs. 17 to 21, Plate XXXVI, are not large enough to pertain to the smallest of the three species of Atlantochelys indicated, but would sufficiently relate in size with the remains of *Cynocercus incisus* to belong to that animal.

The bones appear unusually flat, but this condition, in part at least, is due to compression.

Fig. 17, Plate XXXVI, represents the upper extremity of a humerus extending to the commencement of the distal expansion of the shaft. It resembles nearly the corresponding portion of the humerus of the snapper completely flattened, or a miniature of that of Atlantochelys in the same condition. The greater tuberosity appears to spring from above the top of the head externally, so that its upper anterior border looks like an extension of the articular surface of the latter. A strong muscular impression is situated upon the inner fore part of the shaft. The lesser tuberosity projects posteriorly, and ends in a thick, roughened, convex surface.

The breadth of the specimen between the two tuberosities obliquely measures 33 lines; the breadth of the shaft, where narrowest, is 10 lines.

Fig. 18, represents a complete femur, apparently from the same individual as the former. As in the snapper and Trionyx, it is of proportionately less breadth than the humerus. It is apparently much flattened by pressure, so as to differ considerably from its exact original form. The trochanters appear relatively to have been as well developed as in the snapper, and the distal articulation may be supposed to have had nearly the same form.

The length of the femur is $5\frac{1}{4}$ inches. The breadth of the upper extremity is 20 lines, of the lower extremity 16 lines, and of the middle of the shaft 7 lines.

Several additional bones accompanying the former appear to belong to the shoulder of the same animal.

Fig. 19, represents what appears to be a portion of the left scapula with its upper end and the præ-coracoid prolongation broken away. The specimen appears distorted and flattened from its normal condition as the result of pressure.

Fig. 20, represents what appears to be a portion of the coracoid bone of the same side, also somewhat distorted by pressure.

Fig. 21, represents another bone-fragment, apparently from the same individual, which I cannot determine to my own satisfaction. Like the other specimens, it appears flattened from its normal condition.

Order *Mosasauria*.

Large, extinct, marine saurians, most nearly constructed as in Lacertilians, but having limbs constructed as paddles for swimming. The relations of these reptiles with the serpents, as suggested by Professor Cope, in his Synopsis of the Extinct Batrachia, Reptilia, &c., have been much reduced by the subsequent discoveries of Professor Marsh; and they appear hardly sufficient to justify the name of Pythonomorpha.

The remains of mosasauroid reptiles are comparatively abundant in the Cretaceous formation of the United States. The specimens collected have formed the basis of a multitude of species and genera, the number of which will probably be somewhat reduced on more careful study and comparison of the materials.

In the description of the few mosasauroid remains which have been submitted to my examination, I have referred them to species for the most part as recently named by Professor Marsh, who, with the rich materials in his possession, has the best opportunity of determining their generic and specific characters.

TYLOSAURUS.

TYLOSAURUS DYSPELOR.

Among the fossils submitted to my examination by the Smithsonian Institution, there are some bones of a large mosasauroid animal, collected by Dr. W. B. Lyon, United States Army, in the vicinity of Fort McRae, New Mexico. They consist of vertebræ, mostly more or less crushed and otherwise mutilated, and a few limb-bones, and were obtained from a stratum of soft, yellowish chalk. Specimens from the same collection and skeleton were described by Professor Cope, and referred to a species with the name of *Liodon dyspelor*. This was subsequently referred to a genus, by Professor Marsh, with the name of Rhinosaurus, which, being pre-occupied, Professor Cope proposed that of Rhamphosaurus, and, as this also was previously appropriated, Professor Marsh has now proposed the name of Tylosaurus.

Of the specimens selected by me for examination half a dozen consist of centra and parts of others of posterior dorsal vertebræ, most of which are remarkable for the extent of compression they have undergone with little appearance of fractures. They look as if they had been in a plastic condition, and in this state had been flattened from above downward.

In three of the specimens, consisting of posterior halves of dorsal centra, the articular ball presents a half oval outline below, with slanting sides above, and an emarginate summit. The measurements of the ball, indicating a successive increase in the degree of flattening in the three specimens, are as follows:

	Lines.
Depth of specimen represented in Fig. 1, Plate XXXV	44.2
Breadth of specimen	60.0
Depth of second specimen	42.0
Breadth of second specimen	61.6
Depth of specimen represented in Fig. 2	39.8
Breadth of specimen	63.0

In the other three specimens, consisting of nearly complete dorsal centra, and measuring about 4¼ inches in length, the compression is still greater. In one of the specimens the distal articulation, represented in Fig. 3, is so flattened as to appear transversely lenticular in outline and emarginate above. It measures 30 lines in depth and 62¼ lines in breadth.

Seven selected specimens consist of caudals which have mostly undergone little or no compression. They all present beneath a pair of strong processes projecting obliquely backward from nearer the posterior part, and excavated in a conical pit directed backward and downward for articulation with chevrons. Three have been provided with strong diapophyses projecting in advance of the middle and nearly half way up the sides. A fourth specimen has a small, narrow diapophysis projecting in advance of the middle and about two-thirds up the sides. The remaining two vertebræ have no diapophyses.

The caudals with diapophyses have the articular ends of the body transversely oval, with a slightly hexahedral outline, emarginate above, and in a less degree below. Those without diapophyses have the articular ends of proportionately less width, of less hexahedral outline, and not emarginate below, so that they appear more cordiform than oval.

The largest caudal with diapophyses has measured as follows:

	Lines.
Estimated length of centrum beneath	37.5
Estimated breadth of articular ends	48.0
Depth of articular ends	42.0

A smaller caudal with diapophyses, less mutilated, and represented in Figs. 4, 5, measures as follows:

	Lines
Length of centrum beneath	28.5
Breadth of articular ends	44.0
Depth of articular ends	40.0

The caudal with small diapophyses, represented in Figs. 6, 7, measures as follows:

	Lines.
Length of centrum beneath	30.0
Breadth of articular ends	40.5
Depth of articular ends	36.0

The better preserved of the caudals without diapophyses, represented in Fig. 8, measures as follows:

	Lines.
Length of centrum beneath	24
Breadth of articular ends	38
Depth of articular ends	36

The researches of Professor Marsh have proved the mosasauroid reptiles to have had four limbs constructed as paddles and adapted to swimming. Previous to his discoveries it was supposed that posterior limbs were absent.

Specimens of limb-bones, found in association with the vertebral specimens above described, are supposed to belong to the posterior limbs, from the latter pertaining to the back part of the column.

The femur represented in Fig. 9, Plate XXXV, is a broad bone strikingly different from the humerus of Clidastes. The specimen is probably more flat than in the normal condition, as its many fractures are evidences of its having been crushed.

The distal extremity is much the wider, and the upper extremity is but little wider than the shaft. The head appears as a wide, lenticular, convex, and very rugged surface. The lower extremity ends in a long, narrow, elliptical, rugged surface for articulation with the bones of the fore-arm. From the posterior part of the shaft there projects a thick, convex ridge, which terminates above in an oval, flat, rugged surface, sloping from that of the head of the bone.

The rugged, articular surfaces of the femur would appear to indicate a cartilaginous continuity with the contiguous bones more intimate than in Clidastes.

The measurements of the specimen are as follows:

	Lines.
Length of femur	96
Breadth of head	46½
Breadth of distal extremity	69
Breadth at narrowest part of shaft	34
Thickness of head and trochanter	28
Greatest thickness of lower extremity	17

The remaining two bones I take to be those of the leg, though I am uncertain in regard to their relative position with each other and the femur.

The specimen represented in Fig. 10 I suppose to be a fibula, though it may be an ulna. It is a broad bone, almost as wide as the femur, but not so long. Its flatness has been somewhat increased by pressure. The upper extremity presents a wide, lenticular, uneven, convex, and roughened surface for cartilaginous union with the femur. The lower extremity presents a similar surface, but wider and of less depth or thickness.

The measurements of the specimen are as follows:

	Lines.
Length of fibula	66
Width of upper extremity	54
Thickness of upper extremity	21
Width of lower extremity, partly estimated	65
Thickness of lower extremity	17
Width of shaft near middle	38
Thickness of shaft	13

The supposed tibia, represented in Fig. 11, is a much smaller bone than the fibula. It is clavate, with the lower extremity the more expanded and thinner. The upper part of the shaft is compressed cylindroid, and becomes wider and more compressed below. The upper extremity presents a triangularly oval, slightly convex, articular surface, rugged as in the other bones. The lower articular surface is transversely convex and widely lenticular.

The measurements of the specimen are as follows:

	Lines.
Length of the tibia	45
Width of upper extremity	21
Thickness of upper extremity	14
Width of lower extremity	34
Thickness of lower extremity	11
Width of narrowest part of shaft	12
Thickness of narrowest part of shaft	10

TYLOSAURUS PRORIGER.

Dr. Sternberg's collection of Kansas Cretaceous fossils, preserved in the Museum of the Smithsonian Institution, contains specimens pertaining to several individuals of a large Mosasaurus-like reptile, approximating in size the Maestricht Monitor of Europe, and the *Mosasaurus Mitchelli* of New Jersey. The specimens appear to pertain to the same animal as that described by Professor Cope under the names of *Macrosaurus* and *Liodon proriger*, and afterward, as in the case of the former species, referred to another genus by Professor Marsh, under the name of Rhinosaurus, then by Professor

Cope to Rhamphosaurus, and finally by Professor Marsh to Tylosaurus. A series of specimens belonging to one individual, from the yellow chalk of Kansas, consists of several small fragments of jaws with bases of teeth, a basi-occipital bone, and five vertebræ.

The basi-occipital is obliquely distorted, from pressure. It has attached the diverging processes of the basi-sphenoid. The condyle has approximated 3 inches in transverse diameter, and is about 2 inches in depth. The diverging processes of the basi-sphenoid, at their conjunction with the basi-occipital, are about $3\frac{3}{4}$ inches wide. The vertebræ are all more or less crushed and distorted. One of the specimens, a posterior cervical, has the body below 3 inches in length, and the truncated hypapophysis about $1\frac{1}{5}$ inches in diameter. The articular ball and socket approximate $2\frac{1}{4}$ inches in diameter.

In three of the vertebral specimens, of about the same length as the preceding, the hypapophysis is rudimental. The remaining specimen is a more posterior dorsal, and is of nearly the same size as the other vertebræ.

A second series of specimens, belonging to another individual, consists of several much-mutilated cervical centra, small fragments of jaws with bases of teeth, a coronoid bone, a fragment of a quadrate bone, and the end of the premaxillary.

The latter specimen, represented in Fig. 12, Plate XXXV, exhibits the peculiar character of the extremity of the muzzle in Mosasaurus and its allies. It forms a solid, conical, osseous prominence, with the end obtusely rounded and projecting beyond the anterior teeth. The sides of the premaxillary toward the end are perforated with large vasculo-neural foramina. The projecting end of the bone extends about $1\frac{1}{4}$ inches in advance of the bases of the first pair of teeth. Immediately in front of the latter there is a small conical process.

Several specimens of a third individual consist of a caudal vertebra and two teeth, which, from the adherent matrix, have been obtained from white or cream-colored chalk.

The vertebra is comparatively well preserved, not being crushed nor having its body distorted, as is so frequently the case in the specimens which have come under my observation. It is from the caudal series without diapophyses or transverse processes, and is represented in Figs. 1, 2, Plate XXXVI. The posterior ball, defined from the body by a narrow ledge, and the anterior cup are nearly circular, with a slight hexahedral disposition.

The neural arch exhibits rudiments of zygapophyses. The bottom of the body is provided with a pair of deep, conical pits for the attachment of a chevron-bone. The pits are defined with a prominent margin most projecting anteriorly.

The measurements of the specimen are as follows:

	Lines.
Length of the body inferiorly	20
Diameter of the body at the extremities	30

Of the two specimens of teeth, one is crushed nearly flat; the other, well preserved, is represented in Fig. 3. It presents the usual form more or less characteristic of Mosasaurus. It is curved conical, with the inner and outer surfaces defined by acute ridges. The surfaces are subdivided by longitudinal ridges, becoming obsolete toward the point of the tooth. The intervals of the ridges are feebly concave and faintly rugose. Internally near the base they are delicately striate.

The length of the crown externally is 20 lines; the diameter at base is 10 lines.

Several additional specimens, apparently belonging to another individual, consist of small fragments of jaws and palatine bones with bases of teeth. Among the specimens is a portion of a splenial bone, with its posterior articular surface nearly entire, as represented in Fig. 13, Plate XXXV. The articular surface is a pyriform excavation, with a ridge descending from the upper part internally to near its center.

LESTOSAURUS.

LESTOSAURUS CORYPHÆUS.

Dr. Sternberg's collection of fossils from the Smoky Hill River, of Kansas, belonging to the Smithsonian Institution, contains numerous specimens of dorsal vertebræ of a mosasauroid, which have the appearance as if they had pertained to a single individual. There are about fifty of these vertebræ, but all have been more or less compressed from pressure of the superincumbent beds to that in which they lay, so that not a single specimen preserves the exact original form. They differ but little in size, the more anterior being somewhat shorter than the others.

The specimens appear to belong to the animal described by Professor Cope under the name of *Holcodus coryphæus*, which Professor Marsh has referred to another genus with the name of Lestosaurus.

Fig. 5, Plate XXXVI, represents one of the best preserved of the specimens from the back of the series. In its present condition the centrum beneath is 27 lines long, and the ball and socket ends are about 16 lines in depth and 2 inches in width. The neural arch between the ends of the fore and aft zygapophyses measures 34 lines.

Another similar specimen, represented in Fig. 4, exhibits distinct rudiments of a zygosphenal articulation. The length of its centrum beneath is 33 lines.

The shortest of the series of the dorsals measures beneath about 2 inches in length; the longest from the back of the series measures about 3¾ inches in length.

The same collection contains six specimens of cervical vertebræ, which may perhaps belong to the same species, if not the same individual, as the dorsals above noticed. The specimens are all distorted from pressure. One of them is an axis without the odontoid process and the suturally connected pieces of the atlas. The articular ball of the centrum is transversely hexagonally oval, 1.2 inches wide, and scarcely 1 inch deep.

Another cervical centrum, in some degree compressed from above downward, is represented in Fig. 6, Plate XXXVI. It measures 1.9 inches in length below and is 3 inches wide between the ends of the transverse processes.

Another specimen, represented in Fig. 7, probably a second cervical, is nearly complete, but considerably distorted. Its measurements are as follows:

	Inches.
Length of centrum inferiorly	2.00
Length between fore and back zygapophyses	2.80
Height from hypapophysis to end of spinous process	3.20
Depth of posterior ball of centrum	1.55
Width of posterior ball of centrum	1.05

Dr. Sternberg's collection further contains a number of specimens of caudal vertebræ, probably belonging to the same species as the former, and apparently pertaining to two different individuals. There are twenty-six specimens, all provided with diapophyses or transverse processes, and with hypapophyses for chevron articulation.

Figs. 8, 9, 10, Plate XXXVI, represent the first and last of a consecutive series of four anterior caudals. The body of these has the length nearly as great as the breadth and about equal to the depth. The neural arches are without zygapophyses, or exhibit mere rudiments of them. The articular ball and socket are wider than high, and are widest below the middle. The outline of the articular surfaces is emarginate and sloping at the sides above, and semicircular below. The neural canal is triangular. The transverse processes project obliquely from the lower part of the body, and they become successively narrower. The hypapophyses are excavated into deep conical pits, directed obliquely backward, for movable articulation with chevrons. The pits are small in the first of the series of specimens and become successively larger.

Measurements of the two vertebræ represented are as follows:

	Inches.	Inches.
Length of centrum, including edge of ball	1.65	1.45
Width of ball	1.75	1.65
Depth of ball	1.50	1.45

In a consecutive series of four posterior caudals with small diapophyses, the bodies have nearly the same form as in the preceding, but the articular extremities are of more uniform diameter and of a more hexahedral outline. The transverse processes are small and project just below the center of the sides. The chevron-pits are well developed, and resemble those of the preceding caudal specimens. Two of the caudals are represented in Figs. 11, 12, Plate XXXVI.

The four caudals together measure 5.3 inches in length. The diameters of the cup of the first of the series is 1.4 inches; the diameters of the ball of the last of the series is 1.3 inches.

A mutilated posterior caudal centrum, apparently of the same animal as the preceding, is without diapophyses, but has well-produced chevron-pits. The length of the centrum is less than the depth, and this is greater than the width. The articular ends are hexahedral in outline. The centrum measures 9 inch in length; 1.2 inches wide, and 1.3 inches deep.

The same collection contains the greater part of a palate-bone, with teeth, represented in Fig. 12, Plate XXXIV, which may perhaps belong to the same species as the specimens above described. The specimen contains the remains of seven teeth, which probably is within two or three of the complete series. The teeth are compressed conical, strongly curved backward or hooked, obtuse in front, acute-edged behind, are perfectly smooth, and present no facets or subdivisional planes of the surface.

Two limb-bones, represented in Figs 13, 14, Plate XXXVI, pertaining to the same collection, are supposed also to belong to the same animal as the above. I feel unable to determine their character. The broader one I suppose to be an ulna or a fibula. It resembles in its shape and construction the corresponding bone of the New Mexico mosasauroid, represented in Fig. 10, Plate XXXV, but is much smaller.

Its measurements are as follows:

	Inches.	Lines.
Length at the upper extremity	4	3
Breadth of upper extremity	2	9
Thickness of upper extremity		10
Breadth of lower extremity	3	4
Thickness of lower extremity		7
Width of shaft at middle	2	0
Thickness of shaft at middle	0	11

The smaller bone of Fig. 14 is probably a radius or a tibia.

Its measurements are as follows:

	Inches.	Lines.
Length	3	7
Breadth of upper extremity	1	4½
Thickness of upper extremity	0	9
Breadth of lower extremity	1	10
Thickness of lower extremity	0	9
Width of shaft at middle	0	9
Thickness of shaft at middle	0	7

MOSASAURUS ———— (?)

The cabinet of Swarthmore College, in the vicinity of Philadelphia, contains a number of fossils from the Cretaceous formation of Nebraska, presented by Mr. George S. Truman. They were collected by him from the hills on the Missouri River, near the Santee Agency, in L'Eau qui Court County. They consist of bones and teeth of fishes and reptiles, among which are a number pertaining to the *Polycotylus latipinnis* of Professor Cope, originally described from remains found in Kansas.

An anterior caudal vertebra of a Mosasaurus, in Mr. Truman's collection, is represented in Fig. 15, Plate XXXVI. The vertebra has the form usually assigned to the genus. It retains the neural arch, but has lost its spine. From the lower part of the body project the roots of strong transverse processes. Beneath the body there is a strong pair of eminences projecting just back of the middle and terminating in nearly flat articular facets for a chevron. The surface between these eminences forms a moderately deep concavity.

The measurements of the specimen are as follows:

	Lines.
Length of the body beneath	31
Depth anteriorly	35
Breadth anteriorly	42

Mr. Truman's collection contains several teeth which may probably belong to the same animal as the vertebra just described.

The largest of the teeth is represented in Fig. 18, Plate XXXIV. It presents the usual mosasauroid form, being curved conical, with the inner and outer surfaces unequal in extent and degree of convexity and separated by acute ridges becoming more prominent near the apex of the crown. The enamel is longitudinally striate, especially toward the base of the crown, where more marked ridges show a tendency to divide the surfaces into narrow planes.

The specimen is a shed tooth, and measures a little more than 2 inches in length; and the diameter at base is about 14 lines.

A second tooth, represented in Fig. 21, has nearly the same characters as the former. It is smaller, more compressed, so that its section is more elliptical, and its inner and outer surfaces are more equal. It is also a shed specimen, and measures $1\frac{1}{2}$ inches in length. Its base, an outline of which is seen in Fig. 22, measures 10 lines fore and aft, and 8 lines transversely.

The third specimen, represented in Fig. 19, has nearly the same form as the preceding, but has its surfaces distinctly subdivided into narrow, slightly depressed, smooth planes, of which there are six externally and seven internally. Transverse outlines of the base and of the crown a short distance above are given in Fig. 20. The length of the tooth, also, like the other, a shed specimen, has been about 20 lines. The diameter of the base fore and aft is $10\frac{3}{4}$ lines; transversely $8\frac{3}{4}$ lines.

Fig. 16 represents a small tooth, accompanying the former specimens, which I suppose to be from the back part of the series of the same species as the teeth of Figs. 18 and 21. It is more curved in proportion with its length than in these, but has nearly the same outline in transverse section, and has the enamel striated in the same manner. Its length when complete has been about an inch. Its diameter at base fore and aft is $6\frac{1}{4}$ lines; transversely 6 lines.

CLIDASTES.

The extinct reptilian genus Clidastes, characterized by Professor Cope, is especially distinguished from Mosasaurus and its nearer allies by the possession of an additional mode of articulation to the ordinary one in the vertebræ, such as is found in the living iguanas. The vertebræ are otherwise nearly like those of Mosasaurus. The general form and construction of the skull and the character of the dentition are the same in both genera.

Half a dozen species of Clidastes have been indicated by Professors Cope and Marsh, from remains found in the Cretaceous formations of New Jersey, Alabama, and Kansas.

CLIDASTES INTERMEDIUS.

A species different from those described by the authors just named is indicated by a small collection of remains, presented to the writer by Dr. J. C. Nott, formerly of Mobile. The specimens, consisting of several jaw-fragments and vertebræ, were taken from an excavation 40 feet beneath the surface, imbedded in the rotten limestone, of Cretaceous age, in Pickens County, Alabama.

The remains indicate a species of more robust proportions than *Clidastes propython*, described by Professor Cope, from the great part of a skeleton discovered in the same formation near Uniontown, Alabama. It was a third less in size than the typical species *C. iguanavus*, described by the same author, from an isolated dorsal vertebra obtained from the Cretaceous green sand of New Jersey.

Fig. 1, Plate XXXIV, represents the anterior extremity of a dentary bone, probably more than one-half of the whole. It would appear to have been proportionately of greater depth and thickness in relation with its length than in *C. propython*. It is also of more uniform depth at its fore part and less pointed at the end.

The fragment contains the remains of a series of nine teeth, occupying a space of $5\frac{1}{2}$ inches. The teeth surmount robust osseous pedestals, of which about two-thirds of the length are included within about an equal extent of the depth of the jaw.

The crown of a second tooth, (Fig. 5,) inclosed within a cavity of the pedestal of its predecessor, is 5 lines in length and about $2\frac{1}{2}$ lines in breadth at base. It is curved conical, feebly compressed from without inwardly, and has its inner and outer surfaces well defined by acute borders. The exposed inner surface of the crown exhibits no divisional planes, and has its enamel minutely wrinkled.

The crown of a tooth (Fig. 4) occupying a corresponding cavity of the ninth pedestal, probably not fully produced in its length, in its present condition has a breadth exceeding the latter. The crown is a broad cone about the length of the tooth first described, but with double the width at base. The exposed inner surface is defined in the usual manner from the outer, and exhibits no divisional planes. The enamel is minutely wrinkled.

The depth of the jaw-fragment below the visible base externally of the first dental pedestal is three-fourths of an inch; the depth below the seventh pedestal is 14½ lines.

The splenial bone advanced as far as the back part of the sixth tooth. Beyond it the Meckelian groove is deep and wide compared with that in *C. propython*, and extends to near the end of the jaw.

Fig. 2 represents a posterior fragment of the opposite dentary bone, containing the remains of a series of six teeth. The mutilated crowns of the anterior two teeth retained in the specimen exhibit a swollen base, which may also be seen to be the case in the crowns of the sixth and seventh teeth of the anterior dentary fragment.

From the two fragments of opposite dentary bones I am unable to ascertain the number of teeth which belonged to the complete series, but it seems to me that the seventh tooth of the anterior fragment about corresponds with the first retained tooth of the posterior fragment, which would indicate a series of twelve teeth.

Fig. 10 represents an axis from the same individual as the preceding specimens. It has the same form as that of *C. propython*. The odontoid process, and the elements of the atlas, all of which articulate suturally with the axis, are detached from the specimen and do not accompany it.

The measurements of the axis are as follows:

	Lines.
Length of body through center, devoid of odontoid process	18
Breadth of axis between posterior ends of diapophyses	28
Width of posterior ball	10½
Height of posterior ball	10
Width of hypopophysis	8

Two mutilated dorsal vertebræ exhibit the zygosphenes and zygantra as well developed proportionately as in *Clidastes propython*. Measurements of the better preserved of the specimens are as follows:

	Lines.
Length of the body inferiorly	18
Width of the ball and socket	12
Height of the ball and socket	10

CLIDASTES AFFINIS.

Some remains submitted to my examination by the Smithsonian Institution may perhaps indicate a species of Clidastes distinct from the former. The specimens were discovered by Dr. George M. Sternberg, United States Army, in the Cretaceous formation on the Smoky-Hill River, Kansas.

Fig. 6, Plate XXXIV, represents a nearly complete dentary bone, which is accompanied by that of the opposite side. It contains the remains of a series of twelve teeth, while there is one less in the other bone.

The anterior extremity of the jaw is of rather less depth and slightly greater thickness than in the corresponding part of *Clidastes intermedius*. The splenial bone appears to have reached as far forward as the position of the fourth tooth.

The anterior teeth appear to have been larger, and the intermediate ones smaller, than in *C. intermedius*, though this may have been a variable character in the same species. Portions of the bases of the crowns of several of the back teeth exhibit the enamel strongly striated, and the surfaces of the teeth also present evidences of subdivision into narrow planes.

A fragment from the back part of a maxillary from the same individual contains the bases of four teeth. The last of the series retains part of the crown, which is strongly striated internally, and distinctly subdivided into narrow planes externally. In the remains of the teeth of the specimens referred to *C. intermedius* there is no trace of subdivisional planes to the crowns, but this may have been a variable character in the species.

Fig. 7 represents the back part of the right ramus of the mandible of the same individual, seen on its inner side. It exhibits the same construction as the corresponding part in *C. propython*. The articulation is nearly equally divided between the angular and articular bones.

Measurements of the jaw-specimens are as follows:

	Lines.
Length of dentary bone	126
Length of series of twelve teeth	111
Depth of jaw below first tooth	10

	Lines.
Depth of jaw below fourth tooth	11
Depth of jaw at the outer side of the glenoid articulation	24
Length of projection back of the glenoid articulation	20
Transverse diameter of glenoid articulation	18
Vertical diameter of glenoid articulation	16

An axis and a dorsal vertebra accompanying the former specimens probably pertained to the same individual. They are both considerably distorted from pressure at the sides.

The axis is rather longer than that of *C. intermedius*, while its hypopophysis is considerably smaller at the extremity, and the ball of the centrum is more uniform in diameter, or is less emarginate above. The lower element of the atlas remains in firm sutural connection with the body of the axis, but the odontoid element of the latter and the lateral elements of the atlas are absent.

The dorsal specimen retains the neural arch with its characteristic zygantral articulation.

Measurements of the vertebræ are as follows:

	Lines.
Length of axis through center of the body	22
Width of ball of body of axis	$10\frac{1}{2}$
Height of ball of body of axis	$10\frac{1}{4}$
Length of body of dorsal vertebra inferiorly	22

Accompanying the former specimens there are several others which, if they did not pertain to the same individual, probably belonged to the same species.

Two fragments of the upper part of the cranium represented in Fig. 8 resemble the corresponding portions in *Clidastes propython*, as described and figured by Professor Cope, and differ only in the greater size. Fig. 9 represents an isolated basi-sphenoid bone, probably from the same skull.

These skull-fragments indicate an animal about one-third larger than *C. propython*, as described by Professor Cope.

It is a question of some importance how far difference in size among the mosasauroids may be a test of difference in species. Among the numerous remains of these animals which have been discovered I have never yet observed any which presented any evidence relating to age. In no case have I seen a vertebra in which the neural arch was not continuous with the centrum, so that I have been led to suspect that the former grew out of the latter, as in most fishes, and was never united with it by articulation, as in the crocodiles, &c. In this view of the case, some of the many described species of mosasauroids may have been founded on different ages of the same.

Fig. 11 represents a humerus accompanying the former specimens, and probably belonging to the same species, if not the same individual. In its form and construction it closely resembles the corresponding bone of *C. propython*.

The specimen is somewhat crushed, which perhaps to some extent makes it appear proportionately flatter than the humerus of *C. propython*, described and figured by Professor Cope.

The length of the bone does not exceed the breadth of its distal extremity, which is the wider one.

The measurements of the specimen are as follows:

	Lines.
Length of humerus at middle	35
Breadth of proximal extremity	28
Breadth of distal extremity	35
Breadth at constricted middle of shaft	20
Thickness of head	$8\frac{1}{2}$
Thickness of distal end	$9\frac{1}{4}$

Order *Lacertilia.* (?)

TYLOSTEUS.

Tylosteus ornatus.

The above name has been proposed for a supposed genus of lacertian reptiles, founded on a singular fossil represented in Fig. 14, Plate XIX. The specimen was obtained by Professor Hayden in the "Black Foot" country, at the head of the Missouri River, and was probably derived from the Cretaceous formation. It looks as if it might be an element of the osseous dermal armor of some animal, whether reptile or mammal is by no means certain, though, as before intimated, I suspect the former.

The specimen is imperfect or broken at the borders. Its inner surface is concave; the outer convex, and ornamented with large mammillary bosses. The latter are about fifteen in number and of different sizes. They are porous and of a less dense character than their shield-like basis. The diameter of the fossil is about 2 inches; its thickness an inch.

Accompanying the specimen just described there is an isolated phalanx, represented in Fig. 13. Though suspected to pertain to the same animal, the reference is uncertain. It is a terminal phalanx nearly 2 inches long, and with the expanded extremity nearly circular at the border and 16 lines wide. The upper part of the bone presents a nearly straight slope in its length, and is convex transversely. The under part is likewise straight along the middle, transversely convex posteriorly, and nearly flat at the expanded end. The lower surface of the latter presents near the middle a pair of vascular foramina, and several similar foramina are found near the border. The articular end is transversely elliptical and barely depressed. Its transverse diameter is 15 lines; its vertical diameter 10½ lines.

Order *Sauropterygia*.

OLIGOSIMUS.

OLIGOSIMUS GRANDÆVUS.

A fossil obtained on Henry's Fork of Green River, Wyoming, during Professor Hayden's exploration of 1870, would appear to indicate an extinct reptile allied to Plesiosaurus and Discosaurus. In general aspect, the specimen is different from those in company with it, and I think it doubtful whether it was an associate of the other fossils, which belong to the Bridger Tertiary formation. It was found as a detached specimen, and has no adherent matrix. It probably is of Cretaceous age.

The fossil, represented of natural size in Figs. 18, 19, Plate XVI, consists of the body of a caudal vertebra, apparently from the root of the tail. It was evidently from a mature animal, as the neural arch was firmly co-ossified, leaving no trace of the original separation.

In shape and construction the body resembles the corresponding portion of the vertebræ in Plesiosaurus and Discosaurus, but the proportion of length to the other dimensions is much less, and the depth also is not so great.

The body is biconcave, the concavities being of moderate and nearly equal depth. Deepest at the central half of the area, the peripheral half of the articular surfaces becomes more abruptly shallower, and with the deflexed edges somewhat convex. Near the border, the articular surfaces are defined by a narrow circular groove.

The posterior articular surface of the body at the sides below is deflected in a pair of widely separated facets for a chevron-bone. The facets are sustained on processes extending forward more than a third of the length of the body. Similar facets and processes are absent on the front of the bone.

The sides of the body are comparatively feebly constricted, much less than in Plesiosaurus, and beneath, the constriction is trifling in degree.

Transverse processes or diapophyses project from the sides of the body, just above its middle and below the conjunction of the neural arch. Their bases are broadly conical; wider than high, and appear originally to have had a sutural connection. The ends are broken off in the specimen.

The usual nutritive foramina are visible at the floor of the vertebral canal and beneath the body.

The peculiarities of the fossil appear to justify its reference to a previously undescribed genus and species, and we have therefore attributed it to an animal with the name at the head of this chapter.

The measurements of the specimen are as follows:

	Lines.
Length of body beneath	12
Depth of body in front	19
Width of body in front	23
Width of body at chevron-facets	18
Width of vertebral canal	6
Length of axis of the body	8

NOTHOSAURUS.

NOTHOSAURUS OCCIDUUS.

The above name was appropriated to a saurian indicated by a detached vertebral body or centrum, represented in Figs. 11 to 13, Plate XV. The specimen was obtained by Professor Hayden on the Moreau River, a tributary of the Upper Missouri, and is probably a Cretaceous fossil. In form and construction it resembles the vertebral centra of Nothosaurus, an extinct reptile of the Triassic formation of Europe, and probably it belongs to an animal of the same order if not the same genus. The specimen appears to pertain to a dorsal vertebra, to which the neural arch was attached by broad suture, as usual in the sauropterygians.

The body is nearly cylindric, longer than wide or high, and is moderately narrowed a short distance from the ends. Inferiorly it presents a central roughness, probably for ligamentous attachment. The articular ends are nearly round, but flattened above, and are nearly as wide as high. They are slightly concave and exhibit a slight central protuberance, apparently the ossified notochord.

The sutures for the neural arch extend nearly three-fourths the length of the centrum from its posterior end, and they reach downward to the middle of the sides. The bottom of the spinal canal is narrowest at the middle, grooved on each side, and widens toward the ends of the centrum.

The measurements of the specimen are as follows: Length of centrum, 1 inch; depth in front, 10½ lines; width, 10 lines.

FISHES.

TELEOSTEI.

Order *Acanthopteri.*

SPHYRÆNIDÆ.

CLADOCYCLUS.

CLADOCYCLUS OCCIDENTALIS.

The genus of fishes above named was proposed by Agassiz on some remains consisting of large scales and portions of a vertebral column found in the chalk of Lewes, England. The name was applied on account of the branching of the tube in the scales of the lateral line; and the fish was referred to the sphyrænoids. (Poissons Fossiles, V, 103; Atlas V, Tab. 25 a, Figs. 5, 6.)

Some large scales, found by Dr. John E. Evans, and subsequently by Professor Hayden and Mr. Meek, in ash-colored shales of the Cretaceous-series of Nebraska, I have supposed to belong to the same or a nearly allied genus. The scales vary in form and size, and may probably belong to several species. Mostly they are oval, with the length but little more than half the depth, while others are circular, and these may really pertain to a different species, if not genus, from the former.

A broad oval scale, somewhat distorted and broken at the edges, is represented in Fig. 5, Plate XXX. The inner portion exhibits numerous radiating ridges, while the outer portion, separated from the former by a narrow smooth tract, presents a minutely tubercular or granular aspect. The depth of this scale is estimated to have been nearly 2½ inches, and its length nearly 1½ inches.

Another similar but less perfect specimen appears to have measured about 1¾ inches wide by 1¼ inches long.

A third specimen, represented in Fig. 21, Plate XVII, has measured rather more than 1 inch wide and ¾ inch in length.

Another scale, represented in Fig. 22, has the same structure as the preceding, but is circular in form. Its diameter is about 14 lines. This probably belongs to a different species, and perhaps genus, from the former.

Another specimen is a nearly smooth oval scale, which has been about 13 lines wide and 9 lines long. It exhibits obscure radiant lines on the inner portion, but no granulations are evident on the outer portion. This may belong to another fish than that of the preceding specimens.

ENCHODUS.

Enchodus Shumardi.

The extinct genus above named was inferred by Agassiz from some remains, consisting of jaws and teeth, found in the chalk of Europe, and was by him attributed to the sphyrænoid family. Several species have been since described from similar remains found in the deposits of Cretaceous age in the United States. One of these, under the above specific name, was indicated by a dentary bone with teeth, found by Dr. Benjamin F. Shumard in the same formation in which were discovered the large scales referred to Cladocyclus.

The specimen is rudely represented in a reversed position in Fig. 20, Plate XVII. The dentary margin of the bone is 11 lines long, and contains six long narrow teeth, and in the back intervals a number of minute ones.

The first of the larger teeth is the longest, and is situated a short distance from the end of the bone. Including its thickened base it is 2 lines long by about one-fifth of a line wide. It is a long, narrow, straight cone, laterally compressed, trenchant at the borders, and ends in a point with a slight posterior projection or half barb. The posterior five larger teeth are situated at irregular distances apart, and measure from one to one and one-fifth lines in length by about one-sixth of a line in breadth at base. They are nearly like the largest tooth, but are slightly more curved, and have no projection to the back of the point. The minute teeth in the back intervals of the larger ones and back of these are not over the one-fifth of a line long.

PHASGANODUS.

Phasganodus dirus.

An extinct genus of fishes supposed to belong to the sphyrænoid family, and nearly related with Enchodus, has been described under the above name. It was inferred from a specimen of a mutilated dentary bone with teeth, imbedded in a piece of brown sandstone, obtained by Professor Hayden from a Cretaceous deposit he has indicated as No. 5, on Cannon Ball River.

The specimen with the remains of five large teeth, reduced one-third, is represented in Fig. 24, Plate XVII. The third tooth of the series, preserved entire and separated from the former, is represented in Fig. 23.

The dentary bone exhibits nothing peculiar in the present condition of the fossil, and appears not to have differed in any important point from that of Enchodus.

The teeth differ from those of the latter. They are proportionately shorter, saber-like, and situated on broad bases, with an oblique direction to the edge of the jaw. The thick back border is directed inwardly; the trenchant border forward and outward. The point is cut off in a slanting manner posteriorly. The back part of the crown toward the base and extending on the sides is fluted, but toward the point and trenchant border is smooth. In section the crown is ovate, with the long diameter $2\frac{3}{4}$ lines. The length of the tooth, including its thickened base, is 10 lines; without the base, the crown measures 7 lines.

Order *Malacopteri.*

Siluridæ. (?)

XIPHACTINUS.

Xiphactinus audax.

Under the above name, I described an ichthyodorulite belonging to the collection of the Smithsonian Institution. The specimen was obtained from the Cretacous formation of Kansas, by Dr. George M. Sternberg, United States Army. I supposed it to be the pectoral spine of a large siluroid fish, but according to Professor Cope, who has had the opportunity of examining many remains of fishes from the Cretaceous formation of Kansas, it belongs to a fish of a peculiar family. This he names Saurodontidæ, represented by Saurocephalus and some other genera. At first the spine was referred to the last-named genus, but latterly he appears to be in doubt whether it belongs to this or some other nearly allied genus. The specimen is represented in Figs. 9, 10, Plate XVII, one-third the diameter of nature.

The spine is unsymmetrical, thus rendering it probable that it belonged to one of the lateral pairs of fins rather than to any of the vertical fins. It is a broad saber-shaped weapon, in its present condition about 16 inches in length, which is nearly its entire extent, judging from the thinning and rounding of the broken end. Its breadth the greater part of the length is nearly uniform, and at the middle is nearly 2 inches. Toward the distal end it becomes slightly less wide and thinner; toward the proximal end it undergoes a greater reduction in width, and becomes much thicker.

The upper surface of the spine, represented in Fig. 10, for the most part is nearly flat except toward the rounded borders. It is invested with a thin layer of ossific substance of a more dense character than the compact bone beneath. The surface is striated or ornamented with raised lines, which are longitudinal and parallel, but on portions of the surface are somewhat irregular. Some of the lines branch, and the slightly divergent branches include other commencing lines. At the distal end of the spine, near the anterior border, the lines break up into finer branches which curve outwardly to the edge.

The under surface of the spine (Fig. 9) is uneven. A prominent ridge, commencing at its proximal extremity and occupying more than two-thirds its width, extends outwardly and gradually declines to a point near the center of the inferior surface. A shallow groove commences in front of the ridge, widens outwardly, and extends beyond the former upon the anterior half of the inferior surface of the spine. Back of the commencement of the ridge there is a concave hollow, which narrows outwardly into a deep groove, and this, pursuing the same course, widens and opens downward upon the posterior half of the inferior surface of the spine to its distal end.

The posterior groove for nearly half its length proximally exhibits a row of irregular pits at the bottom. The upper boundary of the groove in advance of the pits is transversely striate, and beyond the position of the pits externally the corresponding surface presents the striæ curling outward to the back edge of the spine. The bottom of the groove, external to the position of the pits, continues as a shallow channel running along the middle of the spine inferiorly to its distal end.

The anterior border of the spine is convex in the length, obtuse internally, and acute externally. The posterior border is concave longitudinally, obtuse internally, and less acute externally than the anterior border.

The inner extremity of the spine appears bent upward into a hook-like eminence with a pyramidal base extending above the general level of the spine. The end of the hook-like process is broken off. Its inner surface forms an ellipsoidal longitudinal convexity, with the lower half more prominent and appearing to be an articular eminence. The outer extremity of the spine is broken, but it appears to have been rounded transversely, though it may have been pointed.

The measurements of the spine are as follows:

	Inches.
Length of the spine in its present condition	16

	Lines.
Breadth of the spine beyond the articular hook	17
Breadth at the inner third	21
Breadth at the middle	23
Breadth near the distal end	20
Thickness of the spine beyond the articular hook	14
Thickness at the inner third	9½
Thickness at the middle	6
Thickness near the distal end	3

The transverse section of the spine near the middle forms an irregular ellipse, as represented in the accompanying figure. The left-hand side beneath represents the posterior groove opening downward.

Since writing the above, I have had the opportunity of examining the proximal half of a similar spine, from L'Eau qui Court County, Nebraska. It was found in association with remains of Mosasaurus, &c., by George S. Truman, and presented by him to Swarthmore College, Pennsylvania.

GANOIDEI.

PYCNODUS.

This genus, typical of an extinct family of fishes, was originally indicated by Agassiz in the Poissons Fossiles. Many species have been described, mainly from teeth and fragments of jaws with teeth, which are comparatively large and stout, and were adapted to crushing hard food, such as mollusks with their shells, crustaceans, &c. The remains have been found in the Triassic, Jurassic, Cretaceous, and early Tertiary formations of Europe.

PYCNODUS FABA.

A specimen, represented in Fig. 16, Plate XIX, indicates a species to which the above name has been given. It was submitted to my examination by Dr. William Spillman, who obtained it from the Cretaceous formation near Columbus, Mississippi.

The specimen consists of a fragment of the ramus of a lower jaw containing a number of teeth. Four principal teeth and part of the attachment of another are retained in the fragment. These teeth are ranged obliquely parallel with one another from within backward and outward. In outline they are elongated-bean shaped, being slightly concave in front and convex behind, and slightly wider externally than internally. The first of the series is $7\frac{1}{2}$ lines wide by $2\frac{3}{4}$ lines fore and aft, and they successively increase in breadth to the last, which measures $8\frac{3}{4}$ lines wide by $2\frac{3}{4}$ lines fore and aft.

At the bottom of the slope, to the inner side of the large teeth, there is a row of three smaller teeth and the traces of attachment of a fourth one. The three teeth, like the others, successively increase in size from before backward. They are ovoid, and situated obliquely nearly opposite the intervals of the large teeth. The first of the series is $2\frac{1}{2}$ lines in diameter fore and aft and $1\frac{1}{2}$ lines transversely; the last one is $3\frac{1}{2}$ lines by 2 lines.

The jaw-bone internal to the teeth just described rises in a ridge toward the symphysis. The slope at the fore part of the ridge exhibits the attachments of two minute teeth, indicating a second row internal to the largest teeth.

To the outer side of the latter the specimen retains evidences of two rows of smaller teeth. Of these, the first row shows remains of seven teeth in the length of space occupied by the five principal teeth, and, like these, they successively increased in size. Only the fourth tooth of the row is preserved, and this is transversely ovoid, with the long diameter 2 lines wide and the short diameter $1\frac{3}{4}$ lines.

Fig. 15, of the same plate, represents a specimen apparently from the same species, belonging to the Museum of the Academy of Natural Sciences of Philadelphia. It was presented by Dr. J. H. Slack, who obtained it from the green sand marl of Crosswicks, Burlington County, New Jersey. It consists of a small jaw-fragment containing three broad teeth similar to the largest ones above described.

An isolated tooth from New Jersey, submitted to my inspection by Professor G. H. Cook, is noticed in the Proceedings of the Academy of Natural Sciences of Philadelphia for 1857, p. 168, under the name of *Pycnodus robustus*. The specimen represented in Figs. 18, 19, Plate XXXVII, has the same shape as in the largest teeth of those referred to *P. faba*, but is much larger. Its long diameter is $14\frac{1}{2}$ lines, and its short diameter nearly 4 lines.

A similar tooth, nearly the same size but slightly more sigmoid, is represented in Fig. 96, page 244, of Professor Emmons's Report of the North Carolina Geological Survey, published in 1858. The specimen is attributed to the Miocene Tertiary, and is referred to a species with the name of *Pycnodus carolinensis*.

HADRODUS.

HADRODUS PRISCUS.

The genus above named is obscure in its relations, and was originally described in 1857, in the Proceedings of the Academy of Natural Sciences of Philadelphia. It was founded on a specimen consisting of a bone with two singular-looking teeth, discovered by Dr. William Spillman, in the Cretaceous formation in the vicinity of Columbus, Mississippi.

The specimen represented in Figs. 17 to 20, Plate XIX, I have supposed to be a premaxillary bone of an animal allied to the extinct genus Placodus, formerly considered to be a pycnodont fish, but now determined to be a sauropterygian reptile.

The bone is unsymmetrical, and supports two strongly co-ossified teeth. Whether the specimen is complete in itself or whether it is part of a larger bone, I have not been able to ascertain.

The bone is quadrate in outline; thicker and longer on one side, and oblique at the upper border. The anterior surface is convex and comparatively smooth. On each side and extending posteriorly, the bone is deeply excavated into large reserve cavities for successional teeth. The back surface between the cavities inclines from each side, forming a median angular groove descending to the interval of the teeth. The bone is more porous and striated posteriorly than anteriorly.

The teeth remind one of the premolars of some pachyderm, rather than the teeth of a fish or reptile. They are not exactly alike, and are co-ossified with the bone by a firm osseous base or root, striated in front. They are quadrate in outline, with the breadth and height nearly the same, and the thickness about half. The crown is convex in front and at the sides, and is bilobed at the triturating border, which slopes off posteriorly. An acute ridge and the conical blunted summits of the lobes define the outer from the inner surface. Smooth enameloid substance invests the crown, extending twice the depth on the outer surface that it does on the inner surface. In transverse section the teeth are ovoid.

The measurements of the fossil are as follows:

	Lines.
Depth of bone	18–20
Breadth of the bone	17
Length of the larger tooth	8
Width of same	8
Thickness of same	5¼
Depth of enamel	6½
Length of smaller tooth	7½
Width of same	7¼
Thickness of same	4½
Depth of enamel	6¼

I have arranged Hadrodus with the Pycnodonts, though, like Placodus, the discovery of additional material may prove it to be a sauropterygian reptile. Of Placodus, Professor Owen remarks that the "teeth are implanted by short simple bases in distinct hollow sockets," (Palæontology, 218;) and Meyer says, "In wircklichen Alveolen stecken eigentlichen nur die Schneidezähne mit gut ausgebildeten Wurzeln, der Wurzeltheil der übrigen Zähne ist mehr mit dem Knochen, dem die Zähne angehören, verbunden." Hadrodus in the relation of the teeth would appear to be different, as they are firmly co-ossified by short bases with the border of the jaw. They exhibit no trace of implantation by sockets, though the successional teeth before being established in a fixed manner in functional position must appear at least to spring from sockets.

ELASMOBRANCHII.

Order *Plagiostomi*.

PTYCHODUS.

PTYCHODUS MORTONI.

The extinct genus of cestraciont fishes above named was inferred by Agassiz, from isolated teeth, the only parts yet found which can be with any certainty referred to the same animal. A number of species have been indicated, mostly by the same authority, from specimens found in the Cretaceous formations of Europe and America.

Teeth of Ptychodus Mortoni have been discovered in the Cretaceous deposits of Alabama, Mississippi, and Kansas, but I have seen none from the corresponding formation of New Jersey or elsewhere.

The Smithsonian Institution has submitted to my examination a collection of fourteen specimens of teeth obtained by Dr. George M. Sternberg, United States Army, from the banks of Chalk Bluff Creek, a branch of Smoky Hill River, about sixty miles east of Fort Wallace, Kansas. The specimens were found in two parcels, each together, as if pertaining to two individuals.

The two largest teeth, of which one is represented in Figs. 1, 2, Plate XVIII, are probably from a median position in the mouth or jaws. They are symmetrical in form, and in outline are transversely quadrate oblong with rounded angles.

The crown is prominently convex, with the front and lateral borders nearly straight, the back border slightly concave, and the angles rounded. Posteriorly it is impressed with a moderately concave crescentoid sinus. The summit is crossed by a short transverse ridge, from which numerous ridges radiate. Descending on the sides of the crown the ridges branch, and about half way down terminate in a fine reticulation which extends to the borders of the tooth. The root is a quadrate plate with the same outline of form as the border of the crown.

Three other specimens of the same parcel as the preceding appear to have been lateral teeth in relation to them in position in the mouth. They are nearly alike in shape and size; one of them being represented in Figs. 3, 4. They are not symmetrical as in the larger teeth, and their outline is more reniform. They are proportionately narrower at one side, and wider and more extended on a base beyond the conical elevation of the crown at the other side. The sinus is of less height, and the ridges of the crown are more convergent at the apex of the cone. The root appears to recede from the narrower side and reaches nearly to the edge of the crown on the opposite side. The remaining two teeth of the same parcel have the same character as those just described, but are considerably smaller.

Measurements of some of the specimens are as follows:

	Figs. 1, 2.			Figs. 3, 4.	
	Lines.	Lines.	Lines.	Lines.	Lines.
Breadth transversely	22	20	18½	17	13
Breadth antero-posteriorly	12½	12½	9	9	7½
Height from bottom of root	12½	12	8	9	7½

Of the two largest teeth, of the second parcel, which are nearly alike, one is represented in Figs. 5 and 6. They are intermediate in character to those previously described, being less symmetrical than the large teeth and more so than the smaller ones, and their crown is proportionately more prominent than in any of them. Of three teeth smaller than the former and successively diminishing, that of intermediate size is represented in Figs. 7 and 8. They have the same form as the unsymmetrical ones of the first parcel, but have their crown proportionately much more prominent.

The remaining two teeth are different in shape from the former. The larger one has the crown proportionately less prominent, with the central conical elevation less strongly radiate. The inner side of the base forms an obtuse angle, and is strongly impressed toward the back border. The front border of the base of the crown is short, nearly straight, and forms with the oblique outer border an obtuse angle.

The smaller tooth is represented in Figs. 9 and 10, and has nearly the same shape as the former, but the crown appears comparatively flat with a central nipple-like eminence, and the anterior and outer borders are more continuous.

The measurements of the teeth are as follows:

		Figs. 5, 6.		Figs. 7, 8.		Figs. 9, 10.	
	Lines.	Lines.	Lines.	Lines.	Lines.	Lines.	Lines.
Breadth of crown transversely	14½	14	11½	8½	7	10	7
Breadth of crown antero-posteriorly	7	8	6	5	4	6	4½
Height of crown from bottom of root	10½	9½	8	6	4	5½	3½

Several specimens of teeth of *Ptychodus Mortoni* have been submitted to my inspection by Dr. William Spillman, who obtained them from the Cretaceous formation near Columbus, Mississippi. One of the teeth, of large size, and considerably worn at the summit of the crown, is represented in Figs. 11 and 12. It is symmetrical in shape, but has a more reniform outline than the large teeth from Kansas. The anterior and lateral borders of the crown nearly form a semicircle, and the posterior border is deeply emarginate. The sinus is deeper than in the Kansas specimens, but the arrangement of the striations of the crown appear to be the same.

Two other specimens, about half the size of the preceding, have nearly the same shape, but have their crown proportionately more convex at the fore part of the base.

The measurements of the Mississippi specimens are as follows:

	Figs. 11, 12.		
	Lines.	Lines.	Lines.
Breadth of crown transversely	20	12	10¼
Breadth of crown fore and aft	10	6¾	7
Height of crown	9½*		7

* To worn summit.

The Museum of the Academy of Natural Sciences of Philadelphia contains nine specimens of teeth of *Ptychodus Mortoni* from the Cretaceous formation of Alabama. These in general resemble the symmetrical and unsymmetrical teeth above described. One of the specimens from Green County, Alabama, is represented in Figs. 13, 14. Its sinus is more sharply triangular than in the previous specimens.

Another tooth, approaching in size the largest Kansas specimens, has a more distinct conical ridge on the summit of the crown from which the other ridges radiate. A third tooth nearly resembles the Kansas specimen represented in Figs. 3, 4, and has the summit of the crown worn away, as in the large Mississippi specimen, represented in Figs. 11, 12.

Measurements of Alabama specimens of teeth are as follows:

		Figs. 13, 14.			
	Lines.	Lines.	Lines.	Lines.	Lines.
Breadth of crown transversely	16	11½	16	13	9
Breadth of crown fore and aft	10½	8½	9	7	5½
Height of crown from bottom of root	10½	9	....	8	6½

PTYCHODUS OCCIDENTALIS.

A peculiar species, to which the above name has been given, is indicated by specimens of teeth discovered by Dr. John L. Leconte in an ash-colored chalk of the Cretaceous formation a few miles east of Fort Hays, Kansas.

The most characteristic, and at the same time the largest specimen, is represented in Figs. 7, 8, Plate XVII, of the natural size.

The shape of the tooth and the arrangement of the ridges of the crown are quite different from what they are in the preceding species. The tooth is symmetrical, as in the largest teeth of *Ptychodus Mortoni*, but it is proportionately of less breadth transversely, and also higher.

The crown forms a prominent cone with evenly sloping sides, and with a transversely oblong square base narrowing a little posteriorly. The posterior sinus of the crown comports in its height and breadth with the proportions of the former. The principal ridges of the surface of the crown cross the summit and posterior slope transversely. Descending, they branch in a divergent manner and anastomose, so as to form a comparatively coarse reticulation, extending to the borders of the crown. The reticulation covers the anterior slope of the crown and the sinus posteriorly. The direction and arrangement of the ridges resemble those in the European *Ptychodus decurrens*, but in this the principal ridges are much coarser and more widely separated. The root is mutilated in the specimen. The transverse diameter of the crown is 14 lines; its fore and aft diameter and its height about 1 inch.

Figs. 15, 16, Plate XVIII, represent two views of a small tooth, which may probably belong to the same species. It is unsymmetrical, and is worn away at the summit of the crown. The latter is proportionately less prominent than in the large tooth, but has its ridges arranged in the same general manner. The root is very thick in comparison with the size of the tooth. The transverse diameter of the crown is $7\frac{1}{2}$ lines; the fore and aft diameter $\frac{1}{2}$ inch.

Of five remaining specimens, one is a smaller and unworn tooth nearly like that last described. Its crown is 5 lines wide and 4 lines from before backward.

The specimen represented in Fig. 17 is more symmetrical, and nearly resembles in shape the smaller symmetrical teeth of *Ptychodus Mortoni*, as represented in Figs. 13, 14. The apex of the crown is not so pointed, but is prolonged fore and aft in an acute ridge, and the rugæ of the surface are not convergent, but cross the summit in the usual transverse manner of the other teeth. The breadth of the crown in this specimen is $4\frac{1}{2}$ lines; the antero-posterior diameter $3\frac{3}{4}$ lines.

The remaining teeth, of which the largest is represented in Fig 18, have a transversely ovoidal crown slightly elevated to one side of the center. The surface is crossed by rugæ in the same manner as in the large teeth.

The measurements of the three specimens are as follows:

	Lines.	Lines.	Lines.
Transverse diameter of the crown	3¼	2½	2
Antero-posterior diameter of the crown	1¾	1½	1½

PTYCHODUS WHIPPLEYI.

Marcou, in his Geology of North America, describes and figures a tooth from the Cretaceous formation near Galisteo, New Mexico, and refers it to a peculiar species under the above name.

A similar tooth submitted to my inspection by Dr. Benjamin F. Shumard, from the Cretaceous rocks of Texas, is represented in Figs. 19, 20, Plate XVIII. It is remarkable for the abrupt nipple-like prolongation of the crown.

The tooth is unsymmetrical, and probably held a lateral position in the series. The base of the crown is quadrate, with the fore and outer borders forming a single curve, while the other borders form a nearly right angle. The nipple-like eminence of the crown inclines, as I suppose, outwardly. The posterior sinus is shallow. The rugæ of the surface of the crown cross the summit transversely and diverge and branch descending upon the sides of the cone. They are comparatively feeble, but this condition may be partially due to friction. The surface of the base of the crown appears rather nodulated than reticulated.

The breadth of the tooth at the base of the crown is 7 lines transversely and fore and aft; its height from the bottom of the root 8 lines.

The tooth resembles that of *Ptychodus altior* of Agassiz, from the chalk of Sussex, England, as represented in Fig. 10, Plate XXX, of Dixon's Geology of Sussex.

ACRODUS.

This extinct genus of cestraciont sharks, first described by Agassiz, was represented in Europe by many species whose remains occur in the various formations from the Permian to the Cretaceous inclusive.

ACRODUS HUMILIS.

A species to which this name has been given is indicated by an isolated tooth represented in Fig. 5, Plate XXXVII, magnified 1½ diameters. The specimen was obtained from the yellow limestone of the Cretaceous series, near Vincentown, Burlington County, New Jersey, and it belongs to the Museum of the Academy. The crown of the tooth is 7½ lines by 2½ lines. The extremities are angular; the sides nearly straight or in the feeblest degree sigmoid. The upper surface is convex; and its median ridge is almost obsolete. The secondary ridges, proceeding transversely from the former, become branched and finely reticulated at the boundaries of the crown. The groove on the inner side of the latter, for co-adaptation to the contiguous tooth, is about three-fourths of a line in width. The fang or contracted base of the tooth is about half the breadth of the crown.

Professor Emmons has represented the tooth of an Acrodus in Fig. 97 of his Report of the North Carolina Geological Survey for 1858, which he attributes to the Miocene Tertiary. If it really pertains to this formation, it indicates the latest known species of the genus. The species has been named *Acrodus Emmonsi*

GALEOCERDO.

Galeocerdo falcatus.

The teeth of Galeocerdo are nearly as broad as they are long, and the root is but moderately notched. The anterior border of the crown is strongly arched and oblique; the posterior border is slightly curved and nearly vertical, but is abruptly prolonged backward at its base. The borders of the crown are serrated; the point is somewhat acuminate.

Teeth from the chalk formations of Europe figured in the "Poissons Fossiles," and ascribed by its illustrious author to half a dozen different species, are, with reason, by Reuss referred to a single one with the name of *Corax heterodon*. As Agassiz, according to Gibbes, does not now consider *Corax* different from *Galeocerdo*, I have used this name, together with the earlier specific one of *falcatus*, to represent the *Corax heterodon* of Reuss.

Many specimens of well-preserved teeth, submitted to my examination, from various localities of the American Cretaceous formation, appear to belong to *Galeocerdo falcatus*. The variations in the form and size of different teeth I think are sufficiently accounted for from the difference of position the teeth occupied in the jaws and upon difference in age.

Figs. 29 to 31, Plate XVIII, represent three of these teeth, obtained with others by Dr. George M. Sternberg, United States Army, from the vicinity of Camp Supply, on the North Canadian River, Indian Territory, probably from a formation of Cretaceous age.

Their apparent specific identity with the teeth of *Galeocerdo falcatus* of Europe is seen by comparing the figures with Fig. 43, taken from a tooth imbedded in a block of chalk from Sussex, England.

Figs. 32 to 36 represent a series of similar teeth obtained, with many others of the same character, by Dr. William Spillman, from near Columbus, Mississippi.

Figs. 37 to 40 represent smaller teeth, which I suspect to belong to the same species, found by Dr. John L. Leconte, about three miles east of Fort Hays, Kansas. Similar specimens were also obtained by Dr. Hayden, in bed No. 2 of the Cretaceous rocks, near the mouth of Vermilion River, Kansas.

Figs. 41, 42 represent small teeth, likewise of the same species, obtained by Dr. Shumard from the Cretaceous formation of Texas.

OXYRHINA.

The teeth of Oxyrhina have a simple, compressed demiconical crown, with sharp borders, and without lateral denticles.

OXYRHINA EXTENTA.

Figs. 21 to 23, Plate XVIII, represent specimens of teeth of an Oxyrhina discovered by Dr. George M. Sternberg, United States Army, in the vicinity of Camp Supply, on the North Canadian River, Indian Territory. Figs. 24, 25 represent similar teeth found by Dr. William Spillman in the Cretaceous formation near Columbus, Mississippi. These teeth differ especially, from those of other species previously described and figured, in the greater proportionate extension laterally of the base of the crown. They most nearly resemble the teeth of the *Oxyrhina Mantelli* of the chalk of Europe.

In the Museum of the Academy of Natural Sciences of Philadelphia there is a specimen of an Oxyrhina tooth in a block of chalk from Sussex County, England, resembling those just described in the unusual extension laterally of the crown. If this specimen pertained to *O. Mantelli*, it is probable that the specimens from Mississippi and the Indian Territory do likewise. It was not until after I had described the latter under the above name that I noticed the specimen from the English chalk. My comparisons had been made with the figures of Agassiz, Dixon, and Reuss, and in none of these do the teeth exhibit so conspicuous a lateral extension of the base of the crown as in the American specimens and the English chalk specimen of our Museum. An exception to this statement may be made in reference to Fig. 26, Plate XXX, in Dixon's Geology of Sussex, representing a tooth, which is referred to *Lamna acuminata.*

Measurements of the specimens referred to *Oxyrhina extenta* are as follows:

Specimens represented in Plate XVIII.	Fig. 21.	Fig. 22.	Fig. 23.	Fig. 24.	Fig. 25.
	Lines.	Lines.	Lines.	Lines.	Lines.
Length from notch of root		12	8½	14	10
Length of crown at middle		8	6½	12	8½
Breadth of crown at base	9	15	12	16½	13
Breadth of root	12	17	13	18	14
Thickness of root	5	4½	4½	5	4

LAMNA s. OXYRHINA.

The teeth of Lamna are in general characterized by the long, narrow crown, with a single denticle on each side of the base, and a strong root with narrow branches separated by a deep notch. Those of Oxyrhina usually have a broader crown without lateral denticles, and also have a broader root with a shallower notch. In both genera, however, the proportion of breadth to length, and most other characters, except the presence or absence of the lateral denticles, vary in different parts of the jaws. In both, the side teeth are wider than those in advance, the disproportion usually being greater in Oxyrhina than in Lamna. Some of the teeth in the two genera so nearly assume the form of one another, that when isolated fossil teeth of either are found without the base it is sometimes difficult to know to which to refer them.

A number of times I have seen specimens of teeth, reputed to have been derived from the Cretaceous formation, which so closely resemble those of certain Tertiary species of Lamna, except in the possession of lateral denticles, that I have suspiciously regarded them as pertaining to the same. The absence of denticles I thought might be accidental or abnormal. The report that the teeth had been found in a Cretaceous formation I suspected might be a mistake; or, if they had, that they were, perhaps, accidental in their occurrence in that formation, and had probably been derived from some contiguous Eocene deposit. The frequent repetition of the same thing has led me to view the specimens as having really pertained to Cretaceous fishes. The absence of the lateral denticles would refer the teeth to the genus Oxyrhina, and the general form and other characters rather to the genus Lamna. May the teeth not be regarded as having belonged to Oxyrhina ancestors of some of the later Lamnæ?

Fig. 44, Plate XVIII, represents a tooth which lies imbedded in a portion of gray rock, obtained by Dr. John L. Leconte from the Cretaceous formation three miles east of Fort Hays, Kansas. The specimen is perfect and unabraded. In all respects it is like the teeth of *Lamna cuspidata* of the early Tertiary deposits, except that it is devoid of lateral denticles, and presents no trace of ever having possessed them.

Fig. 45 represents a tooth, which lies in a block of chalk, from Sussex, England. The specimen is preserved in the Museum of the Academy of Natural Sciences of Philadelphia. Like the former, it closely resembles the teeth of *L. cuspidata*, but exhibits no trace of lateral denticles.

Figs. 46, 47 represent two teeth which the writer found with the skeleton of *Hadrosaurus Foulkii* and shells of *Exogyra costata*, *Ammonites placenta*, &c., in clay near Haddonfield, Camden County, New Jersey. These specimens, unworn and perfect, except in positions having no relation with the point in question, are identical in character with the teeth of *Lamna elegans* of the early Tertiary deposits, except that they exhibit no trace whatever of the existence of lateral denticles.

Figs. 48, 49 represent two teeth selected from eight specimens obtained by Dr. William Spillman from the Cretaceous formation near Columbus, Mississippi. Most of the specimens are complete and well preserved, and in no instance exhibit traces of lateral denticles, while in all other respects they are like the teeth of *L. elegans*.

Seven specimens of teeth from the Cretaceous formation of Green County, Alabama, presented to the Academy of Natural Sciences of Philadelphia by Dr. Joseph Jones, also agree with those of *L. elegans*, except that they have no lateral denticles.

In a collection of similar teeth, presented to the Academy by William M. Gabb, in all the specimens retaining the root, twenty in number, the lateral denticles are absent. Most of the teeth appear slightly water-worn, but the best of them exhibit no trace of the lateral denticles. These specimens were obtained by Mr. Gabb from the Cretaceous green sand of Mullica Hill, Gloucester County, New Jersey.

Fig. 50 represents a tooth which lies partially imbedded in a fragment of gray sandstone, obtained by Professor Hayden from the Cretaceous deposit, indicated by him as No. 2, near the mouth of Vermilion River, Kansas. In the attempt to dislodge the tooth from its matrix the ends of the root were broken off, but it is otherwise complete. It also appears not to have possessed lateral denticles, but otherwise is like the teeth of *L. elegans*.

Roemer describes and figures a tooth, (Kreidebildungen v. Texas, page 29, Plate I, Fig. 7,) under the name of *L. Texana*, from the Cretaceous formation of Texas. The figure represents what appears to be a perfect tooth without lateral denticles, and otherwise resembles those of *L. elegans*.

Dr. B. F. Shumard also submitted to my inspection several teeth from the Cretaceous formation of Texas resembling those of *L. elegans*, but in these the root was broken off, excepting on one side of one specimen, and in this no lateral denticle existed.

Notwithstanding all that has been stated above, I must add that I have noticed among collections of teeth of *L. elegans* from Tertiary formations specimens in which the lateral denticles were feebly developed, and others in which they were entirely absent. In some of the latter, traces of their accidental detachment were perceptible, but in others I could see none.

It would appear, however, from the facts thus given, that during the Cretaceous period there existed two species of sharks in which the teeth resembled those of *L. cuspidata* and *L. elegans* of the Tertiary period, except that the teeth possessed no lateral denticles. The two Cretaceous sharks were probably the ancestors from which the species just named were evolved.

OTODUS.

OTODUS DIVARICATUS.

Among a small collection of fossils submitted to me for examination by Dr. William Spillman, of Columbus, Mississippi, there is a specimen of a shark-tooth of rather peculiar character, which is represented in Figs. 26 to 28, of Plate XVIII. The specimen is labeled "lime formation," Texas, and nothing is further known in relation to its locality, but I suspect it to be of Cretaceous age. Of known species, it bears most resemblance to the teeth of *Otodus semiplicatus*, Ag., of the chalk of Europe. It also has some likeness with a tooth from the chalk of France, represented in Fig. 11, Plate LXXVI, of Gervais's Paleontologie Française.

The crown forms a narrow demicone, with an expanded base supporting a pair of inwardly-diverging denticles. The surface of the principal cone near the base is plicated. The root is thick and deeply notched, and extends posteriorly more than half the length of the tooth. The anterior surface of the crown in the median line is as long as the base is wide, and is about one-fifth greater than the posterior surface.

The measurements of the specimen are as follows :

	Lines.
Length of tooth at middle	16
Length from ends of the root	21
Breadth at ends of the root	15¼
Length of crown in front	13
Length of crown behind	10
Breadth of crown at base	12¾

Holocephali.

EDAPHODON.

In the extinct chimæroid fish Edaphodon the inferior maxillaries are produced anteriorly in a long beak, and the superior maxillaries are provided with three large dental areas. In the allied genus Ischyodus the inferior maxillaries are not prolonged in a beak, and the upper ones are provided with four large dental areas.

EDAPHODON MIRIFICUS.

A species under the above name was indicated by the author in the Proceedings of the Academy for 1856, page 221. It was founded on eight specimens of maxillaries obtained from the Cretaceous green sand of Burlington County, New Jersey, by Professor George H. Cook, during the State geological survey.

The inferior maxillaries, represented in Figs. 6 to 9, Plate XXXVII, are about twice the length of the depth. The two rami converge in a curve, and end together in a long, bird-like beak, (Fig. 6.)

The outer surface (Fig. 7) of each ramus is lozenge-like in outline, defined by a concave upper border, a convex anterior border, a short, oblique, posterior border, and a convex lower border. The surface is concave longitudinally, and is convex transversely in front and behind, and concave in the middle.

The inner surface (Fig. 8) is flat transversely, slightly convex longitudinally, and with the fore and back borders prominent. It is moderately striated in the length, and at its upper part presents a symphysial bevel, extending the length of the beaked portion of the bone.

The oral surface (Fig. 6) on the beak is concave fore and aft, and at the back half of the bone forms a lozenge-like plane sloping inwardly, and having the outer border elevated. The sloping plane exhibits at its fore part internally a large cordiform dental area, with the notch at the base of the beak. Externally to this area, near the fore part of the crest defining the outer part of the sloping plane, there is a second much smaller elliptical dental area. These two areas are separated by a groove, widening forward upon the oral surface of the beak, where it presents a third dental area. This is the third in size, is oval in form, and is situated just in advance of the outer part of the largest dental area.

A fourth area, smaller than the others, occupies the back extremity of the symphysial bevel to the inner side of the anterior part of the largest dental area. The dental column which forms this fourth area produces the prominent ridge defining the inner surface of the ramus mandibuli posteriorly.

Beside these dental areas, two or three others are observed at the end of the beak. One of them curves from the symphysis outward and backward on the outer edge of the point of the beak. Another smaller oval one is situated at the edge of the symphysis behind the commencement of the former. In one specimen a still smaller oval area is situated just behind the outer end of the curved area, but in the other specimens it appears not to be distinct from the latter.

The dental areas in the fossils appear as depressed and decomposed, friable, white, chalky tracts, with harder calcigerous tubules of the vaso-dentine projecting from the surfaces. The tubercular eminences originally occupying the position of the areas and terminating the dental columns have disappeared, leaving depressed surfaces. The vaso-dentinal columns corresponding with the areas on the triturating surface are visible at the posterior-inferior extremity of the mandibles, as seen in Fig. 9.

The upper maxillaries, represented in Figs. 10 to 12, bear a near resemblance to those of *Edaphodon Bucklandi* and *E. leptognathus*, as represented in Tab. 40 d, of the third volume of the Atlas of Agassiz's Poissons Fossiles.

The outer surface (Fig. 11) of each maxilla is a broad, sloping plane, the inner surface a vertical plane. The upper surface is also flat, but is occupied at its inner back part by a wide, deep gutter, ending forward in a pit.

The palatine surface (Fig. 10) of the two bones conjoined at its back part forms a wide, transverse concavity, nearly flat in the middle, but curving downward at the outer part. The palatine surface inclines forward to the anterior subacute termination of the bones. The lateral border of each maxilla at the palatine surface is strongly sigmoid.

Three large dental tubercles occupied the palatine surface of each maxilla, indicated in the fossils, as seen in Fig. 10, by three depressed areas of white, decomposing vaso-dentine. The largest area is posterior and internal. It is broken at its back part in the fossils, but, in the entire condition, appears to have been reniform in outline. Immediately in advance of this area is another with an oblong cordiform outline; and external to the largest one is the third area, about as long as this, but not more than half the breadth, and having a clavate outline.

The dental columns corresponding with the three dental areas are seen at the back of the maxillæ, the largest one below the position of the two smaller ones, as represented in Fig. 12.

The measurements of the specimens are as follows:

Inferior maxillary.

	Lines.
Extreme length of bone	64
Length of beak along the symphysis	40
Length of anterior border of ramus	44
Length of posterior or upper border back of the beak	26
Width of inner surface	26
Width of upper surface back of the beak	24
Width of the large postero-internal dental area	13*
Estimated breadth fore and aft	16
Width of external dental area	7
Estimated breadth fore and aft	3
Diameter of anterior dental area fore and aft	4
Diameter of same transversely	3

* The size of the dental areas is in some measure uncertain, as in some cases they appear to have been more or less extended in the fore and aft diameter by fracture over the position of the dental columns.

Superior maxillary.

	Lines.
Extreme length of bone	58
Length of bone internally	43
Breadth posteriorly	23
Diameter fore and aft of large postero-internal dental area	18
Diameter of same transversely	12
Diameter fore and aft of anterior dental area	11
Diameter of same posteriorly and transversely	5
Diameter fore and aft of external dental area	16
Diameter transversely of the same where widest	5

EUMYLODUS.

EUMYLODUS LAQUEATUS.

Among some fossils from the Cretaceous sandstone near Columbus, Mississippi, submitted to my examination by Dr. William Spillman, there is a specimen of the maxillo-dentary apparatus of a chimæroid fish, related with Ischyodus, but apparently distinct from that genus. The specimen is represented in Figs. 21, 22, Plate XIX, and Figs. 13, 14, Plate XXXVII. It most resembles, in its general form, the mandible of Ischyodus, as represented in Fig. 20, Tab. 40, of *I. Townsendi*, and Fig. 16, Tab. 40 c, of *I. Agassizi*, of the third volume of the Atlas of the Poissons Fossiles.

The bone is of denser character than the corresponding one of Edaphodon, and in this respect and several others is more like that of Leptomylus, described by Professor Cope.

The outer surface (Fig. 14, Plate XXXVII) is nearly flat, but slightly depressed below, and bent outwardly behind from the triturating surface. The inner surface (Fig. 21, Plate XIX) is fluted; the anterior third presents a succession of three curved ridges separated by two grooves; the median third forms a wide, concave groove; and the posterior third forms a nearly square plane, sloping from the triturating surface backward and inward, and defined by a subacute border from the outer surface of the bone.

The anterior border of the mandible appears as curved cylindroid termination of the bone. No appearance of a distinct symphysial surface exists.

The oral surface (Fig. 22, Plate XIX) is uneven, and conforms in its outline with the inner and outer faces of the bone. The anterior most prominent portion is convex, and exhibits some scratches and polish, due to its masticating function. Its posterior two-thirds incline from a median dentary ridge, moderately without and behind, but steeply within.

The dentary ridge is near the median line of the oral surface, extending about half its length, but nearer its posterior than anterior extremity. As seen in Fig. 13, Plate XXXVII, it appears to be composed of the prominent tubercular extremities of three connate columns, of which the back two appear oval, and the anterior one rather clavate in outline.

The measurements of the specimen are as follows:

	Lines.
Length or depth of the anterior border	32
Length or depth of the posterior border	11
Thickness of the anterior column or border	6
Thickness at the second ridge of the inner surface	6½
Thickness at the third ridge of the inner surface	8½
Thickness at the middle concavity of the inner surface	7½
Thickness at the commencing ridge of the posterior slope of the inner surface	10
Fore and aft extent of triturating surface	36
Length of dental tract	19

NOTICE OF SOME REMAINS OF FISHES FROM THE CARBONIFEROUS FORMATIONS OF KANSAS.

The remains described below were obtained by Dr. F. V. Hayden and Mr. F. B. Meek in the summer of 1858, and were originally noticed by the writer in the Proceedings of the Academy of Natural Sciences of Philadelphia in January, 1859.

Plagiostomi.

CLADODUS.

Cladodus occidentalis.

The extinct genus of cartilaginous fishes, Cladodus, was first characterized by Agassiz from isolated teeth from the Coal-formation of Europe. A species of the same genus is indicated by a fragment of a tooth discovered by Messrs. Hayden and Meek in the upper Coal-measures of Manhattan Kansas.

The specimen has lost one-half its base, a large portion of its principal cusp, and the points of the lateral cusps, but sufficient remains to give us a correct idea of the form of the perfect tooth, as represented in Figs. 4 to 6, Plate XVII.

The base of the tooth is oblong in outline, with the inner border somewhat angular and the outer one concave. Its upper inner surface slopes from the cusps, and near its margin, a short distance from the extremities, supports a pair of oval tubercles. Similar protuberances occupy a position beneath the base externally.

The median or principal cusp of the tooth is elongated demiconical, with acute lateral edges. The inner convex surface of the cusp at its base exhibits sharp, oblique folds or striæ, as represented in Fig. 4. The outer less convex or nearly flat surface is smooth, except a few vertical wrinkles at its base.

The lateral denticles on each side of the principal cusp are two, of which the outer is the larger.

In its perfect condition the tooth has approximated 1¼ inches in length, and about 1 inch in breadth at base.

A similar tooth from the coal-measures of Illinois has been described under the name of *Cladodus mortifer* by Professor Newberry in the second volume of Worthen's Geological Survey of Illinois, published in 1866. Mr. Orestes St. John has likewise described some teeth of the same species from the coal measures of Nebraska, in the Proceedings of the American Philosophical Society for 1870, and in Hayden's Report on the Geological Survey of Nebraska, published this year.

XYSTRACANTHUS.

Xystracanthus arcuatus.

A second cartilaginous fish of the Coal-period is indicated by a remarkable dorsal spine, discovered by Messrs. Meek and Hayden in the Upper Carboniferous rocks of Leavenworth City, Kansas. The specimen, represented in Fig. 25, Plate XVII, lies partially imbedded in a piece of yellowish limestone, also containing a few minute crinoid segments. The point of the spine and its root of insertion are destroyed, and the specimen is otherwise mutilated and appears somewhat crushed, but it is sufficiently characteristic to distinguish it from ichthyodorulites previously described.

The spine is strongly curved, appears flattened at the sides, and is rounded at the borders. Its transverse section is narrow ovoid, with the narrower extremity toward the convex border. The spine is longitudinally striated, and in its present condition the bone is brown and quite friable. The sides and concave border of the spine are furnished with white, shining, enamel-like tubercles of various sizes. The smaller ones are half ovoid; larger ones are conical or half conical; and the largest, which occupy the upper and lower part of the concave border, are crescentoid, and embrace the latter. In shape and attachment the larger tubercles remind one of minute polypori projecting from the stem of a tree. They are convex above, and flat, or slightly concave, below.

PETALODUS.

Petalodus alleghaniensis.

Petalodus is another extinct genus of cartilaginous fishes, allied to our living sharks, which was originally characterized by Owen, and was also established on isolated teeth from the Carboniferous formations of Europe.

A species of the same genus, under the above name, was described by the author in the Journal of the Academy of Natural Sciences for 1856, from a specimen found in the Coal-measures of Blair County, Pennsylvania. A similar tooth was also described and referred to the same species in the Proceedings of the Academy for 1859, which was obtained by Messrs. Meek and Hayden from the Upper Carboniferous formation of Fort Riley, Kansas. The specimen is represented in Fig. 3, Plate XVII.

The crown is broad, and somewhat lozenge-shaped in outline. The base is bordered by a thick annulated ridge, arching downward toward the middle and moderately deflected at the extremities. The free border is sharp and somewhat arcuate, and the apex is slightly acuminate. The anterior surface of the crown slopes outwardly. The posterior deeper surface is concave at its lower median portion. The fang is about as long as the crown is externally, but is not so wide. Its extremity is angular and everted.

The measurements of the tooth are as follows:

	Lines.
Length of tooth in the entire condition about	19
Breadth of crown at base	20
Length of crown externally	9½
Length of crown internally	12
Length of fang externally	9½
Breadth of fang	14

Similar teeth from the Coal-measures of Illinois have been described by Professor Newberry, under the name of *Petalodus destructor*, in the work above mentioned. Others have also been described or indicated, from the Coal-measures of Indiana, Iowa, and Nebraska, by Mr. St. John, likewise in the works above named.

ASTERACANTHUS.

ASTERACANTHUS SIDERIUS.

Incidentally, I take the opportunity of describing a fossil submitted to my examination by Professor J. M. Safford through Professor Hayden. It was obtained near Glasgow, Tennessee, and is reputed to be of Sub-carboniferous age. The specimen consists of a fragment of an ichthyodorulite, or fossil-fish spine, and is represented in Fig. 59, Plate XXXII. It appears to indicate a species of the extinct genus Asteracanthus, the remains of which had previously only been found in formations of later age than that above mentioned.

The fragment is from an intermediate position at the junction of the root and shaft, and is a little over 3 inches in length. It looks as if when in a complete condition it had been upwards of a foot in length, approximating that of the dorsal spine of *A. ornatissimus*. Broken off at both extremities, and also posteriorly, so as to leave no portion of the usual groove, it appears as a solid, porous bone-fragment, triangular in transverse section toward the apex, and oblong toward the root.

The sides of the shaft are closely studded with mammillary tubercles, arranged in rows directed upward and forward. The tubercles incline in the same direction, and have their sides longitudinally striated. Their summits are worn away, the extent of abrasion increasing, approaching the anterior border of the spine.

SYNOPSIS OF THE EXTINCT VERTEBRATA DESCRIBED OR NOTICED IN THE PRESENT WORK.

MAMMALIA.

Carnivora.

FELIDÆ.

FELIS.

FELIS AUGUSTUS.
> Leidy: Pr. Ac. Nat. Sc. 1872, 39.

Described page 227 of the present work, and represented by Figs. 18, 19, Plate VII, and Fig. 24, Plate XX. From the Pliocene of the Niobrara River, Nebraska.

FELIS IMPERIALIS.

Founded on an upper-jaw fragment, containing the second premolar tooth, from the Quaternary of California. Described page 228, and represented by Fig. 3, Plate XXXI.

CANIDÆ.

CANIS.

CANIS INDIANENSIS.
> Leidy: Ext. Mam. of N. America 1869, 368.
> *Canis primævus.* Leidy: Pr. Ac. Nat. Sc. 1854, 200; Jour. Ac. Nat. Sc. 1856, III, 167, Plate XVII, Figs. 11, 12.

Founded on an upper maxillary with teeth from the banks of the Ohio, near Evansville, Indiana. Also indicated by the ramus of a lower jaw from California. Quaternary.

See page 230 for description of the latter specimen, represented by Fig. 2, Plate XXXI.

CANIS VAFER.
> Leidy: Pr. Ac. Nat. Sc. 1858, 21; 1870, 109; Ext. Mam. of N. America 1869, 368.

Founded on jaw-fragments with teeth from the Pliocene of the Niobrara River, Nebraska, and Sweetwater River, Wyoming.

FAMILIES UNDETERMINED.

PATRIOFELIS.

PATRIOFELIS ULTA.
    Leidy: Pr. Ac. Nat. Sc. 1870, 10; Hayden's Rep. Geol. Sur. Wyoming 1871, 344; Hayden's Rep. Geol. Sur. Montana 1872, 355.

Founded on the mutilated rami of a lower jaw from the Bridger Eocene Tertiary, Wyoming. Described page 114, and represented by Fig. 10, Plate II.

UINTACYON.

Probably the same as Miacis, described by Professor Cope in the Proc. Am. Phil. Soc. 1872, 470.

UINTACYON EDAX.
    Leidy: Pr. Ac. Nat. Sc. 1872, 277.

Founded on the ramus of a lower jaw from the Bridger Eocene Tertiary of Wyoming. Described page 118, and represented by Figs. 6 to 10, Plate XXVII.

UINTACYON VORAX.
    Leidy: Pr. Ac. Nat. Sc. 1872, 277.

Founded on a lower-jaw fragment from the Bridger Eocene Tertiary of Wyoming. Described page 120, and represented by Figs. 11 to 13, Plate XXVII.

SINOPA.

SINOPA RAPAX.
    Leidy: Pr. Ac. Nat. Sc. 1871, 113; Hayden's Rep. Geol. Sur. Montana 1872, 355.

Founded on a lower-jaw fragment with teeth from the Bridger Eocene Tertiary of Wyoming. Described page 116, and represented in Fig. 44, Plate VI.

SINOPA EXIMIA.
    Indicated by a lower-jaw fragment, described page 118, and represented in Fig. 45, Plate VI. From the Bridger Eocene Tertiary of Wyoming.

MUSTELIDÆ.

LUTRA?

LUTRA PISCINARIA.
    Indicated by a tibia, described page 230, and represented in Fig. 4, Plate XXXI. From the Pliocene Tertiary of Idaho.

ARTIODACTYLA.

Ruminantia.

CAMELIDÆ.

AUCHENIA.

AUCHENIA HESTERNA.

Founded on specimens of teeth described page 255, and represented in Figs. 1 to 3, Plate XXXVII. From the Quaternary of California.

PROCAMELUS. s. *Protocamelus.*

PROCAMELUS OCCIDENTALIS?
 Leidy: Pr. Ac. Nat. Sc. 1858, 23, 89; Ext. Mam. N. America 1869, 382.

See page 258 of the present work, and represented by Figs. 21, 22, Plate XX. Pliocene of Nebraska and Texas?

PROCAMELUS ROBUSTUS?
 Leidy: Pr. Ac. Nat. Sc. 1858, 89; Ext. Mam. N. America 1869, 381.

See page 259 of the present work. Pliocene of Nebraska and Texas.

PROCAMELUS VIRGINIENSIS.
 Leidy: Pr. Ac. Nat. Sc. 1873, 15.

Page 259, and represented by Figs. 26 to 29, Plate XXVII. Founded on teeth from the Miocene of Virginia.

PROCAMELUS? NIOBRARENSIS.
 Megalomeryx niobrarensis? Leidy: Pr. Ac. Nat. Sc. 1858, 24; Ext. Mam. Dakota and Nebraska 1869, 161, Plate XIV, Figs. 12 to 14.

See page 260, under the name of *Megalomeryx niobrarensis?* and represented in Figs. 24, 25, Plate XXVII. Founded on teeth from the Pliocene of the Niobrara River, and from L'Eau qui Court County, Nebraska.

CERVIDÆ.

LEPTOMERYX.

LEPTOMERYX EVANSI.
 Leidy: Pr. Ac. Nat. Sc. 1853, 394; 1870, 112; Ext. Mam. N. America 1869, 383.

Noticed from the Miocene of Oregon, page 216. Originally described from the Miocene of Dakota.

MERYCODUS

MERYCODUS NECATUS.

Leidy: Pr. Ac. Nat. Sc. 1851, 90, 157; 1857, 89; 1858, 24; 1870, 109; Ext. Mam. N. America 1869, 382.

Noticed from the Pliocene of Sweetwater River, Wyoming. Originally described from Bijou Hill and from Little White River, or the South Fork of White Earth River, Dakota.

BOVIDÆ.
BISON.

BISON LATIFRONS.

Leidy: Pr. Ac. Nat. Sc. 1852, 117; Mem. Ext. Sp. American Ox in Smiths. Contrib. 1852, 8; Ext. Mam. N. America 1869, 371.

Noticed from the Quaternary of California and Pennsylvania, page 253, and represented in Figs. 4 to 8, Plate XXVIII.

Found in the Quaternary of Pennsylvania, Georgia, South Carolina, Kentucky, Mississippi, Texas, and California.

OREODONTIDÆ.
OREODON.

Leidy: Pr. Ac. Nat. Sc. 1851, 238.
Merycoidodon. Leidy: Pr. Ac. Nat. Sc. 1848, 47.

OREODON CULBERTSONI.

Leidy: Owens's Rep. Geol. Sur. 1852, 548; Ext. Mam. N. America 1869, 379; Pr. Ac. Nat. Sc. 1870, 67, 112.

Noticed from John Day's River, Oregon, page 211, and represented in Fig. 12, Plate VII.

Professor Marsh has recently described some remains from the Miocene of Oregon, under the name of *Oreodon occidentalis*. (Am. Jour. Sc. May, 1873.) He observes that it resembles *O. Culbertsoni* in most of its cranial characters, but differs materially in the large auditory bullæ. From this, I suspect the remains, together with those I have described from Oregon under the last-mentioned name, belong to the species I have elsewhere named *O. bullatus.*

Professor Marsh observes that, "in comparing the various species of Oreodon, some new points in the structure of the genus were observed." He then gives in the formula of dentition the number of incisors as $\frac{3}{3}$, canines $\frac{1}{1}$, premolars $\frac{4}{4}$, molars $\frac{3}{3}$, and adds: "The caniniform tooth of the lower jaw is clearly the first premolar, as Dr. Gill has stated." As may be seen by referring to pages 84 and 85 of the Extinct Mammalia of Dakota and Nebraska, although giving the formula of dentition of Oreodon as—incisors $\frac{3}{3}$, canines $\frac{1}{1}$, premolars $\frac{4}{4}$, molars $\frac{3}{3}$, I observe that the inferior canine is a transformed premolar, and that the inferior lateral incisor, as in other ruminants, is to be regarded as an incisiform canine.

OREODON SUPERBUS.
> Leidy: Pr. Ac. Nat. Sc. 1870, 111.

Described page 211, and represented by Fig. 1, Plate I; Fig. 16, Plate II; and Figs. 7 to 11, Plate VII. From the Miocene of Oregon.

MERYCOCHŒRUS.

> Leidy: Pr. Ac. Nat. Sc. 1858, 24; Ext. Mam. N. America 1869, 380.

MERYCOCHŒRUS RUSTICUS.
> Leidy: Pr. Ac. Nat. Sc. 1870, 109.

Described page 199, and represented by Figs. 1 to 3, Plate III; Figs. 1 to 5, Plate VII; and Figs. 9 to 11, Plate XX. From the Pliocene of Sweetwater River, Wyoming.

AGRIOCHŒRUS.

AGRIOCHŒRUS ANTIQUUS.
> Leidy: Pr. Ac. Nat. Sc. 1850, 121; Ext. Mam. N. America 1869, 381; Pr. Ac. Nat. Sc. 1870, 112.

Noticed from the Miocene of Oregon, page 216.

AGRIOCHŒRUS LATIFRONS.
> Leidy: Pr. Ac. Nat. Sc. 1867, 32; 1870, 67; Ext. Mam. N. America 1869, 381.

Noticed from the Miocene of Oregon, page 216.

Omnivora.

SUIDÆ.

DICOTYLES.

DICOTYLES PRISTINUS.
> *Peccary.* Leidy: Pr. Ac. Nat. Sc. 1870, 112.

Described page 216, and represented by Figs. 13, 14, Plate VII. From the Miocene of Oregon.

ANTHRACOTHERIDÆ.

ELOTHERIUM.

Pomel: Bibl. Univ. Genève, Archives, 1847, 307.
Entelodon. Aymard: Mem. Soc. Agric., &c., du Puy 1848, 240.
Archæotherium. Leidy: Pr. Ac. Nat. Sc. 1850, 90.

ELOTHERIUM MORTONI?

Leidy: Pr. Ac. Nat. Sc. 1857, 175; Ext. Mam. N. America 1869, 388.

Noticed from Wyoming, page 125, and represented by Figs. 28, 29, Plate VII.

ELOTHERIUM IMPERATOR.

Inferred from several mutilated teeth from the Miocene of Oregon, described page 217, and represented in Figs. 3, 4, Plate II, and Fig. 27, Plate VII. Supposed to be the same as *E. superbum* in Pr. Ac. Nat. Sc. 1870, 112.

ELOTHERIUM INGENS

Leidy: Ext. Mam. N. America 1869, 388; Pr. Ac. Nat. Sc. 1870, 112.

Noticed from the Miocene of Oregon. Originally from the Miocene of White River, Dakota.

FAMILIES UNDETERMINED.

HYOPSODUS.

HYOPSODUS PAULUS.

Leidy: Pr. Ac. Nat. Sc. 1870, 110; 1872, 20; Hayden's Rep. Geol. Sur. Wyoming 1871, 354; Hayden's Rep. Geol. Sur. Montana 1872, 363.

To this species, described page 75, I refer Figs. 1 to 9, 18 to 22, Plate VI. From the Bridger Eocene of Wyoming.

HYOPSODUS MINUSCULUS.

This species, described page 81, is represented by Fig. 5, Plate XXVII. From the Bridger Eocene of Wyoming.

MICROSYOPS.

Leidy: Pr. Ac. Nat. Sc. 1872, 20; Hayden's Prelim. Rep. Geol. Sur. Montana 1873, 363.
Limnotherium. In part of Marsh: Am. Jour. Sc. 1871, II, 42.

MICROSYOPS ELEGANS.

Limnotherium elegans. Marsh: Am. Jour. Sc. 1871, II, 43.
Microsyops gracilis. Leidy: Pr. Ac. Nat. Sc. 1872, 20; Hayden's Prelim. Rep. Geol. Sur. Montana 1872, 363.

In Microsyops, six molar teeth immediately succeed the canine in the lower jaw. In the typical *Limnotherium elegans* seven molars occupy the same position. Described page 82, and represented by Figs. 14, 17, Plate VI. From the Bridger Eocene formation of Wyoming.

MICROSUS.

MICROSUS CUSPIDATUS.
Leidy: Pr. Ac. Nat. Sc. 1870, 113.

See page 81; not positively determined as a distinct species and genus. Represented by Figs. 10, 11, Plate VI. From the Bridger Eocene formation of Wyoming.

HIPPOSYUS.

HIPPOSYUS FORMOSUS.
Leidy: Pr. Ac. Nat. Sc. 1872, 37.

Described from a few isolated teeth, page 90, and represented in Fig. 41, Plate VI, and Figs. 1, 2, Plate XXVII. From the Bridger Eocene of Wyoming.

HIPPOSYUS ROBUSTIOR.
Notharctus robustior. Leidy: Hayden's Rep. Geol. Sur. Montana 1872, 364.

Described page 93, and represented by Fig. 40, Plate VI. From the Bridger Eocene of Wyoming.

HADROHYUS.

HADROHYUS SUPREMUS.
Leidy: Pr. Ac. Nat. Sc. 1871, 248.

Indicated by a mutilated tooth from the Miocene of Oregon. Described page 222, and represented by Fig. 26, Plate XVII.

PERISSODACTYLA.

Solidungula.

EQUIDÆ.

EQUUS.

EQUUS MAJOR.
Dekay: Nat. Hist. New York, Zool. 1842, 108. Leidy: Ext. Mam. N. America 1869, 399.
Equus complicatus. Leidy: Pr. Ac. Nat. Sc. 1858, 11; Ext. Mam. N. America 1869, 399.

Remains described page 244, and represented by Figs. 3 to 18, Plate XXXIII. From the Quaternary of the United States.

EQUUS OCCIDENTALIS.
Leidy: Pr. Ac. Nat. Sc. 1865, 94.
? Equus. Von Meyer: Palæontographica 1867, 70.
Equus excelsus. Leidy: Pr. Ac. Nat. Sc. 1868, 26; Ext. Mam. Dakota and Nebraska 1869, 266, 400, Plate XIX, Fig. 39; XXI, Fig. 31.
Equus pacificus. Leidy: Pr. Ac. Nat. Sc. 1868, 195; Ext. Mam. N. America 1869, 400.

Described page 242, and represented by Figs. 1, 2, Plate XXXIII. From the Quaternary ? of Nebraska, Idaho, California, and Mexico.

HIPPARION.

HIPPARION SPECIOSUM ?
Leidy: Pr. Ac. Nat. Sc. 1858, 27; Ext. Mam. N. America 1869, 101.

See pages 247, 248, and Figs. 14, 15, Plate XX. From the Tertiary of Texas.

PROTOHIPPUS, s. MERYCHIPPUS.

PROTOHIPPUS PERDITUS (?) s. MERYCHIPPUS MIRABILIS ?
Protohippus perditus. Leidy: Pr. Ac. Nat. Sc. 1858, 26; Ext. Mam. N. America 1869, 401.
Merychippus mirabilis. Leidy: Pr. Ac. Nat. Sc. 1858, 27.

See pages 248, 249, 250, and Figs. 16, 20, Plate XX. From the Tertiary of Texas and Utah.

PROTOHIPPUS PLACIDUS.
Leidy: Ext. Mam. N. America 1869, 401.

See pages 249, 250, and Figs. 17, 18, Plate XX. From the Tertiary of Texas.

ANCHITHERIDÆ.

ANCHITHERIUM.

Meyer: Jahrbuch Mineralogie 1844, 298.

ANCHITHERIUM BAIRDI.
Leidy: Owen's Rep. Geol. Sur. Wisconsin, &c., 1852, 572; Ext. Mam. N. America 1869, 402; Pr. Ac. Nat. Sc. 1870, 112.

A full account of the remains of the species from the Mauvaises Terres of White River, Dakota, is given in the Extinct Mammalia of Dakota and Nebraska, page 303. A notice of remains from Oregon is given page 218 of the present work, and a tooth representing the species is given in Fig. 15, Plate VII. Miocene.

ANCHITHERIUM CONDONI.
Leidy: Pr. Ac. Nat. Sc. 1870, 112.

Described page 218, and represented by Fig. 5, Plate II. From the Miocene of Oregon.

ANCHITHERIUM AGRESTE.
Anchitherium. Leidy: Pr. Ac. Nat. Sc. 1871, 199.

Described page 251, and represented by Figs. 16, 17, Plate VII. From the Miocene ? of Montana.

ANCHITHERIUM ? AUSTRALE.

Described page 250, and represented by Fig. 19, Plate XX. From the Tertiary of Texas.

? ANCHITHERIUM.
Equus. Leidy: Pr. Ac. Nat. Sc. 1868, 195.
Equus parvulus. Marsh: Am. Jour. Sc. 1868.

Noticed page 252, and represented by Fig. 23, Plate XX. From the Tertiary of Nebraska.

PALÆOSYOPS.

Leidy: Pr. Ac. Nat. Sc. 1870, 113; 1871, 114, 118, 197, 229; 1872, 168, 241. Hayden's Prelim. Rep. Geol. Sur. Wyoming 1871, 355. Hayden's Prelim. Rep. Geol. Sur. Montana 1872, 358, published April, 1772.
Telmatherium. Marsh: Am. Jour. Sc. 1872, IV, 123, published in advance July 22, 1872.
Limnohyus. Marsh: Am. Jour. Sc. 1872, IV, 124, published in advance July 22, 1872. Cope: Pr. Am. Phil. Soc. 1873.

Remains referable to the genus Palæosyops are the most common of those of the larger extinct mammals occurring in the Bridger Eocene formation of Wyoming. The genus was originally indicated by characteristic specimens of teeth represented in Figs. 4, 5, Plate V, and Figs. 3 to 6, Plate XXIII. Subsequently a number of specimens were received from time to time and indicated in the Proceedings of the Academy from 1870 to 1872, and in Professor Hayden's Preliminary Report of the Geological Surveys of Wyoming and Montana. In the report on Montana, published in April, 1872, the characters of the genus are succinctly stated. Palæosyops is described "as an odd-toed pachyderm, with the skeleton constructed nearly as in the tapir. The thigh-bone possesses a third trochanter. The hind feet nearly repeat the construction of those of the tapir. The skull, with its large temporal fossæ, high and thick sagittal crest, concave occiput, broad, convex face, resembled that of the related Palæotherium. The teeth also agree in number and nearly in constitution with those of that animal. The number of teeth altogether appear to have been 44, consisting of 3 incisors, 1 canine, 4 premolars, and 3 molars to the series on each side, above and below. The teeth in each jaw form a nearly unbroken arch, intervals existing only sufficient to accommodate the passing of the points of the large and bear-like canines.

"The true molars have a resemblance to those of Palæotherium. In the crowns of the upper true molars the inner constituent lobes are more completely isolated from the outer ones than in that genus, and the bottoms of the transverse valleys are proportionately of less depth. *The last upper molar of Palæosyops has but a single lobe to the inner part of the crown.*

"In Palæotherium, the large premolars have the same form as the true molars, but are quite different in this respect in Palæosyops. In the former the crown of the upper premolars, except the first, is composed of four lobes, as in the succeeding molars. In Palæosyops the first premolar has a conical crown, the second a bilobed crown, and the third and fourth have trilobed crowns.

"The canines of Palæosyops are proportionately as large and of the same form as in the bears."

In an article in the American Journal of Science, 1872, V, published in advance July 22, 1872, Professor Marsh, after remarking that the type of the genus Palæosyops is too imperfectly known to determine its more important characters, adds that, "in some specimens which agree best with the original description of paludosus, *the last upper molar has two inner cones,* and to this group the name Palæosyops may in future be restricted. The other specimens have but *a single internal cone on the last upper molar,* and for the genus thus represented the name Limnohyus is proposed."

In this view Professor Cope has recently described some remains of Palæosyops under the name of *Limnohyus lævidens*.

Teeth such as I have attributed to Palæosyops are comparatively abundant, but I have not yet had the opportunity of inspecting a specimen of a last upper molar, such as Professor Marsh ascribes to Palæosyops, in which the inner side of the crown possesses two internal cones. That such exist there can be no question, as proved by Professor Marsh's description of *Palæosyops laticeps*.

Professor Marsh has described some remains which he refers to a genus with the name of Telmatherium. Of this, he observes: "The dentition of this genus, so far as is known, appears to be similar to that of Palæosyops. The upper molar teeth have the inner cones more elevated and more pointed than in Palæosyops, and the basal ridge is well developed. The last upper molar has but a single internal cone." He also remarks that "the two may be readily distinguished by the anterior portion of the skull, which in Telmatherium has the premaxillaries compressed, with an elongated median suture. The zygomatic arch is also much less strongly developed, and the squamosal portion of it is comparatively slender." Such differences are more likely to be of a sexual or individual character than of either specific or generic value.

Since writing the preceding chapters we have attempted to give a restoration of the skull of Palæosyops in Fig. 1, Plate XXXI, built up from a number of specimens. The cranium and face were mainly reconstructed from the specimens of Fig. 51, Plate XVIII, and Figs. 1, 2, Plate XXIV; the lower jaw from the specimen of Fig. 52, of the latter plate, and Fig. 4, of the former plate.

1. PALÆOSYOPS PALUDOSUS.

    Leidy: Pr. Ac. Nat. Sc. 1870, 113; 1871, 114, 197, 229; 1872, 168; Hayden's Rep. U. S. Geol. Sur. Wyoming 1871, 355; Hayden's Rep. U. S. Geol. Sur. Montana 1872, 359.

    Palæosyops. Leidy: Pr. Ac. Nat. Sc. 1871, 118.

    Limnohyus lævidens. Cope: Pr. Am. Phil. Soc. 1873, article published in advance January 31, 1873.

The description of the species is given on page 28 of the present work. The specimens represented in Figs. 1 to 8, Plate IV; Figs. 4 to 11, Plate V; Figs. 1 to 4, Plate XIX; Figs. 1 to 7, Plate XX; Figs. 3 to

6, Plate XXIII; Figs. 6, 7, Plate XXIV; and Fig. 5, Plate XXIX, are considered as pertaining to *Palæosyops paludosus*.

A fine specimen, consisting of the greater part of a skull, exhibited by Professor Cope to the Academy, and described by him under the name of *Limnohyus laevidens*, appeared to me to be the same as *Palæosyops paludosus*. From the Bridger Eocene formation of Wyoming.

2. PALÆOSYOPS MAJOR.
> Leidy: Hayden's Prelim. Rep. Geol. Sur. Montana, April, 1872, 359; Pr. Ac. Nat. Sc. 1872, 168, 241.
> *Lymnohyus robustus*. Marsh: Am. Jour. Sc. 1872, IV, 124, published in advance July 22, 1872.

I am not convinced that this is a really distinct species from *Palæosyops paludosus*. A large number of specimens referable to the genus would indicate a considerable variation in the size of individuals, of which the more robust forms may have been males. The species is described on page 45. The specimens regarded as pertaining to it are represented in Fig. 8, Plate XX; Figs. 1, 2, 7 to 12, 14 to 16, Plate XXIII; and Figs. 1 to 5, Plate XXIV. From the Bridger Eocene formation of Wyoming.

3. PALÆOSYOPS HUMILIS.
> Leidy: Pr. Ac. Nat. Sc. 1872, 168, 277.

Probably a small species indicated by an upper molar, represented in Fig. 8, Plate XXIV, and noticed page 58. From the Bridger Eocene of Wyoming.

4. PALÆOSYOPS JUNIUS.
> *Palæosyops junior*. Leidy: Pr. Ac. Nat. Sc. 1872, 277.

Described page 57. From the Bridger Eocene of Wyoming.

LIMNOHYUS.

LIMNOHYUS LATICEPS.
> *Palæosyops laticeps*. Marsh: Am. Jour. Sc. 1872, 122.

Indicated page 58, and represented by Fig. 13, Plate XXIII. From the Bridger Eocene of Wyoming.

HYRACHYUS.

HYRACHYUS AGRARIUS.
>Leidy: Hayden's Prelim. Rep. Geol. Sur. Wyoming 1871, 357; Pr. Ac. Nat. Sc. 1871, 229; 1872, 19, 168; Hayden's Rep. Geol. Sur. Montana 1872, 361.
>*Hyrachyus agrestis.* Leidy: Hayden's Rep. Geol. Sur. Wyoming 1871, 357.
>*Lophiodon Bairdianus.* Marsh: Am. Jour. Sc. 1871, II, 3.

The species is described page 60 of the present work, and specimens attributed to it are represented in Figs. 11, 12, Plate II; Figs. 9 to 18, Plate IV; and Figs. 25, 26, Plate XX. From the Bridger Eocene of Wyoming.

HYRACHYUS EXIMIUS.
>Leidy: Pr. Ac. Nat. Sc. 1871, 229; 1872, 168; Hayden's Rep. Geol. Sur. Montana 1872, 361.

Described page 66, and represented by Figs. 19, 20, Plate IV; Fig. 5, Plate XIX; and Figs. 9, 10, Plate XXVI. From the Bridger Eocene of Wyoming.

HYRACHYUS MODESTUS.
>Leidy: Pr. Ac. Nat. Sc. 1872, 20; Hayden's Rep. Geol. Sur. Montana 1872, 361.
>*Lophiodon modestus.* Leidy: Pr. Ac. Nat. Sc. 1870, 109.

Described page 67, and represented by Fig. 13, Plate II. From the Bridger Eocene of Wyoming.

HYRACHYUS NANUS.
>Leidy: Pr. Ac. Nat. Sc. 1872, 20; Hayden's Rep. Geol. Sur. Montana 1872, 361.
>? *Lophiodon nanus.* Marsh: Am. Jour. Sc. 1871, II, 37.

Described page 67, and represented by Fig. 14, Plate II; Fig. 42, Plate VI; Fig. 11, Plate XXVI; and Figs. 21, 22, Plate XXVII. From the Bridger Eocene of Wyoming.

LOPHIODON?

LOPHIODON OCCIDENTALIS.
>Leidy: Pr. Ac. Nat. Sc. 1868, 232; Ext. Mam. N. America 1869, 391.

Noticed as probably found in the Miocene of Oregon, page 218, and represented in Fig. 1, Plate II.

LOPHIOTHERIUM.

LOPHIOTHERIUM SYLVATICUM.
>Leidy: Pr. Ac. Nat. Sc. 1870, 126.

Described page 69, and represented by Figs. 33 to 35, Plate VI. From the Bridger Eocene of Wyoming.

RHINOCEROTIDÆ.

RHINOCEROS.

RHINOCEROS PACIFICUS.
Leidy: Pr. Ac. Nat. Sc. 1871, 248.
Rhinoceros occidentalis. Leidy: Pr. Ac. Nat. Sc. 1870, 112.

Described from teeth on page 221, and represented by Figs. 6, 7, Plate II, and Figs. 24, 25, Plate VII. From the Miocene of Oregon.

RHINOCEROS HESPERIUS?
Leidy: Pr. Ac. Nat. Sc. 1865, 176; 1870, 112; Ext. Mam. N. America, 1869, 390.

Originally described from the ramus of a lower jaw from the Miocene? of California. Also supposed to be indicated by teeth described page 220, and represented by Figs. 8, 9, Plate II, from the Miocene of Oregon.

In the May number of the American Journal of Science for 1873, Professor Marsh has noticed remains of rhinoceros, which he refers to two additional species. One named *R. annectens* is founded on remains from the same formation as those of the preceding species. The other, named *R. oregonensis,* is reputed to have pertained to the Pliocene deposits of Oregon.

FAMILIES UNDETERMINED.

ANCHIPPODUS.

ANCHIPPODUS RIPARIUS
Leidy: Pr. Ac. Nat. Sc. 1868, 232; Ext. Mam. N. America, in Jour. Ac. Nat. Sc. 1869, VII, 403, Figs. 45, 46, Plate XXX.
Palæosyops minor. Marsh: Am. Jour. Sc. 1871, II, 36.
Trogosus castoridens. Leidy: Pr. Ac. Nat. Sc. 1871, 113; Hayden's Rep. Geol. Sur. Montana 1872, 360.

Described page 71, under the name of *Trogosus castoridens,* and also represented as such in Figs. 1 to 3, Plate V.

The genus Anchippodus was originally named from an isolated tooth from a Tertiary formation of Monmouth County, New Jersey. The specimen is represented in Figs. 45, 46, Plate XXX, of the seventh volume of the Journal of the Academy for 1869, and is described on page 403 of that work. It was not until after the description of the lower jaw referred to *Trogosus castoridens*, on page 71 of the present work, and represented under the same name in Figs. 1 to 3, Plate V, that I noticed the identity in character of the corresponding tooth. Previous to the description of the jaw referred to Trogosus, Professor Marsh had published a notice of a similar tooth under the name of *Palæosyops minor*.

It is not improbable, after all, that Trogosus may be distinct from Anchippodus, for there are several genera which, while they have the inferior true molars alike, have the premolars and upper true molars quite different. While regarding Trogosus the same as Anchippodus, for the same reason I have considered *Trogosus castoridens* the same as *Anchippodus riparius*, for the specimen upon which the latter was originally made known is identical in form and size with the corresponding tooth in the jaw of the former. Nor is it improbable that they are the same, for they were probably of contemporaneous age, and perhaps extended throughout the continent, as the American mastodon did at a later period. Specimen from the Bridger Eocene of Wyoming.

ANCHIPPODUS VETULUS.

Trogosus vetulus. Leidy: Pr. Ac. Nat. Sc. 1871, 229; Hayden's Rep. Geol. Sur. Montana 1872, 360.

Noticed on page 75, under the name of *Trogosus vetulus*, and represented, with the name of *Anchippodus vetulus*, in Fig. 43, Plate VI. From the Bridger Eocene of Wyoming.

NOTHARCTUS.

NOTHARCTUS TENEBROSUS.

Leidy: Pr. Ac. Nat. Sc. 1870, 114.

Described page 86, and represented by Figs. 36, 37, Plate VI. From the Bridger Eocene of Wyoming.

Proboscidea.

ELEPHAS.

ELEPHAS AMERICANUS.

Dekay: Nat. Hist. New York, Zool., 1842, I, 101. Leidy: Ext. Mam. N. America 1869, 398.
Elephas Columbi. Falconer: Quart. Jour. Geol. Soc. 1857, 319, &c.
Elephas Texianus. Owen: Rep. Brit. Asso. 1858, 84, &c.
Elephas imperator. Leidy: Pr. Ac. Nat. Sc. 1858, 10.

Euelephas Jacksoni. Briggs and Foster: Canad. Nat. and Geol. 1863, 135, 147.
Euelephas Columbi. Falconer: Palæont. Mem. 1868, II, 211 to 251.
Elephas. Von Meyer: Palæontographica, 1867, 70, Plate VII, Figs. 7, 8.

See page 238. Remains noticed from New Mexico and Texas.

MASTODON.

MASTODON AMERICANUS.
> Leidy: Pr. Ac. Nat. Sc. 1868, 175. For synonymy, see Extinct Mammalia of North America 1869, 392.

Some remains described or noticed page 237, and represented in Figs. 5, 6, Plate XXII, and Fig. 9, Plate XXVIII.

Remains of the common American mastodon are found in the Quaternary formation throughout the United States.

MASTODON MIRIFICUS.
> Leidy: Pr. Ac. Nat. Sc. 1858, 10; 1870, 67; Ext. Mam. Fauna of Dakota and Nebraska 1869, 249, 396.
> *Mastodon (Tetralophodon) mirificus.* Leidy: Pr. Ac. Nat. Sc. 1858, 10.

Remains originally described from the Pliocene of the Loup Fork of Platte River. Also reported to occur on the Niobrara River, Nebraska. Noticed page 237. From the Pliocene of Sinker Creek, Idaho.

MASTODON OBSCURUS.
> Leidy: Ext. Mam. N. America 1869, 396. For earlier synonymy, see the same work. Pr. Ac. Nat. Sc. 1870, 99; 1871, 199; 1872, 142.
> *Mastodon Shepardi.* Leidy: Pr. Ac. Nat. Sc. 1870, 98; 1871, 199.
> *Rhynchotherium?* See Falconer: Palæontological Memoirs, 1868, II, 74.

Originally named from remains found in Maryland, North Carolina, and Georgia. See Extinct Mammalian Fauna of Dakota and Nebraska, 1869, 244, 396. Remains from California and New Mexico described page 231 and represented in Figs. 1 to 4, Plate XXI and Figs. 1 to 4, Plate XXII, of the present work, are supposed in whole or part to belong to the same species. If they do not, they would represent another species, which might retain the name of *M. Shepardi.*

In the Palæontographica for 1867, page 64, Von Meyer has given a description of the right ramus of the lower jaw of a Mastodon, from Mechoacan, Mexico. The specimen is represented in Plate VI of the same work, and it contains the last molar and the one in advance, both entire. The portion of the last molar tooth in the jaw-fragment from New Mexico, described page 235, and represented in Figs. 1, 4, Plate XXII, bears a very near resemblance with the corresponding part of the same tooth in the Mechoacan specimen. Notwithstanding this likeness, it would appear that the fore part of the jaw differs so much that the two may be supposed not to pertain to the same species. As stated in the account of the New Mexico Mastodon, the anterior extremity of the jaw is enormously prolonged and provided with a pair of incisors. Von Meyer observes of the Mechoacan specimen, "Too little of the symphysis is preserved to speak with any certainty of its constitution; but it appears not to have contained incisors and rather ended in front in a short beak, as in the elephant." The jaw he refers with doubt to the *Mastodon Humboldti*.

I have said that the New Mexican and Mechoacan specimens may be supposed not to pertain to the same species. However, when we consider the difference in the fore part of the lower jaw of the sexes in the *Mastodon americanus*, it is not improbable that the male of the *Mastodon Shepardi* may have had the lower jaw provided with a long beak and incisors which might have been absent in the female.

UINTATHERIUM.

Titanotherium. Marsh: Am. Jour. Sc. 1871, II, 35; the article published in advance June 21, 1871; *ibid*. 1872, IV, 123, published in advance July 22, 1872.

Mastodon. Marsh: Am. Jour. Sc. 1872, note to p. 123, published in advance July 22, 1872.

UINTATHERIUM.

Leidy: Pr. Ac. Nat. Sc. 1872, 169; in a letter addressed to the Academy and published in advance of the proceedings August 1, 1872. Reprinted in Am. Jour. Sc. September, 1872, 239. Pr. Ac. Nat. Sc. 1872, 241. Marsh: Pr. Am. Phil. Soc. 1872, 578; Am. Jour. Sc. 1873, V, 118; American Naturalist 1873, 147. Cope: Pr. Ac. Nat. Sc. 1873, 10, 102; Pr. Am. Phil. Soc. Feb., 1873. Nature: March 13, 1873, 366.

Uintamastix. Leidy: Pr. Ac. Nat. Sc. 1872, 169.

Tinoceras. Marsh: Am. Jour. Sc. 1872, IV, in errata of Sept. No.; do., p. 504, published in advance August 19, 1872; *ibid*. 1872, IV, 322, published in advance August 24, 1872; *ibid*. 1872, IV, 323, published in advance September 21, 1872; *ibid*. 1872, IV, 343, published in advance September 27, 1872; *ibid*. 1873, V, 117, published in advance January 28, 1873; *ibid*. 1873, V, 293, published in advance March 18, 1873; American Naturalist Jan., 1873, 52.

Eobasileus. Cope:* Pr. Am. Phil. Soc. 1872, 185, published in advance August 20, 1872; *ibid.* 1872, 512; *ibid.* 1873, published as a separate pamphlet, "On the Short-Footed Ungulata of the Eocene of Wyoming," March 14, 1873; Pr. Ac. Nat. Sc. 1873, 10, 102; American Naturalist, March 1873, 180.

Loxolophodon. Cope: Pr. Am. Phil. Soc. 1872, 487, 488, published in advance August 22, 1872. Here regarded as the same genus first named in the Proceedings of February 16, 1872, 420, and founded on the tooth of an animal about the size of the American tapir, referred to *Bathmodon semicinctus* and then to Loxolophodon. Pr. Am. Phil. Soc. 1872, 580; Pr. Ac. Nat. Sc. 1873, 102.

Lefalophodon. Typographical error? Cope: Pr. Am. Phil. Soc. 1872, 515.

Dinoceras. Marsh: Am. Jour. Sc. 1872, IV, 344, published in advance September 27, 1872; *ibid.* 1873, V, 117–122, Plates I, II, published in advance January 28, 1873; *ibid.* April, 1873, published in advance March 18, 1873; American Naturalist, March 1873, 146. Nature, March 13, 1873, 366.

Loxolophodon. Cope: "On the Short-Footed Ungulata of the Eocene of Wyoming," read before the Am. Phil. Soc., Feb. 21, 1873, and published in advance of the Proceedings, March 14, 1873. The name is here used as that of a genus recognized as distinct from the one originally described under the same name, which the author now regards as a synonym of Bathmodon.

All the above names I suspect to have been applied to members of the same genus, and in this view have regarded them as synonyms to the first characteristic generic name employed. Of this, however, I am by no means positive, as I have had no opportunity of examining the different fossils upon which the genera were founded, except those described by myself under the name of *Uintatherium robustum*, and the skull described by Professor Cope under the name of *Loxolophodon cornutus*.

In addition, we have the description and figures of the skull described by Professor Marsh under the name of *Dinoceras mirabilis*.

As far as I am able to estimate the differences which have been indicated by the authors just named and those observed by myself, they appear to be rather of specific value, and perhaps in part of sexual character, than of generic importance. We hope, however, that all obscurity in relation to the matter will be cleared away when Professor Marsh and Professor Cope present to us full descriptions with characteristic figures of the fossils in their possession. I may add it is not improbable that the names of Uintatherium, Tinoceras, Eobasileus, Dinoceras, and Loxolophodon, may be

---

* The dates given as those of Professor Cope's publications in advance of the different periodicals named are taken from the publications themselves; but they are, in some instances, contested by Professor Marsh. See an article read before the Philadelphia Academy of Sciences April 8, 1873, and published by Professor Marsh under the title "On the Dates of Professor Cope's Recent Publications."

expressive of more than one genus, in the light that Cariacus, Capreolus, Blastocerus, Axis, Elaphus, &c., are distinct from Cervus. Future comparisons and discoveries will perhaps reduce the nine species of the five genera which have been indicated to the number of two or three species of one or two genera.

Professor Marsh has referred the remarkable animals above indicated to a new order with the name Dinocerata. In the uncertainty as to the true ordinal position of Uintatherium, I have allowed it to remain, according to my first impression, with the Proboscidea.

UINTATHERIUM ROBUSTUM.

> Leidy : Pr. Ac. Nat. Sc. 1872, 169, in a letter addressed to the Academy and published in advance of the proceedings, August 1, 1872. Reprint of the letter in Am. Jour. Sc. September, 1872, 239. Pr. Ac. Nat. Sc. 1872, 241. Cope: Pr. Ac. Nat. Sc. 1873, 102; Pr. Am. Phil. Soc. 1873. Marsh: Am. Jour. Sc. 1873, V, 296; American Naturalist, January, 1873.
>
> *Uintamastix atrox.* Leidy : Pr. Ac. Nat. Sc. 1872, 169; Am. Jour. Sc. 1872, 239.
>
> *Dinoceras mirabilis.* Marsh : Am. Jour. Sc. 1872, IV, 344, published in advance September 27, 1872; *ibid.* 1873, V, 117-122, Plates I, II, published in advance January 28, 1873; *Ibid.* April, 1873, published in advance March 18, 1873. American Naturalist, March, 1873, 146. Nature, March 13, 1873, 366.
>
> *Uintatherium mirabile.* Cope: Pr. Ac. Nat. Sc. 1873, 102; Pr. Am. Phil. Soc. 1873, published in advance "On the Short-Footed Ungulata of the Eocene of Wyoming, March 14, 1873, 28."

The Figs. 6 to 12, Plate XXV, Figs. 1 to 3, Plate XXVI, and Figs. 30 to 34, Plate XXVII, of the present work, represent the chief type-specimens upon which the genus Uintatherium was founded and the species *U. robustum* named. Descriptions of these occur on pages 93 and 96.

The large canine tooth represented in Figs. 1 to 5, Plate XXV, was, on discovery, supposed to belong to a Drepanodon-like carnivore. The discovery of the nearly complete skulls described by Professor Marsh under the name of *Dinoceras mirabilis,* and Professor Cope under the name of *Loxolophodon cornutus,* leaves no doubt that the remarkable tooth belongs to the same kind of an animal, which, from the proportions of the specimen, I suppose to be *Uintatherium robustum.*

The fine skull discovered and described by Professor Cope under the name of *Loxolophodon cornutus,* I had the opportunity of seeing on the occasion when it was exhibited at a meeting of the Academy of Natural Sciences. So far as I could judge from the cursory examination, and from the more recent description and figures of the skull, it appears to me to be a larger species of Uintatherium than the *U. robustum*, but not of a distinct genus.

The remains, which were first noticed by Professor Marsh and referred to *Titanotherium* (!) *anceps*, subsequently to *Mastodon anceps*, and finally to *Tinoceras anceps*, I have not seen. I have suspected that perhaps they might pertain to the same animal as that I have described as *Uintatherium robustum*. Should this prove to be the case, as the specific name of *anceps* is of earliest date, the latter would be correctly designated as *Uintatherium anceps*.

Professor Marsh regards the *Eobasileus* s. *Loxolophodon cornutus*, Cope, as pertaining to Tinoceras, probably *T. grandis*, Marsh, (Am. Jour. Sc. April, 1873.) On the other hand Professor Cope refers Dinoceras to Uintatherium, and also includes as synonyms *Titanotherium* (!) *anceps*, and therefore Tinoceras, Marsh, (Pr. Am. Phil. Soc. 1873.) Thus the conjoint views of these authors rather favor the idea that all are probably of the same genus.

Since the article on *Uintatherium robustum*, page 96, was printed, I have attempted a restoration of the skull in Fig. 1, Plate XXVIII, on an enlarged outline taken from Professor Marsh's Fig. 1, Plate II, of *Dinoceras mirabilis*, published in the American Journal of Science for February, 1873. The cranial fragment and that of the upper jaw with the last molar tooth are taken from the same skull as the specimens of Fig. 8, Plate XXV, and Fig. 1, Plate XXVI. The canine is from the same specimen as Fig. 1, Plate XXV.

In the May number of the American Journal of Science for 1873, Professor Marsh has indicated what he considers to be a new species of Dinoceras with the name of *D. lucaris*. In the account he observes, "From Uintatherium, so far as that genus is at present known, Dinoceras differs in the position of the occipital condyles, in the more anterior position of the posterior horns, and in the last molar, which lacks the external cone between the two transverse ridges, and has a second small tubercle behind the posterior ridge." These characters may, perhaps, together with others more important, point to a different species, but appear hardly sufficient to distinguish a genus. The differences are also more apparent than real; for instance, the so-called "external cone between the two transverse ridges" of the last molar, as seen in Fig. 7, Plate XXV, is nothing more than a tubercle produced from the basal ridge, might be absent in another individual, and is actually so in the molar in advance, as seen in Fig. 12 of the same Plate.

MEGACEROPS* s. *Megaceratops*.

MEGACEROPS COLORADENSIS.
>Leidy: Pr. Ac. Nat. Sc. 1870, 1; Hayden's Rep. Geol. Sur. Wyoming, 1871, 352.
>*Megaceratops coloradoensis*. Cope: Pr. Ac. Nat. Sc. 1873, 102; Pr. Am. Phil. Soc. 1873.

Described page 239, and represented by Figs. 2, 3, Plate I, and Fig. 2, Plate II.

Before the discovery of the more characteristic specimens of the skulls of species of Uintatherium, from the nearer resemblance of the fossil described under the name of Megacerops to the corresponding part of Sivatherium, the animal to which it belonged was supposed to be a ruminant. It now appears probable that Megacerops forms a member of the same order, whatever that may be, with Uintatherium.

Rodentia.

SCIURIDÆ.

PARAMYS.

PARAMYS DELICATUS.
>Leidy: Pr. Ac. Nat. Sc. 1871, 231; Hayden's Rep. Geol. Sur. Montana, 1872, 357.

Described page 110, and represented by Figs. 23 to 25, Plate VI. From the Bridger Eocene of Wyoming.

PARAMYS DELICATIOR.
>Leidy: Pr. Ac. Nat. Sc. 1871, 231; Hayden's Rep. Geol. Sur. Montana, 1872, 357.

Described page 110, and represented by Figs. 26, 27, Plate VI, and Figs. 16 to 18, Plate XXVII. From the Bridger Eocene of Wyoming.

PARAMYS DELICATISSIMUS.
>Leidy: Pr. Ac. Nat. Sc. 1871, 231; Hayden's Rep. Geol. Sur. Montana, 1872, 357.

Described page 111, and represented by Figs. 28, 29, Plate VI. From the Bridger Eocene of Wyoming.

SCIURAVUS?

>Marsh: Am. Jour. Sc. 1871, 122.

A tooth supposed to pertain to this genus is described on page 113, and represented in Fig. 30, Plate VI. From the Bridger Eocene of Wyoming.

---

* For the sake of both brevity and euphony, I have preferred to use Megacerops instead of Megaceratops, just as Megatherium is preferred to Megalotherium, &c.

MURIDÆ. (?)

MYSOPS.

MYSOPS MINIMUS.
  Leidy: Pr. Ac. Nat. Sc. 1871, 232; Hayden's Rep. Geol. Sur. Montana, 1872, 357.

  Described page 111, and represented by Figs. 31, 32, Plate VI. From the Bridger Eocene of Wyoming.

MYSOPS FRATERNUS.

  Described page 112, and represented by Figs. 14, 15, Plate XXVII. From the Bridger Eocene of Wyoming.

Insectivora.

FAMILIES UNDETERMINED.

OMOMYS.

OMOMYS CARTERI.
  Leidy: Pr. Ac. Nat. Sc. 1869, 63.

  Originally described in the Extinct Mammalia of North America, 1869, 408, and represented in Figs. 13, 14, Plate XXIX of the same work. Redescribed in the present work, 120. From the Bridger Eocene of Wyoming.

PALÆACODON.

PALÆACODON VERUS.
  Leidy: Pr. Ac. Nat. Sc. 1872, 21; Hayden's Rep. Geol. Sur. Montana, 1872, 356.

  Described from specimens of teeth page 122, and represented by Fig. 46, Plate VI. From the Bridger Eocene of Wyoming.

WASHAKIUS.

WASHAKIUS INSIGNIS.

  Described page 123 from a small jaw-fragment containing the last two molars, and represented in Figs. 3, 4, Plate XXVII. From the Bridger Eocene of Wyoming.

Sirenia.

MANATUS.

MANATUS INORNATUS.

  Figs. 16, 17, Plate XXXVII, represent the crown of a tooth from the "phosphate beds" of the Ashley River, South Carolina. It most nearly resembles the corresponding part of the lower teeth of the living Manatee of the Florida coast, and it indicates an animal of about the same size. The constituent lobes of the crown are less contracted approaching the summits, and the intervening valleys are wider than in the teeth of the living Manatee. The summits of the lobes being less contracted, are also sharper and not so wrinkled. The summit of the anterior lobe presents a wider and deeper oval pit, and the posterior heel is less mammillary, not wrinkled at the summit, and is broadly sloping at its fore part. The crown measures half an inch four and aft and $4\frac{1}{2}$ lines where widest.

Zeuglodontia.

PONTOBASILEUS.

PONTOBASILEUS TUBERCULATUS.

Fig. 15, Plate XXXVII, represents a fragment of a remarkable tooth, apparently belonging to an animal of the same order as the Basilosaurus. The specimen pertains to the Museum of the Academy of Natural Sciences of Philadelphia. It is without label, and was associated with some Basilosaurus remains from Alabama. I suppose it to have been derived from some Eocene or Miocene formation of the Atlantic States. Upon the fang there are the remains of two white disks, apparently the basal attachment of barnacles.

The fragment consists of the back portion of the crown and the corresponding fang of a double-fanged tooth. The crown has been very unlike that of any known animal of the order. The conical summit occupied a position over the separation of the fangs, including at most the anterior one. The back part of the crown forms a wide, thick heel, extending over more than half the width of the corresponding fang. The enamel is exceedingly tuberculate, and near the most prominent portion of the heel outwardly it is worn off over a small oval space from attrition of an opposed tooth. The fang is widely divergent, and is depressed along the middle externally and internally, and also more deeply on the surface opposed to the absent fang.

Cetacea.

DELPHINIDÆ.

GRAPHIODON.

GRAPHIODON VINEARIUS.

Leidy: Pr. Ac. Nat. Sc. 1870, 122.

An extinct genus and species of cetacean animals, apparently different from any previously described, is indicated by a fossil submitted to my examination by the Smithsonian Institution. The specimen was found by Mr. Pearce in the Miocene formation of Gay Head, Martha's Vineyard. It consists of a tooth represented in Fig. 7, Plate XXII, of the natural size.

The form of the tooth, with its huge gibbous fang, led me at first to mistake it for that of a mosasauroid reptile, nor did I observe my error until it was suggested by Professor Marsh.

The crown of the tooth is curved, conical, and without subdivisional planes upon the surface. The inner and outer surfaces are barely defined posterointernally by a feeble and interrupted ridge. The enamel is singularly wrinkled, the wrinkles being short, vermicular, somewhat branched and crowded, and they remind one of Arabic letters. At the base of the crown the enamel is nearly smooth. The transverse section of the crown is circular, and measures 8 lines in diameter. The length of the crown when complete appears to have been about twice the latter.

The fang of the tooth, broken at the extremity, exposes to view a large interior pulp cavity. It is longer than the crown and very gibbous. In its relation of size and form, it is wonderfully like the corresponding part in the teeth of Mosasaurus. It is ovoidal in form and is curved in the direction of the crown. It is abruptly thickened at the base of the latter, and on one side, near the extremity, exhibits a deep groove. The texture of the fang, as seen at its broken part, appears as dense as ordinary dentine. In the entire condition, the fang has approximated 2 inches in length; its diameter is about half the length.

REPTILIA.

Dinosauria.

POICILOPLEURON.

Deslongchamps: Mem. Soc. Lin. de Normandie VI, 1838, 37.

POICILOPLEURON VALENS.

Leidy: Pr. Ac. Nat. Sc. 1870, 3.
Antrodemus. Leidy: Ibidem, 4.

Founded on several fragments of vertebræ described page 267, and represented by Figs. 16 to 18, Plate XV, under the name of Antrodemus. From Colorado, and supposed to have been derived from the Cretaceous formation.

Chelonia.

TESTUDINIDÆ.

TESTUDO.

TESTUDO CORSONI.
> Leidy: Pr. Ac. Nat. Sc. 1871, 154; 1872, 268; Hayden's Rep. Geol. Sur. Montana, 1872, 366.
>
> *Emys Carteri.* Leidy: Pr. Ac. Nat. Sc. 1871, 228; Hayden's Rep. Geol. Sur. Montana, 1872, 367.

Described page 132, and represented in Figs. 1, 2, Plate XI, under the name of Emys Carteri; in Fig. 7, Plate XV; Figs. 2 to 4, Plate XXIX, and Figs. 1 to 4, Plate XXX. From the Bridger Eocene of Wyoming.

TESTUDO NEBRASCENSIS.
> Leidy: Pr. Ac. Nat. Sc. 1852, 59; Owen's Rep. Geol. Sur. Wisconsin, &c. 1852, 567; Ancient Fauna of Nebraska, 1853, 105, Plate XIX; Ext. Mam. Fauna of Dakota and Nebraska, 1869, 26.
>
> *Stylemys nebrascensis.* Leidy: Pr. Ac. Nat. Sc. 1851, 172; Ancient Fauna of Nebraska, 1853, 103; Ext. Mam. Fauna of Dakota and Nebraska, 1869, 26. See page 223. Cope: Ext. Batrachia, &c. 1870, 124.
>
> *Emys s. Testudo hemispherica, Oweni, Culbertsoni, et lata.* Leidy: Pr. Ac. Nat. Sc. 1851, 173, 327; 1852, 34, 59. Owen's Rep. Geol. Sur. Wisconsin, &c. 1852, 568 to 572; Ancient Fauna of Nebraska, 1853, 105 to 110, Plates XX to XXIV.
>
> *Stylemys Culbertsonii.* Cope: Ext. Batrachia, &c. 1870, 124.

Noticed page 224 under the name of *Stylemys nebrascensis*, and further represented by Figs. 7, 9, 10, Plate XIX.

All the turtle remains from the Mauvaises Terres of White River, Dakota, which have come under my inspection, and which have been described under the various names above indicated, I regard as having pertained to a single species. This agrees so closely in the usual characters of living species of the land tortoises, that I have placed it in the same genus, though it is subgenerically distinct. A mature and nearly perfect specimen of the shell in the Museum of the Academy of Natural Sciences, obtained by Professor Hayden in 1866, has the following dimensions:

	Inches.
Length of carapace in the curve	27
Breadth of carapace in the curve	26
Length of plastron	20
Breadth of plastron	15
Height of shell above the level	8

TESTUDO NIOBRARENSIS.
> *Testudo (Stylemys) niobrarensis.* Leidy: Pr. Ac. Nat. Sc. 1858, 29; Ext. Mam. Fauna Dakota and Nebraska, 1869, 26. See page 224.
> *Stylemys niobrarensis.* Cope: Ext. Batrachia, &c. 1870, 124.

Described page 225, under the name of *Stylemys niobrarensis*, and represented by Figs. 4 to 6, Plate III, and Figs. 6, 8, Plate XIX. From the Pliocene of the Niobrara River.

TESTUDO OREGONENSIS.
> *Stylemys oregonensis.* Leidy: Pr. Ac. Nat. Sc. 1871, 218. See page 225.

Noticed page 226, under the name of *Stylemys oregonensis*, and represented by Fig. 10, Plate XV. From the Miocene of Oregon. I suspect I have been too hasty in regarding this as a species distinct from *Testudo nebrascensis*.

EMYDIDÆ.

EMYS.

EMYS WYOMINGENSIS.
> Leidy: Pr. Ac. Nat. Sc. 1869, 66; Hayden's Rep. Geol. Sur. Montana, 1872, 367.
> *Emys Stevensonianus.* Leidy: Pr. Ac. Nat. Sc. 1870, 5; Hayden's Rep. Geol. Sur. Wyoming, 1871, 366.
> *Emys Jeanesi.* Leidy: Pr. Ac. Nat. Sc. 1870, 123; Hayden's Rep. Geol. Sur. Wyoming, 1871, 366.
> *Emys Haydeni.* Leidy: Pr. Ac. Nat. Sc. 1870, 123; Hayden's Rep. Geol. Sur. Wyoming, 1871, 366.

The species described page 140, and represented by Figs. 2 to 6, Plate IX, Figs. 1, 2, Plate X. From the Bridger Eocene of Wyoming.

EMYS PETROLEI.
> Leidy: Pr. Ac. Nat. Sc. 1868, 176. Cope: Ext. Batrachia, &c. 1870, 128.

Species described page 260, and represented by Fig. 7, Plate IX. From the Quaternary of Texas.

HYBEMYS.

HYBEMYS ARENARIUS.
> Leidy: Pr. Ac. Nat. Sc. 1871, 103.

Noticed page 174, and represented by Fig. 9, Plate XV. From the Bridger Eocene of Wyoming.

FAMILIES UNDETERMINED, APPARENTLY INTERMEDIATE TO THE PLEURODIRIDÆ AND THE CHELYDRIDÆ.

BAPTEMYS.

BAPTEMYS WYOMINGENSIS.

 Leidy: Pr. Ac. Nat. Sc. 1870, 4; Hayden's Rep. Geol. Sur. Wyoming, 1871, 367; Hayden's Rep. Geol. Sur. Montana, 1872, 367.
 Adocus wyomingensis. Cope: Pr. Am. Phil. Soc. 1870, 297; Ext. Batrachia, Reptilia N. Am. in Trans. Am. Phil. Soc. 1870, 233.

Described page 157, and represented by Figs. 1, 2, Plate XII, and Fig. 6, Plate XV. From the Bridger Eocene of Wyoming.

CHISTERNON s. *Chisternum.*

 Leidy: Pr. Ac. Nat. Sc. 1872, 162.

CHISTERNON UNDATUM.

 Baena undata. Leidy: Pr. Ac. Nat. Sc. 1871, 228; Hayden's Rep. Geol. Sur. Montana, 1871, 369.

Described page 169, and represented by Figs. 1, 2, Plate XIV, under the name of *Baena undata.* From the Bridger Eocene of Wyoming.

Chisternon undatum, in the presence of an additional pair of plates to the plastron, resembles the existing Sternothaerus.

BAENA.

BAENA ARENOSA.

 Leidy: Pr. Ac. Nat. Sc. 1870, 123; 1871, 228; Hayden's Rep. Geol. Sur. Wyoming, 1871, 367; Hayden's Rep. Geol. Sur. Montana, 1872, 368.
 Baena affinis. Leidy: Hayden's Rep. Geol. Sur. Wyoming, 1871, 367.

Species described page 161, and represented by Figs. 1 to 3, Plate XIII, under the names of *Baena arenosa* and *Baena affinis,* Figs. 1 to 5, Plate XV, and Figs. 8, 9, Plate XVI.

ANOSTEIRA.

ANOSTEIRA ORNATA.

 Leidy: Pr. Ac. Nat. Sc. 1871, 102, 114; Hayden's Rep. Geol. Sur. Wyoming, 1872, 370.

Described page 174, and represented by Figs. 1 to 6, Plate XVI. From the Bridger Eocene of Wyoming.

TRIONYCHIDÆ.

TRIONYX.

TRIONYX GUTTATUS.

 Leidy: Pr. Ac. Nat. Sc. 1869, 66; 1870, 5; 1871, 228; Hayden's Rep. Geol. Sur. Wyoming, 1871, 367; Hayden's Rep. Geol. Sur. Montana, 1872, 370. Cope: Ext. Batrachia, &c., 1870, 152.

Described page 176, and represented by Fig. 1, Plate IX. From the Bridger Eocene of Wyoming.

TRIONYX UINTAENSIS.
Leidy: Pr. Ac. Nat. Sc. 1872, 267.

Described page 178, and represented by Fig. 1, Plate XXIX. From the Bridger Eocene of Wyoming.

TRIONYX ——— ———?
Fragments described page 180, and represented in Figs. 11, 12, Plate XVI. From the Bridger Eocene of Wyoming.

SPHARGIDIDÆ?

ATLANTOCHELYS.

ATLANTOCHELYS MORTONI.*
Agassiz: Pr. Ac. Nat. Sc. 1849, 169.
Mosasaurus Mitchelli. Leidy: Cret. Rept. in Smith. Contrib. 1865, 43, 116. Determination admitted by Cope: Pr. Bost. Soc. Nat. Hist. 1869, 253.
Protostega neptunia. Cope: Pr. Am. Phil. Soc. 1872, 433.

Founded on the fragment of a large humerus described in the "Cretaceous Reptiles of the United States," 1865, 43, and represented in Figs. 3, 4, 5, Plate VIII of that work. From the Cretaceous green sand of New Jersey. See page 270.

ATLANTOCHELYS TUBEROSUS.
Holcodus acutidens. In part of Leidy: Cret. Rept. in Smiths. Contrib. 1865, 42, 118. Determination admitted by Cope: Pr. Bost. Soc. Nat. Hist. 1869, 253.
Platecarpus tympaniticus. In part of Cope: Pr. Bost. Soc. Nat. Hist. 1869, 265; Synop. Ext. Batrachia, Reptilia, &c. 1870, 199.
Protostega. Cope: Pr. Am. Phil. Soc. 1872, 433.
Platecarpus tuberosus. Cope: Pr. Am. Phil. 1872, 433.
Protostega tuberosa. Cope: Hayden's Rep. Geol. Sur. Montana, 1872, 330, 334.

Founded on a humerus described in the "Cretaceous Reptiles of the United States," 1865, 42, and represented in Figs. 1, 2, Plate VIII of that work. From the Cretaceous formation near Columbus, Mississippi. See page 270.

It was the association of this specimen with several cervical vertebræ, and

---

* Professor Cope observes that "this name was unaccompanied with the necessary description, and is hence useless to science." (Pr. Am. Phil. Soc. 1872, 433.) As the specimen on which it was founded was described and figured in my paper on the Cretaceous Reptiles, so as to be recognized by every student, I have preferred to employ the original name instead of the proposed substitute.

a palatine bone with teeth of an undoubted mosasauroid, that led me into the error of supposing it belonged to the same animal. This suggested the idea that the specimen originally referred to *Atlantochelys Mortoni* likewise belonged to a Mosasaurus. The error was easy at a time when the limb-bones of none of the mosasauroids were known, and when it was even doubted whether these reptiles possessed hinder limbs. My determination was concurred in, not only by Professor Cope, but also by Professor Agassiz, after I had exhibited to him the different specimens and their associates.* It was only after I had had the opportunity of seeing the nearly complete fore-limbs in the skeleton of *Clidastes propython*, described by Professor Cope, that I suspected my reference of the specimens of humeri above indicated was incorrect.

CYNOCERCUS !

CYNOCERCUS INCISUS. ? .
    Cope: Pr. Am. Phil. Soc. 1872, 309.

Remains probably belonging to this species described page 269, and represented by Figs. 17 to 21, Plate XXXVI. From the Cretaceous of Kansas.

Mosasauria.

MOSASAURUS !

MOSASAURUS ———— ?
    See page 279. Represented by Fig. 15, Plate XXXVI. From the Cretaceous of Nebraska.

TYLOSAURUS.

TYLOSAURUS DYSPELOR.
    Marsh: Am. Jour. Sc. 1872, 147.
        Liodon dyspelor. Cope: Pr. Am. Phil. Soc. 1870, 572, 574; 1871, 169, 280; Hayden's Rep. Geol. Sur. Wyoming, 1871, 110; Hayden's Rep. Geol. Sur. Montana, 1872, 333.
        Rhinosaurus dyspelor. Marsh: Am. Jour. Sc. 1872.
        Rhamphosaurus. Cope: Pr. Ac. Nat. Sc. 1872, 141.

See page 271. Represented by Figs. 1 to 11, Plate XXXV. From the Cretaceous of New Mexico and Kansas.

---

* I do not introduce the names of these naturalists as an apology for my error, but rather to show that able authorities are liable to the same mistakes under the same circumstances.

TYLOSAURUS PRORIGER.
>Marsh: Am. Jour. Sc. 1872, 147.
>*Macrosaurus proriger.* Cope: Pr. Ac. Nat. Sc. 1869, 123.
>*Liodon proriger.* Cope: Trans. Am. Phil. Soc. 1870, 202; 1871, 279; Hayden's Rep. Geol. Sur. Wyoming, 1871, 401; Hayden's Rep. Geol. Sur. Montana, 1872, 333.
>*Rhinosaurus proriger.* Marsh: Am. Jour. Sc. 1872.
>*Rhamphosaurus.* Cope: Pr. Ac. Nat. Sc. 1872, 141.

See page 274. Represented by Figs. 12, 13, Plate XXXV, and Figs. 1 to 3, Plate XXXVI. From the Cretaceous of Kansas.

LESTOSAURUS.

LESTOSAURUS CORYPHÆUS.
>Marsh: Am. Jour. Sc. 1872.
>*Holcodus coryphæus.* Cope: Pr. Am. Phil. Soc. 1871, 269; Hayden's Rep. Geol. Sur. Montana, 1872, 331.
>*? Platecarpus.* Cope: Pr. Ac. Nat. Sc. 1872, 141.

See page 276. Represented by Figs. 12 to 14, Plate XXXIV, and Figs. 4 to 14, Plate XXXVI. From the Cretaceous of Kansas.

CLIDASTES.

Cope: Pr. Ac. Nat. Sc. 1868, 233.

CLIDASTES INTERMEDIUS.
>Leidy: Pr. Ac. Nat. Sc. 1870, 4. Cope: Syn. Ext. Batrachia, Reptilia, &c. 1870, 221; Hayden's Rep. Geol. Sur. Wyoming, 1871, 412.

Described page 281, and represented by Figs. 1 to 5, Plate XXXIV. From the Upper Cretaceous of Alabama.

CLIDASTES AFFINIS.
>*C. intermedius.* In part, Leidy: Pr. Ac. Nat. Sc. 1870, 4.

Described page 283, and represented by Figs. 6 to 11, Plate XXXIV. From the Cretaceous of Smoky Hill River, Kansas.

Lacertilia.

SANIWA s. *Sanica.*

SANIWA ENSIDENS.
>Leidy: Pr. Ac. Nat. Sc. 1870, 124; Hayden's Rep. Geol. Sur. Wyoming, 1871, 368; Hayden's Rep. Geol. Sur. Montana, 1872, 370.

Described page 181, and represented by Fig. 15, Plate XV, and Fig. 35, Plate XXVII. From the Bridger Eocene of Wyoming.

SANIWA MAJOR.

Described page 182, and represented by Fig. 14, Plate XV, Figs. 36, 37, Plate XXVII. From the Bridger Eocene of Wyoming.

CHAMELEO.

CHAMELEO PRISTINUS.
Leidy: Pr. Ac. Nat. Sc. 1872, 277.

Described page 184, and represented by Figs. 38, 39, Plate XXVII. From the Bridger Eocene of Wyoming.

GLYPTOSAURUS.

Marsh: Am. Jour. Sc. 1871; Pr. Ac. Nat. Sc. 1871, 105.

GLYPTOSAURUS ————?

Noticed page 182, and represented by Fig. 13 to 17, Plate XVI. From the Bridger Eocene of Wyoming.

TYLOSTEUS.

TYLOSTEUS ORNATUS.
Leidy: Pr. Ac. Nat. Sc. 1872, 40.

Noticed page 285, and represented by Figs. 14, Plate XIX. From the Upper Missouri; probably Cretaceous.

Sauropterygia.

NOTHOSAURUS.

NOTHOSAURUS OCCIDUUS.
Nothosaurops occiduus. Leidy: Pr. Ac. Nat. Sc. 1870, 71.

Noticed page 287, and represented by Figs. 11 to 13, Plate XV. From the Cretaceous? of Moreau River, Dakota.

OLIGOSIMUS.

OLIGOSIMUS GRANDÆVUS.
Leidy: Pr. Ac. Nat. Sc. 1872, 39.

Described page 286, and represented by Figs. 18, 19, Plate XVI. From the Cretaceous (?) of Wyoming.

FISHES.

Teleostei.

LABRIDÆ.

PROTAUTOGA.

PROTAUTOGA CONIDENS.

Tautoga (Protautoga) conidens. Leidy: Pr. Ac. Nat. Sc. 1873, 15; Am. Jour. Sc. 1873, 312.

A short time since, Mr. C. M. Smith, engineer, of Richmond, Virginia, submitted to the writer, for examination, a small collection of fossil bones, which had been discovered by him during the construction of a tunnel beneath the city. Mr. Smith informs me that the material penetrated by the tunnel, in which the bones were found, consists of a stiff blue clay containing remains of infusoria. On examining a portion of the substance with the microscope, I observed an abundance of well-preserved frustules of Coscinodiscus, besides many other less conspicuous diatomes.

The fossil bones consist mainly of vertebræ and teeth of Cetaceans, the teeth of *Procamelus virginiensis*, previously described, a portion of a humerus of a bird, and a number of remains of fishes.

Among the latter there are two specimens which consist of portions of the premaxillaries, with teeth, represented in Figs. 56, 57, Plate XXXII, of a species of Tautoga larger than the living black-fish, *Tautoga americana*.

The better-preserved specimen exhibits the base of attachment of the first large tooth, and succeeding it a row of seven teeth. These are separated by wider intervals than the fewer teeth of the same kind of the recent black-fish. The points of the teeth are more regularly conical than in the latter. Within the position of the larger teeth there is a row of small teeth.

The second specimen contains the first large tooth alone. This tooth is not longer than in the recent black-fish, but is more robust, and its enameloid-covered extremity is more perfectly conical or is less flattened from without inwardly.

The premaxillary bone is flatter externally than in the black-fish, and looks as if it had not turned down in a hook-like end as in the latter. The specimens I have supposed to indicate a genus closely related with Tautoga, and have named it Protautoga.

The more complete specimen contained a row of eight teeth in a space of an inch and a quarter from the symphysis. The first large tooth is 5½ lines long; the crown-like portion is 3½ lines long, with the breadth at base 2½ lines. The second tooth is 4.4 lines long; the crown-like portion is 2 lines long and 1.6 lines in diameter at base. The other teeth range from 2 lines to a line in length.

SPHYRÆNIDÆ.

ENCHODUS.

ENCHODUS SHUMARDI.
Leidy: Pr. Ac. Nat. Sc. 1856, 257.
Described page 289, and represented by Fig. 20, Plate XVII. From the Cretaceous of Sage Creek, Dakota.

PHASGANODUS.

PHASGANODUS DIRUS.
Leidy: Pr. Ac. Nat. Sc. 1857, 167.
Described page 289, and represented by Figs. 23, 24, Plate XVII. From the Cretaceous of Cannon Ball River, Dakota.

CLADOCYCLUS.?

CLADOCYCLUS? OCCIDENTALIS.
Leidy: Pr. Ac. Nat. Sc. 1856, 256.
Noticed page 288, and represented by Figs. 21, 22, Plate XVII, and Fig. 5, Plate XXX. From the Cretaceous of Sage Creek, Dakota.

CLUPEIDÆ.

CLUPEA.

CLUPEA HUMILIS.
Leidy: Pr. Ac. Nat. Sc. 1856, 256.
Page 195, and represented by Fig. 1, Plate XVII. From the Eocene shales of Green River, Wyoming.

CLUPEA ALTA.
Described page 196, and represented by Fig. 2, Plate XVII. From the Eocene shales of Green River, Wyoming.

CYPRINIDÆ.
MYLOCYPRINUS.

MYLOCYPRINUS ROBUSTUS.
Leidy: Pr. Ac. Nat. Sc. 1870, 70.

Described page 262, and represented by Figs. 11 to 17, Plate XVII. From the Pliocene of Idaho.

SILURIDÆ.
PIMELODUS.

PIMELODUS ANTIQUUS.
Leidy: Pr. Ac. Nat. Sc. 1873, 99.

Page 193, and represented by Figs. 44 to 46, Plate XXXII. From the Tertiary of Big Sandy and Green Rivers, Wyoming.

FAMILY UNDETERMINED.
XIPHACTINUS.

XIPHACTINUS AUDAX.
Leidy: Pr. Ac. Nat. Sc. 1870, 12.

Described page 290, and represented by Figs. 9, 10, Plate XVII. From the Cretaceous of Smoky Hill River, Kansas, and L'eau qui Court County, Nebraska.

GANOIDEI.
Cycloganoidei.
AMIA.

AMIA UINTAENSIS.
Amia (Protamia) uintaensis. Leidy: Pr. Ac. Nat. Sc. 1873, 98.

Page 185, and represented by Figs. 1 to 6, Plate XXXII. From the Bridger Eocene of Wyoming.

AMIA MEDIA.
Amia (Protamia) media. Leidy: Pr. Ac. Nat. Sc. 1873, 98.

Page 188, and represented by Figs. 7 to 11, Plate XXXII. From the Bridger Eocene of Wyoming.

AMIA GRACILIS.
Amia (Protamia) gracilis. Leidy: Pr. Ac. Nat. Sc. 1873, 98.

Page 188, and represented by Figs. 23, 24, Plate XXXII. From the Bridger Eocene of Wyoming.

HYPAMIA.

HYPAMIA ELEGANS.
Leidy: Pr. Ac. Nat. Sc. 1873, 98.
Page 189, and represented by Figs. 19 to 22, Plate XXXII. From the Bridger Eocene of Wyoming.

FAMILY UNDETERMINED.

PHAREODUS.

PHAREODUS ACUTUS.
Leidy: Pr. Ac. Nat. Sc. 1873, 99.
Page 193, and represented by Figs. 47 to 51, Plate 193. From the Bridger Eocene of Wyoming.

Rhomboganoidei.

LEPIDOSTEUS.

LEPIDOSTEUS ATROX.
Leidy: Pr. Ac. Nat. Sc. 1873, 97.
Page 189, and represented by Figs. 14, 15, Plate XXXII. From the Bridger Eocene of Wyoming.

LEPIDOSTEUS ————?
Leidy: Pr. Ac. Nat. Sc. 1873, 98.
Page 190, and represented by Figs. 16, 17, 25, 27 to 30, Plate XXXII. From the Bridger Eocene of Wyoming.

LEPIDOSTEUS SIMPLEX.
Leidy: Pr. Ac. Nat. Sc. 1873, 98.
Page 191, and represented by Figs. 18, 26, 31 to 45, Plate XXXII. From the Bridger Eocene of Wyoming.

LEPIDOSTEUS NOTABILIS.
Leidy: Pr. Ac. Nat. Sc. 1873, 98.
Page 192, and represented by Figs. 12, 13, Plate XXXII. From the Bridger Eocene of Wyoming.

PYCNODUS.

PYCNODUS FABA.
Leidy: Pr. Ac. Nat. Sc. 1872, 163.
Described page 292, and represented by Figs. 15, 16 Plate XIX. From the Cretaceous of Mississippi and New Jersey.

Pycnodus robustus.

Leidy: Pr. Ac. Nat. Sc. 1857, 168.

Noticed page 293, and represented by Figs. 18, 19, Plate XXXVII. From the Cretaceous of New Jersey.

Pycnodus carolinensis.

Emmons: North Carolina Geol. Sur. 1858, 244, Fig. 96.

Noticed page 294. From the Miocene of North Carolina.

HADRODUS.

Hadrodus priscus.

Leidy: Pr. Ac. Nat. Sc. 1857, 167.

Described page 294, and represented by Figs. 17 to 20, Plate XIX. From the Cretaceous of Mississippi. Specimen discovered by Dr. William Spillman.

Since the determination of the reptilian character of the genus Placodus, I have suspected that this one may also belong to the same order.

Placoganoidei.

ACIPENSER.

Acipenser ornatus.

Leidy: Pr. Ac. Nat. Sc. 1873, 15; Am. Jour. Sc. 1873, 312.

Among the fossils in Mr. C. M. Smith's collection from the Miocene formation of Virginia, previously mentioned, there is a dermal plate of a sturgeon, especially interesting on account of the rarity of the remains of fishes of the same family.

The specimen is represented of the natural size in Fig. 58, Plate XXXII, and is nearly entire. It appears to have been one of the lateral plates, and indicates a species about the size of our common sturgeon of the Delaware River. Though exhibiting no positive distinctive character, it most probably pertained to a species now extinct.

ELASMOBRANCHI.

Holocephali.

Edaphodontidæ.

EDAPHODON.

Edaphodon mirificus.

Leidy: Pr. Ac. Nat. Sc. 1856, 221.

Described page 306, and represented by Figs. 6 to 12, Plate XXXVII. From the Cretaceous of New Jersey.

EUMYLODUS.

EUMYLODUS LAQUEATUS.

Described page 309, and represented by Figs. 21, 22, Plate XIX, and Figs. 13, 14, Plate XXXVII. From the Cretaceous of Mississippi.

Plagiostomi.

SQUALIDÆ.

LAMNA.

LAMNA ——— ?

Described page 304, and represented by Figs. 44, 45, Plate XVIII. From the Cretaceous of Kansas and the Chalk of England.

LAMNA ——— ?

Described page 304, and represented by Figs. 46 to 50, Plate XVIII. From the Cretaceous of New Jersey, Mississippi, and Kansas.

OTODUS.

OTODUS DIVARICATUS.

Leidy: Pr. Ac. Nat. Sc. 1872, 162.

Described page 305, and represented by Figs. 26 to 28, Plate XVIII. From the Cretaceous of Mississippi.

OXYRHINA.

OXYRHINA EXTENTA.

Leidy: Pr. Ac. Nat. Sc. 1872, 162.

Described page 302, and represented by Figs. 21 to 25, Plate XVIII. From the Cretaceous of Kansas and Mississippi.

GALEOCERDO.

GALEOCERDO FALCATUS.

Described page 301, and represented by Figs. 29 to 43, Plate XVIII. From the Cretaceous of Kansas, Mississippi, Texas, and England.

HYBODONTIDÆ.

CLADODUS.

CLADODUS OCCIDENTALIS.

Leidy: Pr. Ac. Nat. Sc. 1859, 3.

Cladodus mortifer. Newberry and Worthen: Geol. Sur. Illinois, vol. ii, Palæontology 22, Plate I, Fig. 5. St. John: Hayden's Rep. Geol. Sur. Nebraska, 1872, 239, Plate III, Fig. 6; Plate VI, Fig. 13.

Described page 311, and represented by Figs. 4 to 6, Plate XVII. From the Carboniferous formation of Kansas, Nebraska, and Illinois.

CESTRACIONTIDÆ.

ACRODUS.

ACRODUS HUMILIS.
Leidy: Pr. Ac. Nat. Sc. 1872, 163.

Described page 300, and represented by Fig. 5, Plate XXXVII. From the Cretaceous limestone of New Jersey.

ACRODUS EMMONSI.
Leidy: Pr. Ac. Nat. Sc. 1872, 163.
Acrodus. Emmons: North Carolina Geol. Sur. 1858, 244, Fig. 97.

Attributed by Professor Emmons to the Miocene of North Carolina.

PTYCHODUS.

PTYCHODUS MORTONI.
Agassiz: Poissons Fossiles III, 1833–'43, 158, Tab. 25, Figs. 1 to 3; copied in Figs. 773, 773a, of Dana's Manual of Geology. Leidy: Pr. Ac. Nat. Sc. 1868, 205.
Palate-bone of a fish? Morton: Syn. Org. Rem. Cret. Group, 1834, Plate XVIII, Figs. 1, 2.

Described page 295, and represented by Figs. 1 to 14, Plate XVIII. From the Cretaceous of Kansas, Mississippi, and Alabama.

PTYCHODUS OCCIDENTALIS.
Leidy: Pr. Ac. Nat. Sc. 1868, 207.

Described page 298, and represented by Figs. 7, 8, Plate XVII, and Figs. 15 to 18, Plate XVIII. From the Cretaceous of Kansas.

PTYCHODUS WHIPPLEYI.
Marcou: Geology North America, 1858, 33, Plate I, Fig. 4.

Described page 300, and represented by Figs. 19, 20, Plate XVIII. From the Cretaceous of Texas.

PTYCHODUS POLYGYRUS.
Agassiz: Poissons Fossiles III, 1833–'43, 156. Dixon: Geol. Sussex, 1850, 363. Gibbes: Jour. Ac. Nat. Sc. 1849, 299, Plate XLI, Figs. 5, 6. Leidy: Pr. Ac. Nat. Sc. 1868, 208.

From the Cretaceous of Alabama.

PETALODUS.

PETALODUS ALLEGHANIENSIS.
> Leidy: Jour. Ac. Nat. Sc. 1856, 161, Plate XVI, Figs. 4 to 6; Pr. Ac. Nat. Sc. 1859, 3.
> *Sicarius extinctus.* Leidy: Pr. Ac. Nat. Sc. 1855, 414.
> *Petalodus destructor.* Newberry and Worthen: Geol. Sur. Illinois, vol. ii, Palæontology 35, Plate II, Figs. 1 to 3. St. John: Hayden's Rep. Geol. Sur. Nebraska, 1872, 241, Plate III, Fig. 5.

Described page 312, and represented by Fig. 3, Plate XVII. From the Carboniferous formation of Kansas, Nebraska, Iowa, Illinois, and Indiana.

Ichthyodorulites.

XYSTRACANTHUS.

XYSTRACANTHUS ARCUATUS.
> Leidy: Pr. Ac. Nat. Sc. 1859, 3.

Page 312, and represented by Fig. 25, Plate XVII. From the Carboniferous formation of Kansas.

ASTERACANTHUS.

ASTERACANTHUS SIDERIUS.
> Leidy: Pr. Ac. Nat. Sc. 1870, 13.

Described page 313, and represented by Fig. 59, Plate XXXII. From the sub-Carboniferous formation of Tennessee.

RAIE.

ONCOBATIS.

ONCOBATIS PENTAGONUS.
> Leidy: Pr. Ac. Nat. Sc. 1870, 70.

Page 264, and represented by Figs. 18, 19, Plate XVII. From the Pliocene of Sinker Creek, Idaho.

TRYGON.

TRYGON ———.

Indicated by the basal portion of a caudal spine, represented in Figs. 54, 55, Plate XXXII. It resembles the corresponding portion of the caudal spines of our common whip-sting ray, *Pastinaca hastata*, and would appear to have pertained to a species of about the same size. The anterior, shining, enameloid surface is strongly wrinkled longitudinally, and the lateral denticles are directed downward.

From the Miocene formation of Virginia. Specimen discovered by Mr. C. M. Smith in the blue clay beneath the city of Richmond.

MYLIOBATES.

MYLIOBATES ———.

Indicated by the basal portion of a caudal spine, represented in Figs. 52, 53, Plate XXXII. In its relation of breadth to length, in comparison with the spines of ordinary rays, it would appear in the complete condition to have been upward of 8 inches in length. The specimen, however, becomes rather more abruptly narrowed at its upper broken extremity than appears in ordinary spines, so that it may have been proportionately shorter than usual.

The transverse section has almost the Greek ε form. In front the spine is concave along the middle and convex at the sides; behind it has the reverse arrangement. The lateral denticles are directed downward and backward. The anterior enameloid surface is strongly wrinkled along the middle groove, but not so much at the sides, except at the base of the spine. The posterior surface is moderately ridged.

Specimen found with the preceding in the blue clay of the Miocene formation of Virginia. From Mr. C. M. Smith.

INDEX.

[Synonyms are in *italic*.]

A.

	Page.
Acanthopteri	266
Acipenser ornatus	350
Aerodus	300, 352
Emmonsi	301, 352
humilis	300, 352
Adocus wyomingensis	341
Agriochoerus antiquus	216, 319
latifrons	216, 319
Amia	185, 348
gracilis	188, 348
media	188, 348
uintaensis	185, 348
Anchippodus	328
riparius	72, 328
vetulus	329
Anchitheriidæ	322
Anchitherium	218, 250, 322, 323
agreste	251, 323
australe	250, 323
Bairdi	218, 252, 322
Condoni	218
Anosteira ornata	174, 341
Anthracotheriidæ	320
Antrodemus	338
Apatornis	266
Archæotherium	320
Artiodactyla	216, 317
Asineops squamifrons	195
viridensis	195
Asteracanthus siderius	313, 353
Atlantochelys Mortoni	269, 342
tuberosus	342
Auchenia	255, 317
californica	255
hesterna	255, 317

B.

Baëna	160, 341
affinis	161, 341
arenosa	161, 341
undata	162, 341
Baptemys	154, 341
wyomingensis	157, 341
Bison latifrons	253, 318
Bovidæ	318

C.

Canidæ	315
Canis indianensis	230, 315
Canis *primævus*	315
vafer	315
Camelidæ	317
Carnivora	114, 227, 315
Cervidæ	317
Cestraciontidæ	352
Cetacea	337
Chamæleo pristinus	184, 345
Chelonia	132, 223, 260, 269, 339
Chisternon undatum	162, 341
Chisternum	341
Cladocyclus occidentalis	288, 347
Clastes occidentalis	311, 354
mortifer	312, 354
Clidastes	284, 344
affinis	283, 344
intermedius	284, 344
Clupea	195, 347
alta	195, 347
humilis	195, 347
pusilla	195
Clupeidæ	347
Corax heterodon	304
Crocodilia	125
Crocodilus	125
aptus	126
Elliotti	143
Cycloganoidei	348
Cynodictis	313
lacustris	263, 313
Cyprinidæ	262, 348

D.

Delphinidæ	337
Dicotyles	216, 319
hesperius	217
pristinus	216, 319
Dinoceras	232
lacustris	95
lucaris	331
mirabilis	95, 108*, 332, 333
Dinosauria	267, 338

E.

Elaphodon mirificus	306, 350
Elaphodontidæ	350
Elasmobranchi	235, 350
Elephas	238, 329
americanus	238, 329
Columbi	238, 329

	Page.
Elephas imperator	329
Texanus	329
Elotherium	124, 217, 320
imperator	217, 320
ingens	320
lentis	124
Mortoni	125, 320
superbum	218
Emys	140, 260
Carteri	137, 339
Haydeni	140, 145, 340
Jeanesi	140, 143, 340
petrolei	260, 340
Stevensonianus	140, 141, 340
wyomingensis	140, 141, 340
Euchodus Shumardi	289, 347
Entelodon	320
Eobasileus	332, 334
Equidae	321
Equus	242, 321
complicatus	244, 321
excelsus	243, 322
major	241, 321
occidentalis	242, 321
pacificus	322
parvulus	252, 323
Erismatopterus levatus	195
Rickseckeri	195
Eucrotaphus	212
Euelephas Columbi	330
Jacksoni	330
Eumylodus laqueatus	309, 351

F.

Felidae	315
Felis angustus	227, 315
imperialis	228, 315
Fishes	184, 261, 288, 346

G.

Galeocerdo falcatus	304, 351
Ganoidei	292, 348
Glyptosaurus	182, 345
ocellatus	183
Graphiodon vinearius	337

H.

Hadrodus priscus	294, 350
Hadrohyus supremus	222, 321
Hipparion speciosum	247, 322
Hippsyus formosus	91, 321
robustior	93, 321
Holodus acutidens	342
coryphaeus	276, 344
Holocephali	306, 350
Hybemys arenarius	174, 340
Hybodontidae	351
Hyopsodus	75, 320
minusculus	81, 320
paulus	75, 320

	Page.
Hypamia elegans	189, 349
Hyrachyus	59, 327
agrarius	66, 327
agrestis	66, 327
eximius	65, 327
modestus	67, 327
nanus	67, 327

I.

Ichthyodorulites	353
Ichthyornis	266
Insectivora	120, 336

L.

Labridae	346
Lacertilia	160, 285, 344
Lamna	303, 351
cuspidata	305
elegans	305
Texana	305
Lefalofodon	332
Lepidosteus	189, 190, 349
atrox	189, 349
notabilis	192, 349
simplex	191, 349
Leptomeryx Evansi	216, 317
Lestosaurus coryphaeus	276, 344
Limnohyus	57, 323
laevidens	323, 325
laticeps	326
robustus	326
Limnotherium elegans	84, 320
Limnotherium tyrannus	83, 89
Liodon dyspelor	271, 313
proriger	344
Lophiodon	219, 327
Bairdianus	60, 327
modestus	67, 327
nanus	68, 327
occidentalis	220, 327
parisiense	98
Lophiotherium	69, 327
Ballardi	71
sylvaticum	69, 327
Loxolophodon	332, 333
Lutra	230, 316
piscinaria	316

M.

Macrosaurus proriger	340
Malacopteri	294
Mammalia	27, 199, 211, 227, 315
Manatus inornatus	336
Mastodon	231, 330
americanus	237, 330
anceps	94, 331
mirificus	237, 330
obscurus	231, 330
Shepardi	235, 330

	Page		Page
Megacerops	259, 335	Palæosyops	27, 323
coloradensis	230, 335	humilis	58, 326
Megaceratops coloradoensis	335	junius	57, 326
Megalomenyx	260	junior	326
niobrarensis	260, 317	laticeps	325, 326
Merychippus	248, 322	major	45, 326
mirabilis	250, 322	minor	72, 328
Merychyus	202	paludosus	28, 325
elegans	201	Palancheuia magna	255
major	201	Paramys	109, 335
medius	201	delicatior	110, 335
Merycodus necatus	318	delicatissimus	111, 335
Merycochœrus	199, 202, 208, 319	delicatus	110, 335
proprius	201	Patriofelis ulta	114, 316
rusticus	199, 319	Perissodactyla	27, 219, 321
Miacis	316	Petalodus alleghaniensis	312, 353
Microsus cuspidatus	81, 322	destructor	313, 353
Microsyops	82, 320	Phareodus acutus	193, 349
elegans	84, 320	Phasganodus dirus	289, 347
gracilis	83, 320	Pimelodus antiquus	193, 348
Mosasauria	270, 343	Placoganoidei	350
Mosasaurus	279, 343	Plagiostomi	295, 311, 351
Mitchelli	342	Platecarpus	342, 344
Muridæ	336	tuberosus	342, 344
Mustelidæ	316	tympaniticus	342
Myliobates	353	Platygonus Coulderi	217
Mylocyprinus robustus	262, 348	Pœcilopleuron Bucklandi	268
Mysops	111, 335	valens	267, 338
fraternus	112, 336	Polycotylus latipinnis	279
minimus	111, 336	Pontobasileus tuberculatus	337
		Proboscidea	93, 231, 329
N.		Procamelus occidentalis	258, 317
Notharctus robustior	93, 321	niobrarensis	317
Notharctus tenebrosus	86, 320	robustus	258, 317
Nothosaurops occidens	287, 345	virginiensis	259, 317
Nothosaurus occidens	287, 345	*Protania*	185, 348
		Protautoga conidens	346
O.		*Protocamelus*	317
Omnivora	319	Protohippus	248, 322
Omomys Carteri	120, 336	perditus	249, 250, 322
Oncobatis pentagonus	264, 353	placidus	248, 322
Oligosimus grandævus	280, 345	Protostega gigas	289
Oreodontidæ	318	neptunia	342
Oreodon	201, 211, 318	tuberosa	260, 342
affinis	212	Ptychodus Mortoni	295, 352
bullatus	212, 318	occidentalis	298, 352
Culbertsoni	211, 318	polygyrus	352
gracilis	211	Whippleyi	300, 352
hybridus	212	Pycnodus	294, 349
major	211	carolinensis	294, 350
occidentalis	318	faba	294, 349
superbus	211, 319	robustus	294, 350
Osteoglossum encaustum	195		
Otodus divaricatus	305, 351	**R.**	
Osyrhina	302, 303, 351	Raia	264, 353
extenta	302, 351	Reptilia	125, 338
		Reptiles	267
P.		Rhamphosaurus	271, 343, 344
Palæocolon	122, 336	Rhinocerus	220, 328
verus	122, 336	annectans	223

	Page
Rhinoceros hesperius	230, 328
occidentalis	328
oregonensis	328
pacificus	231, 328
Rhinosaurus	271
dyspelor	343
proriger	344
Rhombogauoidei	349
Rhynchotherium	237, 330
Rodentia	109, 335
Ruminantia	199, 211, 253, 317

S.

Sauria	181, 344
Saniwa ensidens	181, 344
major	181, 345
Sauropterygia	286, 345
Sciuravus	113, 335
nitidus	113
undans	113
Sciuridæ	335
Sicarius extinctus	353
Siluridæ	290, 348
Sinopa	116, 316
eximia	118, 316
rapax	116, 316
Sirenia	336
Solidungula	208, 218, 242, 321
Sphargididæ	342
Sphyrænidæ	288, 347
Squalidæ	351
Stylemys	223
Culbertsoni	339
oregonensis	226, 340
nebrascensis	224, 339
niobrarensis	225, 340
Suidæ	319

T.

Tautoga	346
Teleostei	288, 346
Telmatherium	323
Testudinidæ	339
Testudo Corsoni	132, 339

	Page
Testudo *Culbertsoni*	339
hemisphærica	339
lata	339
nebrascensis	339
niobrarensis	340
oregonensis	340
Tetralophodon mirificus	330
Tylosaurus dyspelor	271, 343
proriger	274, 344
Tylosteus ornatus	285, 345
Tinoceras	331
anceps	94, 334
grandis	94, 334
Titanotherium anceps	94, 334
Trincodon falax	123
Trionychidæ	341
Trionyx	176, 180, 341
guttatus	176, 341
uintaensis	178, 342
Trogosus castoridens	71, 328
vetulus	75, 329
Trygon	353

U.

Uintacyon edax	118, 316
vorax	120, 316
Uintamastix	334
atrox	94, 107, 333
Uintatherium	93, 331
anceps	334
mirabile	333
robustum	93, 96, 333

V.

Vulpavus palustris	118

W.

Washakius insignis	123, 336

X.

Xiphactinus audax	290, 348
Xystracanthus arcuatus	312, 353

Z.

Zeuglodontia	337

EXPLANATION OF PLATE I.

Fig. 1. OREODON SUPERBUS:

    A side view of a skull, with the base of the cranium invested in the matrix. Specimen obtained by Rev. Thomas Condon on John Day's River, Oregon. One-half the natural size.

Figs. 2, 3. MEGACEROPS COLORADENSIS:

    Fig. 2. Upper view of the nasal extremity of the face with a pair of horn-cores. One-half the natural size.

    Fig. 3. Front view of the same specimen

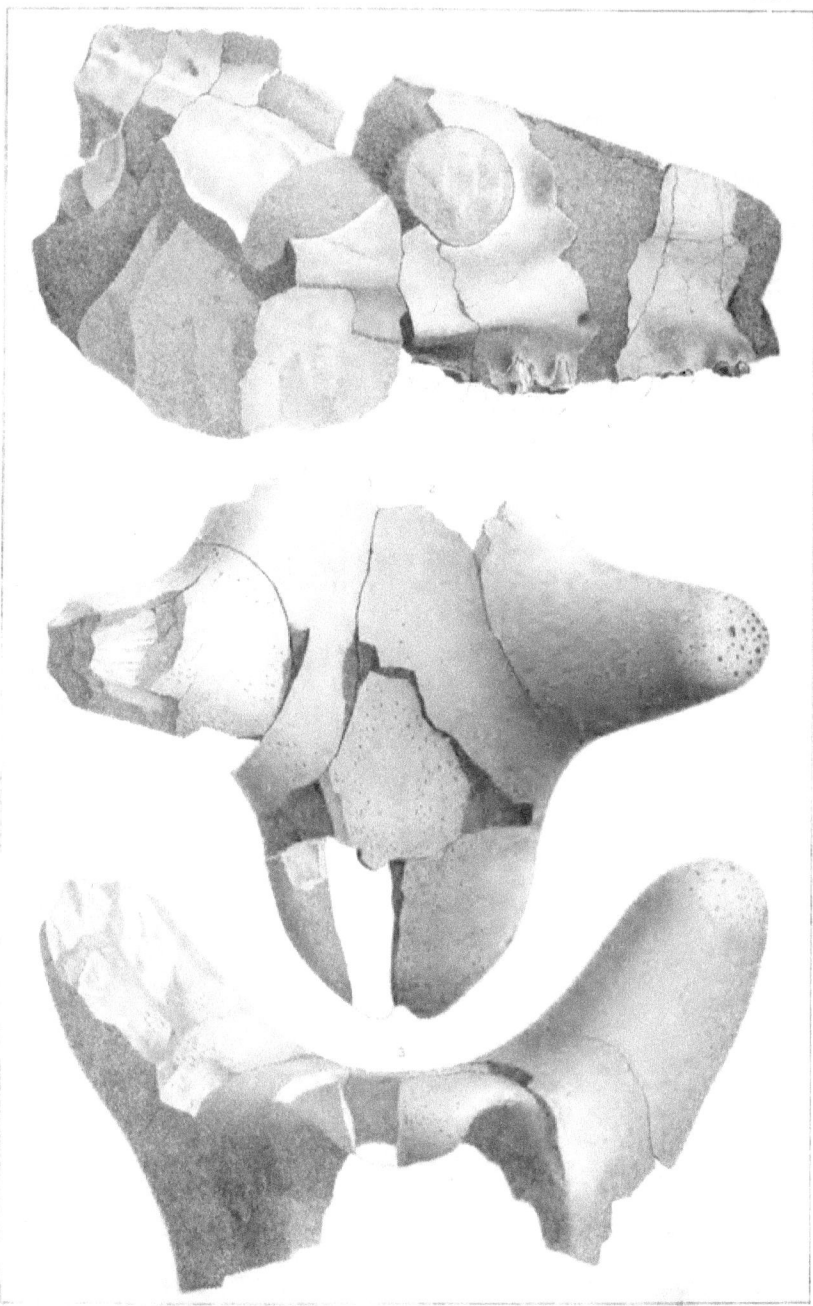

1. TELEODON SUPERBUS ½   2. 3. MEGACEROPS COLORADENSIS ½

EXPLANATION OF PLATE II.

Fig. 1. LOPHIODON OREGONENSIS:
    Two upper molar teeth, much worn and seen on their triturating surfaces. Specimen from Bridge Creek, Oregon. Natural size.

Fig. 2. MEGACEROPS COLORADENSIS:
    Side view of the same specimen as that of figures 2, 3, of Plate I. One-half the natural size.

Figs. 3, 4. ELOTHERIUM SUPERBUM:
    Fig. 3. Portion of a lower canine tooth, natural size. From Bridge Creek, Oregon.
    Fig. 4. Crown of an anterior premolar, natural size. From John Day's River, Oregon.

Fig. 5. ANCHITHERIUM CONDONI:
    A mutilated upper molar tooth, natural size. From Oregon.

Figs. 6, 7. RHINOCEROS PACIFICUS:
    Fig. 6. An upper molar seen on the triturating surface, natural size. From Alkali Flats, Oregon.
    Fig. 7. An upper last premolar, seen on the triturating surface, natural size. From Alkali Flats, Oregon.

Figs. 8, 9. RHINOCEROS HESPERIUS (?):
    Fig. 8. An upper last molar, seen on the triturating surface, natural size. From the Condon collection of Oregon.
    Fig. 9. An inferior molar, seen on the triturating surface.

Fig. 10. PATRIOFELIS ULTA:
    Portion of the right ramus of the lower jaw, half the natural size. It contains the remains of five teeth behind the position of the canine. From near Fort Bridger, Wyoming. See page 114.

Figs. 11, 12. HYRACHYUS AGRARIUS:
    Fig. 11. Left ramus of the lower jaw, one-half the natural size. Specimen obtained by Professor Hayden on Smith's Fork of Green River, Wyoming.
    Fig. 12. Portion of the left ramus of the lower jaw of a young animal, natural size. It contains the temporary series of teeth behind which the first of the true molars is inclosed within the jaw. From Black's Fork of Green River. Hayden's collection.

Fig. 13. HYRACHYUS MODESTUS:
    A first or second upper molar of the left side, slightly larger than natural. From Smith's Fork of Green River. Hayden's collection.

Fig. 14. HYRACHYUS NANUS:
    Portion of left ramus of the lower jaw, with two premolars and the three molars, natural size. Obtained by Dr. Joseph K. Corson from Grizzly Buttes.

Fig. 15. Diseased calcaneum (hyperostosis) of MERYCOCHOERUS RUSTICUS. From Sweetwater River. Hayden's collection of 1870.

Fig. 16. OREODON SUPERBUS:
    Portion of right ramus of lower jaw, with the three premolars and first molar; natural size. Condon collection of Oregon fossils.

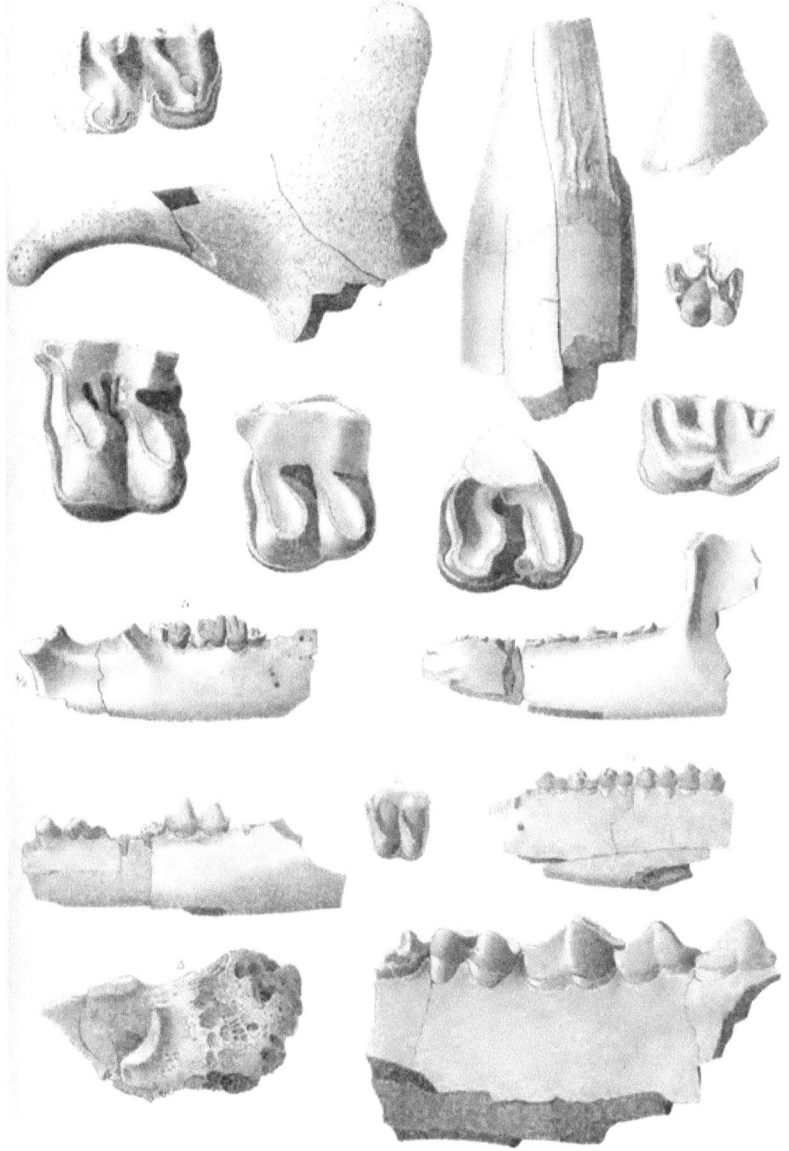

EXPLANATION OF PLATE III.

Figs. 1–3. MERYCOCHŒRUS RUSTICUS. Figures of the natural size. Specimens from Sweetwater River, Wyoming. Hayden's collection of 1870.

Fig. 1. Upper jaw, with a nearly complete series of teeth, the last molar introduced from another specimen.
Fig. 2. Front view of the same specimen, exhibiting the high alveolar border and the narrow nasal orifice.
Fig. 3. Lower jaw, with a full series of molar teeth.

Fig. 4–6. TESTUDO OR STYLEMYS NIOBRARENSIS. Figures of the natural size, except Figure 6, which is one-half the size of nature. From the Niobrara River. Hayden's collection of 1857.

Fig. 4. Internal view of the fore-part of the plastron.
Fig. 5. The last vertebral and the pygal plates.
Fig. 6. Internal view of a posterior portion of the carapace, exhibiting the costal capitula, and the processes for conjunction with the pelvic girdle.

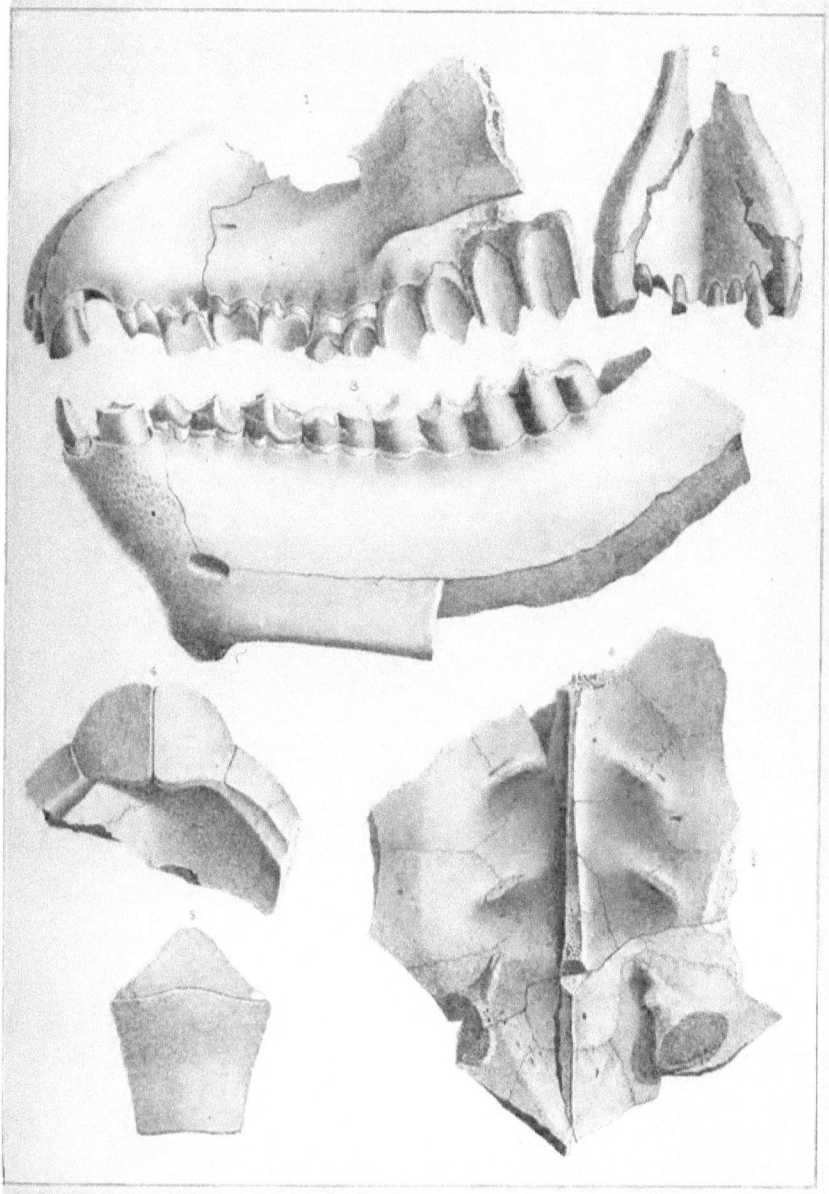

1, 2. MERYCOCHOERUS RUSTICUS. 4–6. PTYLEMYS NIOBRARENSIS.

EXPLANATION OF PLATE IV.

Figures all of the natural size. Specimens all from the Bridger tertiary formation of Wyoming.

Figs. 1–8. PALÆOSYOPS PALUDOSUS:

Fig. 1. A mutilated upper canine of the supposed female, from the same individual as the specimens of figures 5–8.
Fig. 2. Mutilated canine of the supposed male, from the specimen of the following.
Fig. 3. A complete series of molar teeth and the mutilated canine of the left side of a fine specimen discovered at Grizzly Buttes by Dr. J. Van A. Carter. View of the triturating surfaces, partially worn, of the molar teeth; from a supposed male.
Fig. 4. Outer view of the crowns of the same molar series.
Fig. 5. A complete series of molar teeth, discovered by Dr. Carter on Henry's Fork of Green River. View of the triturating surfaces; more worn than in the preceding specimen. From a supposed female.
Fig. 6. Outer view of the anterior two premolars of the same specimen.
Fig. 7. A third upper premolar, left side. Specimen from Henry's Fork. Hayden's collection of 1870.
Fig. 8. Lateral view of an upper incisor. Specimen probably from the same individual as that of Fig. 5.

Figs. 9–18. HYRACHYUS AGRARIUS:

Fig. 9. Outer view of the crowns of an upper series of molar teeth.
Fig. 10. View of the triturating surfaces of the same teeth. From a specimen discovered by Dr. Carter near the Lodge-pole trail, eleven miles from Fort Bridger. All the teeth considerably worn.
Fig. 11. An upper second true molar, left side. Found by Dr. Carter on Henry's Fork of Green River.
Fig. 12. An upper last premolar, left side, but little worn. Specimen found by Dr. Joseph K. Corson at Grizzly Buttes.
Fig. 13. A portion of the lower jaw; from the same individual as Figs. 9, 10. It contains part of the lateral incisor, the canine, and the premolars.
Fig. 14. View of the triturating surfaces of the premolars, from the same specimen.
Fig. 15. Outer view of a second lower molar, from the same individual.
Fig. 16. Triturating surface of the same specimen.
Fig. 17. An upper canine, found at Grizzly Buttes by Dr. Corson.
Fig. 18. A lower incisor, from the same individual as Fig. 13.

Figs. 19, 20. HYRACHYUS EXIMIUS. Specimen found by Dr. Carter on Henry's Fork of Green River.

Fig. 19. Fragment of the left side of the lower jaw, containing the last premolar and the greater part of the first molar.
Fig. 20. View of the triturating surfaces, much worn, of the same teeth.

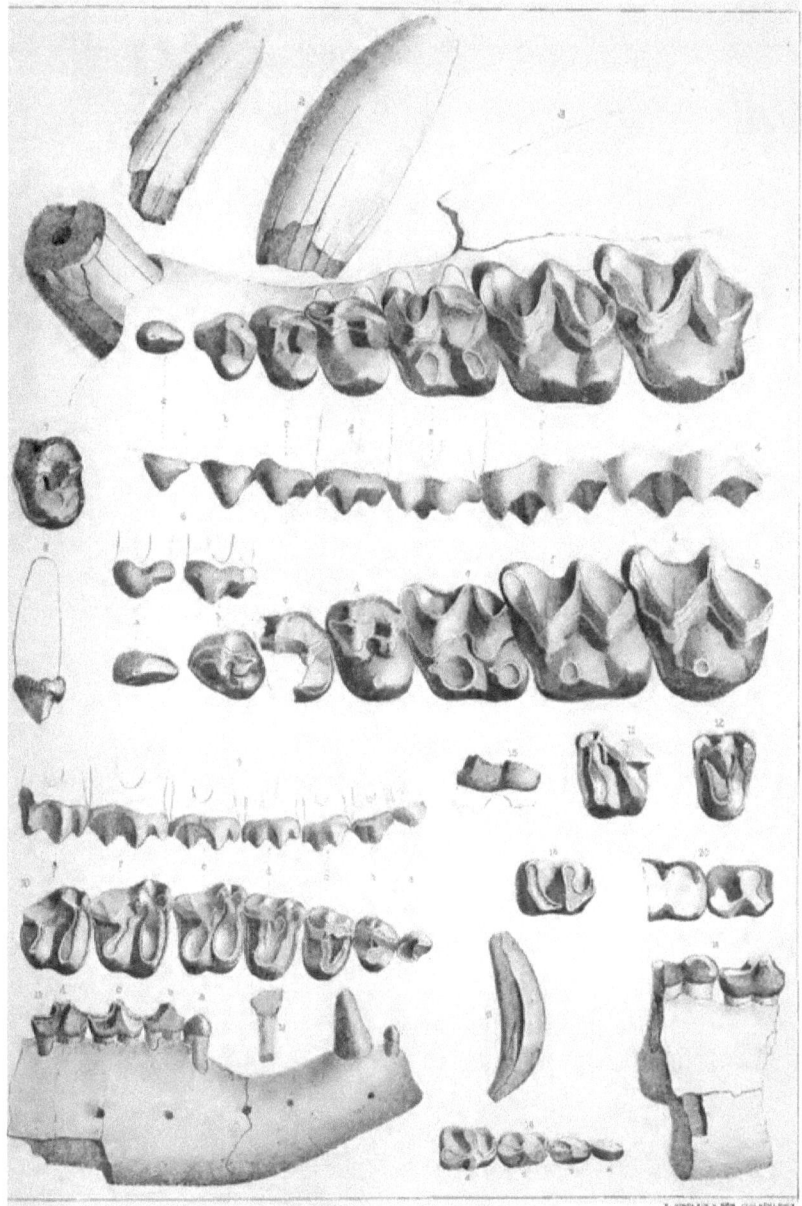

1-8 PALAEOSYOPS PALUDOSUS   9-18 HYRACHYUS AGRARIUS
19-21 HYRACHYUS EXIMIUS

EXPLANATION OF PLATE V.

All the figures of the natural size except Fig. 11, which is one-half the size.

Figs. 1–3. TROGOSUS CASTORIDENS. A lower jaw, discovered in the vicinity of Fort Bridger by Dr. Carter.
    Fig. 1. View of the left ramus of the jaw.
    Fig. 2. Triturating surface of the second true molar, much worn. The other molars are too much injured to be characteristic.
    Fig. 3. Front view of the jaw, exhibiting the large rodent-like incisors.

Figs. 4–11. PALÆOSYOPS PALUDOSUS:
    Fig. 4. An upper last premolar, the triturating surface much worn. From Henry's Fork. Hayden's collection.
    Fig. 5. An upper last premolar, nearly unworn. This is one of the original specimens upon which the genus and species were established. From Church Buttes. Hayden's collection of 1870.
    Fig. 6. Outer view of a last upper molar, left side. Henry's Fork. Hayden's collection of 1870.
    Fig. 7. Triturating surface of the same specimen; the outer fore-part much fissured, with the portions displaced and the single inner lobe partially broken away.
    Fig. 8. Outer view of a second upper molar, from the opposite side of the same individual.
    Fig. 9. The triturating surface, with the outer lobes much worn. Figs. 6–9 are from specimens, which were attributed to the same species at the time of the original notice of it in the Proceedings of the Academy of Natural Sciences, Philadelphia, 1870, p. 113.
    Fig. 10. View of the triturating surfaces of the last two premolars and the molars from the specimen represented in the next figure.
    Fig. 11. Left ramus of a lower jaw, containing the teeth just indicated. This fine specimen was discovered by Dr. Carter thirteen miles southeast of Fort Bridger.

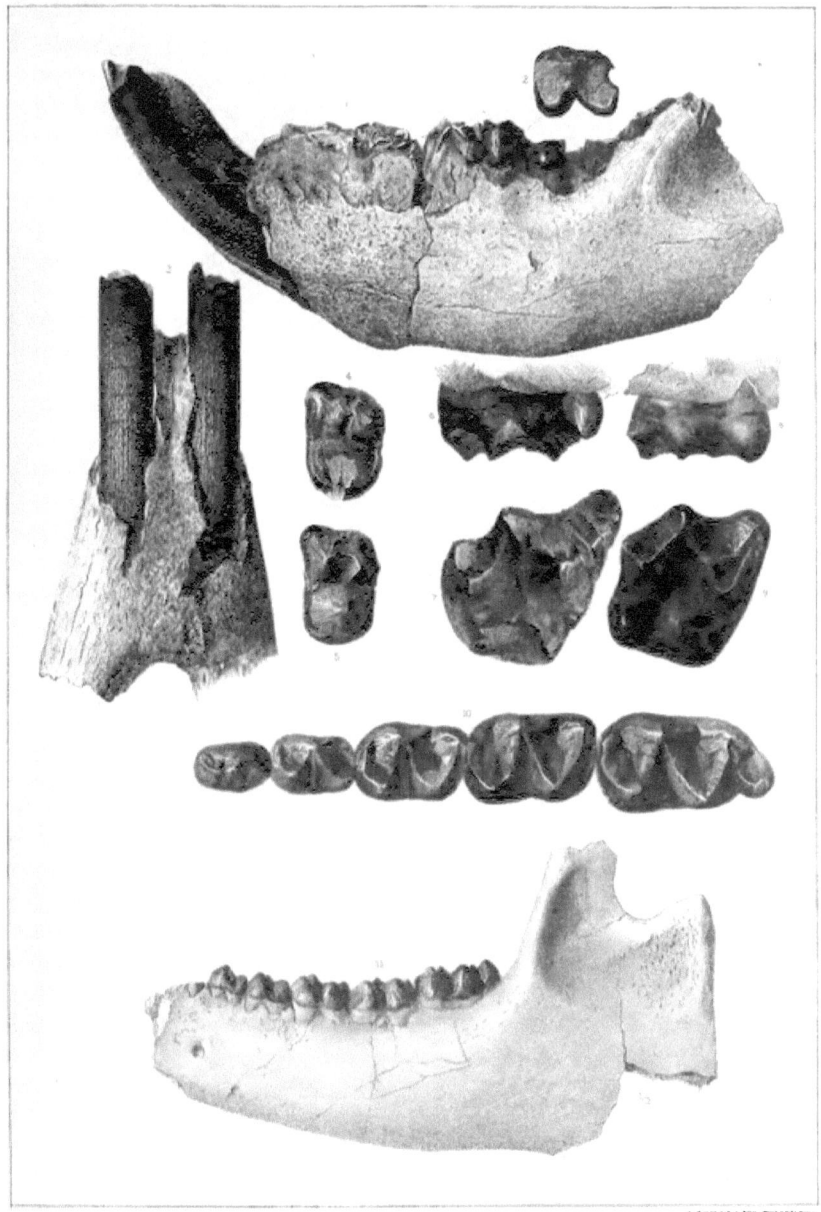

1–3 TROGOSUS CASTORIDENS   4–11 PALAEOSYOPS PALUDOSUS

EXPLANATION OF PLATE VI.

Figs. 1-9. HYOPSODUS PAULUS. All of the natural size except Figs. 2, 5, 8, 9, which are magnified four diameters.

    Fig. 1. Right side of lower jaw, with the three molars. From an individual past maturity. Specimen from which the genus and species were first noticed.
    Fig. 2. Triturating surfaces of the molars of the same specimen.
    Fig. 3. Left side of lower jaw, with last premolar and the three molars. Specimen obtained by Dr. Corson at Grizzly Buttes.
    Fig. 4. Right side of lower jaw, with last premolar and the molars, but slightly worn. Specimen obtained by Dr. Carter.
    Fig. 5. Triturating surfaces of the teeth from the same.
    Figs. 6, 7. Left side of two lower jaws containing the molars. From mature but comparatively young individuals. Dr. Carter.
    Fig. 8. Series of the back two premolars and the molars of the right side. From a specimen loaned by Dr. Carter.
    Fig. 9. First and second lower molars of the right side. From another specimen loaned by Dr. Carter.

Figs. 10, 11. MICROSUS CUSPIDATUS:

    Fig. 10. Portion of left side of lower jaw, with back two molars, natural size. Specimen from Black's Fork of Green River.
    Fig. 11. Triturating surfaces of the two molars, magnified four diameters.
    Fig. 12. Portion of right side of lower jaw, probably pertaining to the last-named animal. It contains the roots of the molars and the last premolar, the triturating surface of which is represented in Fig. 13, magnified four diameters. Specimen obtained by Dr. Carter near Fort Bridger.

Figs. 14-17. MICROSYOPS GRACILIS:

    Fig. 14. Left side of lower jaw with the molars, natural size. Fig. 15. Triturating surfaces of the molars, magnified four diameters. Specimen obtained at Grizzly Buttes by Dr. Carter.
    Fig. 16. Left side of lower jaw, with the second molar and portions of the others, natural size. Fig. 17. The triturating surface of the second molar magnified four diameters. Specimen obtained by Dr. Carter at Grizzly Buttes.

Figs. 18-22. HYOPSODUS PAULUS:

    Fig. 18. Right upper jaw, with three premolars and the molars, magnified two diameters.
    Fig. 19. Triturating surfaces of the teeth magnified four diameters. Specimen obtained by Dr. Carter at Grizzly Buttes, and apparently pertaining to the same individual as that of Fig. 14.
    Fig. 20. Triturating surfaces of the right upper molars, magnified four diameters, from a second specimen. Dr. Carter.
    Fig. 21. Triturating surfaces of back two premolars and first molar of the left side, magnified four diameters. Obtained by Dr. Carter at Lodge-pole trail.
    Fig. 22. Triturating surfaces of upper second and third premolars of right side, magnified four diameters. Dr. Carter.

Figs. 23-25. PARAMYS DELICATUS:

    Fig. 23. Right side of lower jaw, with all the molars, natural size. Fig. 24. Triturating surfaces of the molars except the last, which is broken away excepting the outer portion, magnified three diameters. Grizzly Buttes. Dr. Carter.
    Fig. 25. Triturating surfaces of the molar series, except the last, of the lower right side, magnified three diameters. From another specimen loaned by Dr. Carter.

PLATE VI.

Figs. 26, 27. PARAMYS DELICATIOR:
  Fig. 26. Left side lower jaw, with the second molar, natural size. Grizzly Buttes. Dr. Carter. Fig. 27. Triturating surface of the second molar, magnified three diameters.

Figs. 28, 29. PARAMYS DELICATISSIMUS:
  Fig. 28. Right side of lower jaw, with all the molars, natural size. Grizzly Buttes. Dr. Carter. Fig. 29. Triturating surfaces of the molar series, magnified three diameters.

Fig. 30. SCIURAVUS (?) Triturating surface of a lower left third molar, magnified eight diameters. From a portion of the lower jaw obtained at Grizzly Buttes by Dr. Carter.

Figs. 31, 32. MYSOPS MINIMUS:
  Fig. 31. Right side of lower jaw, with third and fourth molars, magnified two diameters. Fig. 32. Triturating surfaces of the teeth, magnified eight diameters. Dr. Carter.

Figs. 33–35. LOPHIOTHERIUM SYLVATICUM:
  Fig. 33. Portion of left side of lower jaw, with last premolar and first and last molars, natural size. Fig. 34. Triturating surfaces of the last premolar and first molar. Fig. 35. Triturating surface of the last molar. Specimen from Henry's Fork of Green River.

Figs. 36, 37. NOTHARCTUS TENEBROSUS:
  Fig. 36. Right side of lower jaw, with canine and all the molar series except the first premolar, natural size. Fig. 37. The triturating surfaces of the molars, magnified two diameters. Specimen from Black's Fork of Green River.

Figs. 38, 39. HIPPOSYUS FONOSUS (?)
  Fig. 38. Triturating surface of a lower right second molar, magnified two diameters. From a jaw-fragment from near Fort Bridger. Dr. Carter.
  Fig. 39. Triturating surface of a left lower first molar, magnified two diameters. From a jaw-fragment obtained by Dr. Carter near Fort Bridger.

Fig. 40. HIPPOSYUS ROBUSTUS:
  Triturating surface of a left lower second molar, magnified two diameters. From Henry's Fork of Green River. Professor Hayden.

Fig. 41. HIPPOSYUS FORMOSUS:
  Triturating surfaces of the upper left first and second molars, magnified three diameters. Specimen from near Fort Bridger. Dr. Carter.

Fig. 42. HYRACHYUS NANUS:
  Triturating surfaces of the back two premolars, and the molars, magnified one and a half diameters. Taken from the left side of the lower jaw of the same specimen represented in Fig. 14, of Plate II. Specimen obtained by Dr. Corson at Grizzly Buttes.

Fig. 43. TROGOSUS VETULUS, probably *Anchippodus*:
  Right lower incisor, natural size. From near Fort Bridger. Dr. Carter.

Fig. 44. SINOPA RAPAX:
  Portion of left side of lower jaw, with last premolar and first molar, natural size. From Grizzly Buttes. Dr. Carter.

Fig. 45. SINOPA EXIMIA:
  Portion of left side of the lower jaw, supposed to belong to a smaller species of the former, natural size. From Grizzly Buttes. Dr. Carter.

Fig. 46. PALÆACODON VERUS:
  Penultimate molar of the upper left side, magnified four diameters. From Lodge-pole trail. Dr. Carter.

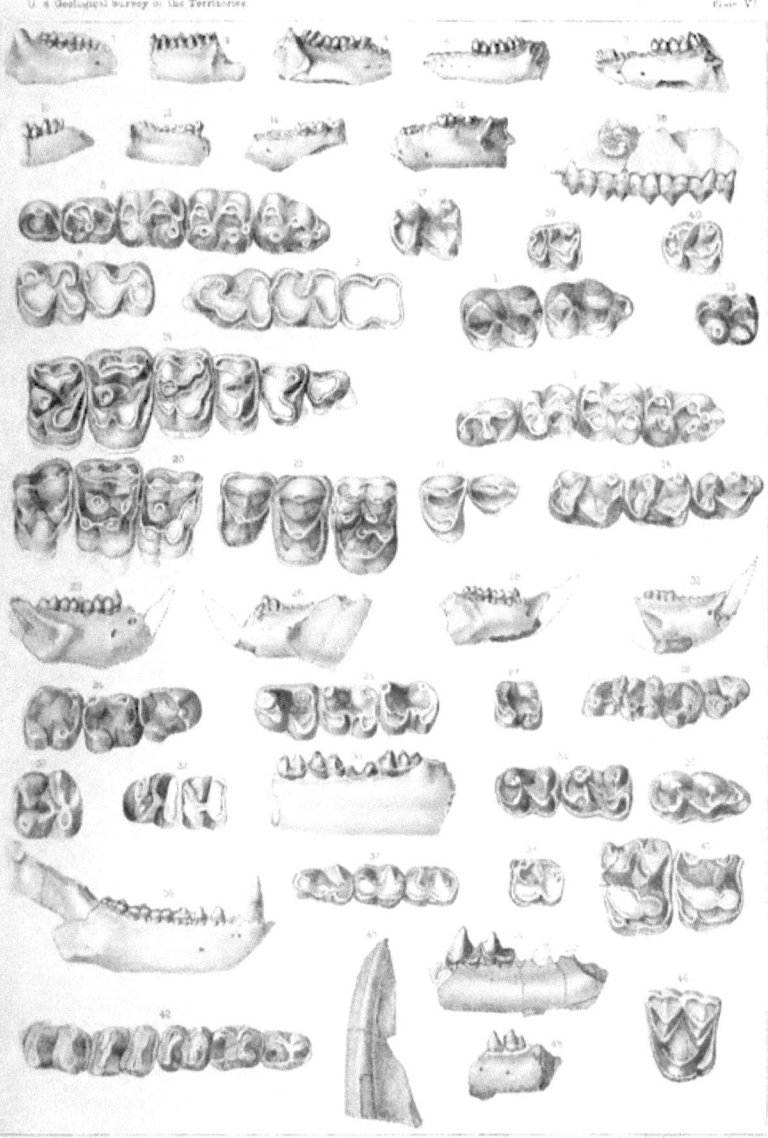

1-9 HYOPSODUS PAULUS
10, 11 MICRMUS CUSPIDATUS
12, 13
14-17 MICROSYOPS GRACILIS
18-22
23-25 PARANYS DELICATUS
26, 27 P DELICATIOR
28, 29 P DELICATISSIMUS
30
31, 32 MYSOPS MINIMUS
33-35 LOPHIOTHERIUM SYLVATICUM
36, 37 NOTHARCTUS TENEBROSUS
38, 39
40.
41
42 HYRACHYUS NANUS
43 ANCHIPPODUS VETULUS
44 SINOPA RAPAX
45
46 PALÆACODON VERUS

EXPLANATION OF PLATE VII.

All the figures are of the natural size, except Fig. 10, which is reduced to one-half the diameter of the original.

Figs. 1–5. MERYCOCHŒRUS RUSTICUS. From specimens obtained on Sweetwater River, Wyoming, by Professor Hayden's party in 1870.
    Fig. 1. Series of upper molars of the right side, viewed on their triturating surfaces. The last tooth had not entirely protruded, and in the first one the median enamel-pits are nearly obliterated.
    Fig. 2. Upper last premolar and molar of the left side, of the temporary series.
    Fig. 3. Upper second and third premolars of the left side, of the permanent series. The triturating surfaces but slightly worn.
    Fig. 4. Outer view of the same teeth, in a small jaw-fragment.
    Fig. 5. Symphysis of the lower jaw, with the four incisors on each side.

Fig. 6. MERYCOCHŒRUS PROPRIUS. First and second upper molars of the right side. From a specimen obtained on the Niobrara River, by Professor Hayden, in 1857.

Figs. 7–11. OREODON SUPERBUS. From specimens discovered in Oregon by the Rev. Thomas Condon.
    Fig. 7. Last lower molar of the right side, viewed on the triturating surface.
    Fig. 8. First and part of the second molars, from the same jaw-fragment as the preceding figure.
    Fig. 9. The three lower premolars of the right side, viewed on their triturating surfaces. From the same specimen as Fig. 16, Plate II.
    Fig. 10. Upper view of the intermediate portion of the face, one-half the natural size.
    Fig. 11. View of the inner surface of a lower canine, from the left side of a specimen of a jaw, which lies with its outer face imbedded in a hard mass of rock.

Fig. 12. OREODON CULBERTSONI. A series of upper true molars of the left side. Specimen discovered by Mr. Condon on John Day's River, Oregon.

Figs. 13, 14. DICOTYLES PRISTINUS. Specimens in the Condon collection of Oregon fossils.
    Fig. 13. Triturating surface of a lower penultimate molar.
    Fig. 14. Outer view and view of the triturating surface of a lower last molar.

Fig. 15. ANCHITHERIUM BAIRDI. An upper right molar. From the Condon collection.

Figs. 16, 17. ANCHITHERIUM AGRESTE. From a specimen found on Red Rock Creek, one of the head streams of the Jefferson Fork of the Missouri. Obtained by Professor Hayden in 1871.
    Fig. 16. Lower last premolar and first molar of the left side. Triturating surface much worn.
    Fig. 17. Last molar, from the same specimen of the jaw as the former.

Figs. 18, 19. FELIS AUGUSTUS. Specimens discovered by Professor Hayden on the Loup Fork of the Niobrara River, Nebraska.
    Fig. 18. Portion of the right premaxillary, containing the second incisor, viewed in front.
    Fig. 19. Upper sectorial molar of the left side, viewed externally.

Fig. 20. PATRIOFELIS ULTA (?) A premolar, probably of the upper jaw. Specimen found by Dr. Carter in the vicinity of Fort Bridger, Wyoming.

Figs. 21–23. Tooth of a carnivore, undetermined. Obtained by Professor Hayden's party on Henry's Fork of Green River, Wyoming.
    Fig. 21. Outer view of the crown of an anterior premolar.   Fig. 22. Upper view of the same.
    Fig. 23. Outer view of the crown of a canine tooth.

PLATE VII.

2

Figs. 24, 25. RHINOCEROS PACIFICUS. A left inferior molar tooth, from Bridger Creek, Oregon, belonging to the Condon collection.
Fig. 24. View of the outer part of the crown. Fig. 25. Triturating surface of the same specimen.

Fig. 26. A canine tooth of an undetermined animal, probably of a large carnivore, but it may be of an Elotherium-like pachyderm. The specimen belongs to the Condon collection of Oregon fossils, and is labeled "Alkali Flats."

Fig. 27. ELOTHERIUM IMPERATOR. A supposed incisor tooth, inner view. Specimen labeled "Bridge Creek," and belonging to the Condon collection of Oregon fossils.

Figs. 28, 29. ELOTHERIUM MORTONI? An incisor tooth, obtained by Mr. Peirce, of Denver, twenty miles southeast from Cheyenne City, Wyoming.
Fig. 28. Inner view of the tooth. Fig. 29. Outer view of the same.

Fig. 30. Canine of an undetermined carnivore. It resembles the inferior canines of a bear, but is more compressed. Specimen discovered by Professor Hayden on White River, Dakota, in 1866. The crown is compressed conical, with the inner surface defined in the usual manner by acute borders. The fang exhibits a gibbous character. Length of crown 11 lines; breadth at base, 8 lines; thickness, 4½ lines.

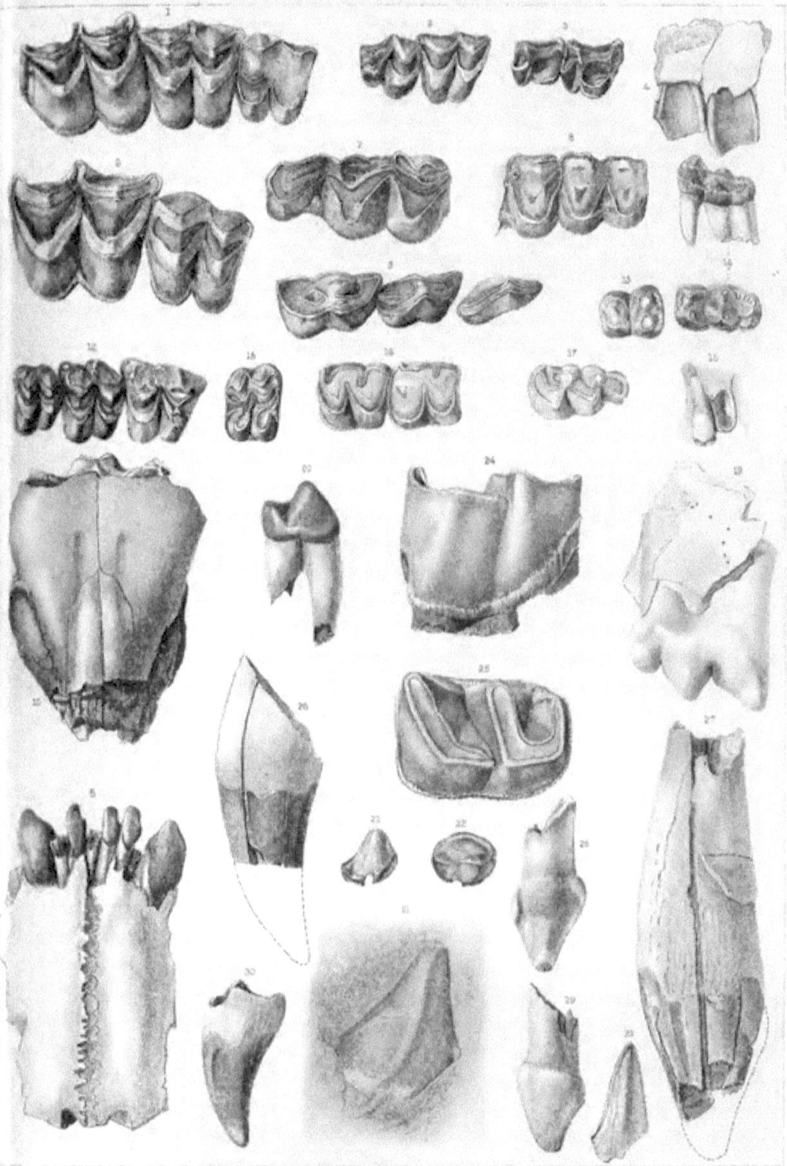

EXPLANATION OF PLATE VIII.

Figures all one-half the diameter of nature.

Fig. 1. A lumbar vertebra of a crocodile. From Little Sandy River. Hayden's collection of 1870. (*Crocodilus Elliotti.*)

Fig. 2. CROCODILUS APTUS:

A cervical vertebra, found on South Bitter Creek, Wyoming.

Fig. 3. A first caudal vertebra of a crocodile. From Little Sandy River. Hayden's collection of 1870.

Figs. 4–6. CROCODILUS ELLIOTTI. Hayden's collection of 1870.

Fig. 4. Portion of the left maxillary, containing the fourth and fifth teeth of that bone. From the junction of Big Sandy and Green Rivers.

Fig. 5. Upper extremity of a left femur. From near Little Sandy River.

Figs. 6, 7. Upper view of a large portion of the skull. Found by H. W. Elliott, on Little Sandy River.

Fig. 8. Left ramus of the lower jaw of a larger individual, or perhaps of a larger species. Discovered in the vicinity of Fort Bridger by Dr. Joseph K. Corson, and presented by him to the Academy of Natural Sciences of Philadelphia.

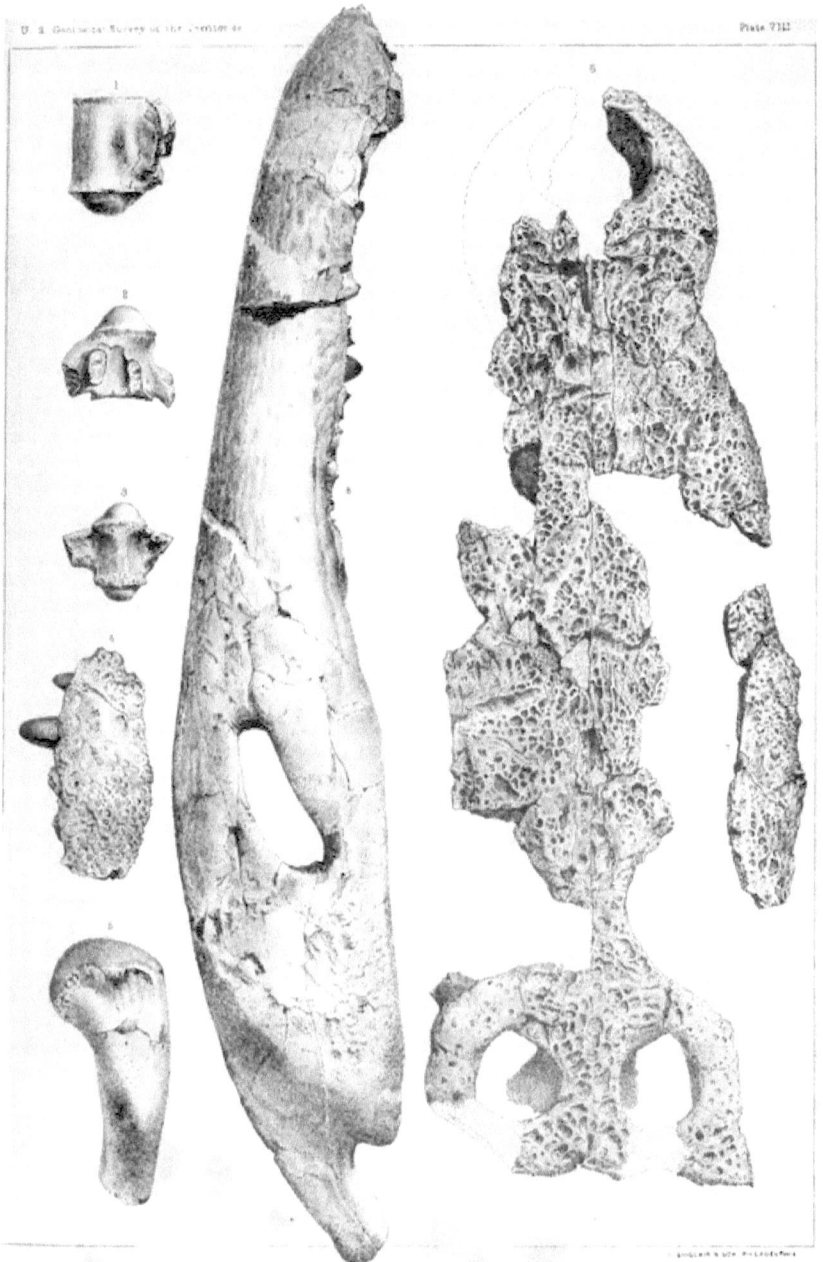

CROCODILES. Pl. 22.

EXPLANATION OF PLATE IX.

All the figures half the natural size.

Fig. 1. TRIONYX GUTTATUS:

Portion of a carapace, consisting of the third to the sixth vertebral plates, inclusively, together with parts of the contiguous costal plates. Specimen obtained at Church Buttes during Professor Hayden's exploration of 1868.

Figs. 2–6. EMYS WYOMINGENSIS:

Fig. 2. Portion of a carapace comprising the vertebral plates from the first to the eighth inclusively, together with small portions of some of the contiguous costal plates. Specimen, originally referred to *Emys Stevensonianus*, obtained by Dr. Carter in the vicinity of Fort Bridger, and presented by him to the Smithsonian Institution.

Fig. 3. Portion of a plastron, which accompanied the preceding specimen and was originally referred to *E. Stevensonianus*.

Fig. 4. Anterior fragment of another plastron, accompanying the former two specimens, and likewise referred to *E. Stevensonianus*.

Fig. 5. An episternal, upon which the species *Emys wyomingensis* was first noticed. Specimen found by Dr. Carter near Fort Bridger.

Fig. 6. Central portion of a carapace, originally attributed to *Emys Haydeni*. Specimen obtained near Fort Bridger. Hayden's collection.

Fig. 7 EMYS PETROLEI:

Two episternals from different individuals. Specimens from Hardin County, Texas.

1 TRIONYX GUTTATUS   2-6 EMYS WYOMINGENSIS
7 EMYS PETROLEI   ⅓ SIZE

EXPLANATION OF PLATE X.

Represents the nearly complete shell of EMYS WYOMINGENSIS, one-half the natural size. It was originally referred to a species with the name of *Emys Jeanesi*. Specimen obtained from the vicinity of Fort Bridger, during Professor Hayden's exploration of 1870.

Fig. 1. View of the plastron.
Fig. 2. View of the carapace.

EMYS WYOMINGENSIS

EXPLANATION OF PLATE XI.

TESTUDO CORSONI:

    Both specimens pertained to the same shell, and were originally described under the name of *Emys Carteri*. They were discovered near Fort Bridger by Dr. Carter, and presented to the Academy of Philadelphia.

    Fig. 1. The greater part of the plastron, its anterior extremity to the right, one-third the natural size.

    Fig. 2. The anterior intermediate portion of the carapace, its front to the left, one-half the natural size.

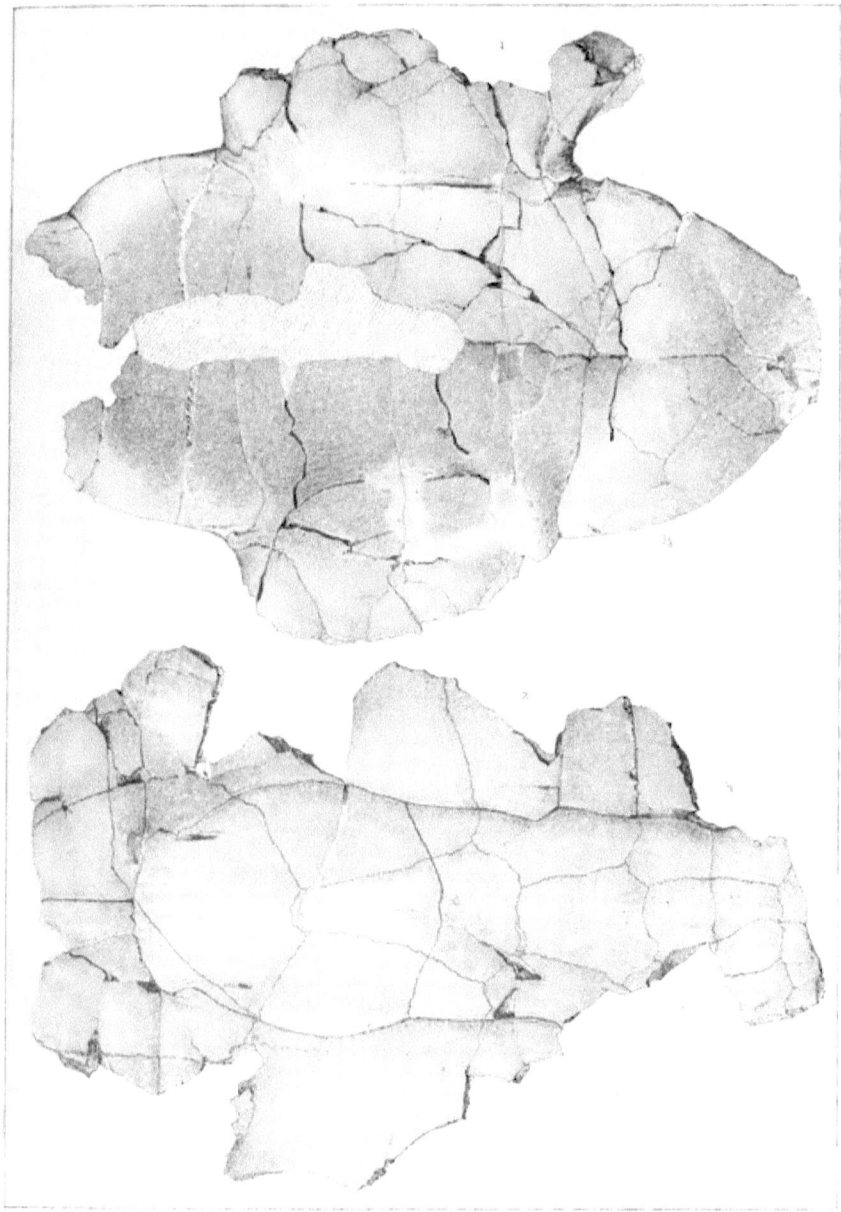

EXPLANATION OF PLATE XII.

BAPTEMYS WYOMINGENSIS:

Figures one-third the natural size. Specimen discovered at Church Buttes, Wyoming, by Mr. O. C. Smith, of Leverett, Massachusetts, while engaged in service of the Union Pacific Railroad. It now belongs to the museum of the Academy of Natural Sciences of Philadelphia.

Fig. 1. View of the carapace.
Fig. 2. View of the sternum.

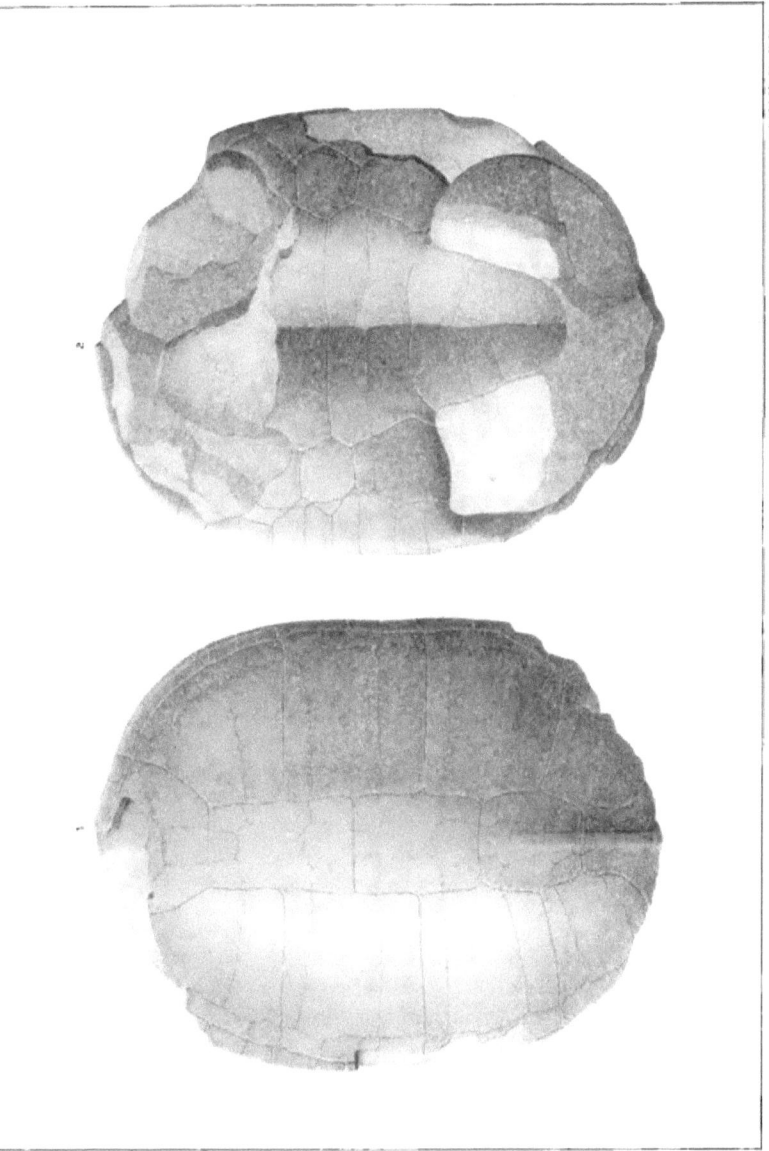

BAPTEMYS WYOMINGENSIS. ½

EXPLANATION OF PLATE XIII.

BAËNA ARENOSA:
>   Figures one-half the natural size.
>   Figs. 1, 2. Specimen on which the genus and species were originally established. Discovered at the junction of the Big Sandy and Green Rivers, Wyoming, during Professor Hayden's exploration of 1870.
>   Fig. 1. View of the carapace.
>   Fig. 2. View of the plastron, its anterior extremity lost.
>   Fig. 3. View of the plastron of another specimen, originally referred to a species with the name of *Baëna affinis*. It was discovered by Dr. Carter at Church Buttes, and was presented by him to the Academy of Natural Sciences of Philadelphia.

1, 2. BAENA ARENOSA ½

3. BAENA AFFINIS ½

EXPLANATION OF PLATE XIV.

CHISTERNON UNDATUM, originally referred to *Baëna undata*. Figures one-half the natural size. Specimen discovered in the vicinity of Fort Bridger by Dr. Carter, and presented by him to the Academy of Natural Sciences.

Fig. 1. View of the carapace; the sutures scarcely visible.

Fig. 2. View of the greater portion of the plastron, with the left border of the carapace. The crucial suture of the plastron is visible, from which the genus received its name.

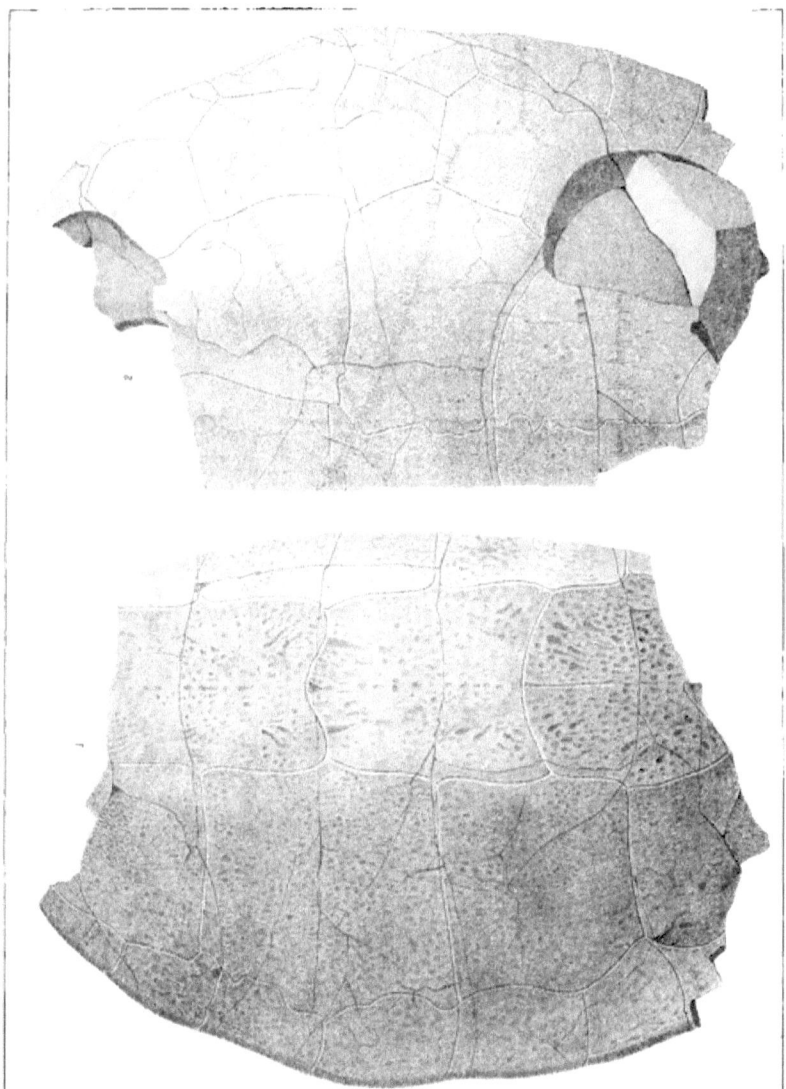

BAENA UNDATA ⅔

EXPLANATION OF PLATE XV.

Figs. 1–5. BAENA ARENOSA?

    Fig. 1. Anterior extremity of the plastron, exhibiting the two pairs of gular scute areas. From the same specimen as Fig. 3, of Plate XIII. One-half the natural size.

    Fig. 2. Anterior extremity of the plastron, from another specimen found by Dr. Carter on Henry's Fork of Green River. The gular scute areas are larger, and the surface of the plates is comparatively smooth. One-half the natural size.

    Fig. 3. From a specimen found by Dr. Corson at Grizzly Buttes. It is of greater proportionate breadth than the former, and presents a want of symmetry in the gular scute areas. One-half the natural size.

    Figs. 4, 5. Of the natural size. From a young specimen obtained by Professor Hayden's party at the junction of Big Sandy and Green Rivers. It retains the sutures, which are obliterated in the preceding mature specimens.

    Fig. 4. Inferior view.

    Fig. 5. Superior view, exhibiting the trident form of the entosternal bone.

Fig. 6. BAPTEMYS WYOMINGENSIS. One-half the natural size. A portion of the anterior extremity of the plastron, from a specimen obtained by Professor Hayden's party at Church Buttes. It presents no distinction between gular and humeral scute areas.

Fig. 7. TESTUDO CORSONI. Anterior extremity of a plastron, one-half the natural size. From a specimen discovered by Dr. Corson at Grizzly Buttes.

Fig. 8. Supposed turtle egg, natural size. A frequent fossil of the indurated clays of the Bridger beds. They are usually about the size of the specimen represented, though quite small ones are also found, like that represented in Fig. 61, Plate XXXII. They have an outer calcareous crust, and are filled with the same material as the imbedding matrix. Usually one end is truncated and rough, as if the shell had been originally broken. Sometimes the truncated end appears covered with a low conical disk, resembling an operculum, as represented in Figs. 60, 61, Plate XXXII.

Fig. 9. HYBEMYS ARENARIUS. A marginal plate, exhibiting the bosses on its outer extension. From a specimen found by Professor Hayden's party on Little Sandy Creek. Natural size.

Fig. 10. STYLEMYS OREGONENSIS. A vertebral plate, one-half the natural size. From Crooked River, Oregon.

Figs. 11–13. NOTHOSAUROPS OCCIDUUS. Three views of a vertebra, natural size, from a specimen obtained by Professor Hayden on Moreau River.

    Fig. 11. Side view of the centrum, exhibiting the sutural surface of the neural arch.

    Fig. 12. Upper view of the same.

    Fig. 13. View of the anterior end.

Figs. 14, 15. SANIWA. Natural size.

    Fig. 14. SANIWA MAJOR. Distal extremity of a humerus, from a specimen found by Dr. Carter at the Lodge-pole trail, on Dry Creek, Wyoming.

    Fig. 15. SANIWA ENSIDENS. Two dorsal vertebræ as they lie in the matrix, inferior view, from a specimen obtained near Granger, Wyoming, during Professor Hayden's exploration.

Figs. 16–18. ANTRODEMUS. In the text, page 267, under the name of POICILOPLEURON VALENS. Figures one-half the natural size. Three views of one-half of a vertebra, from Middle Park, Colorado.

    Fig. 16. End view, exhibiting the articular surface of the centrum.

    Fig. 17. Side view.

    Fig. 18. View of the broken surface of the vertebra, exhibiting the large areolæ of the interior of the centrum, inclosed by thick walls of compact substance.

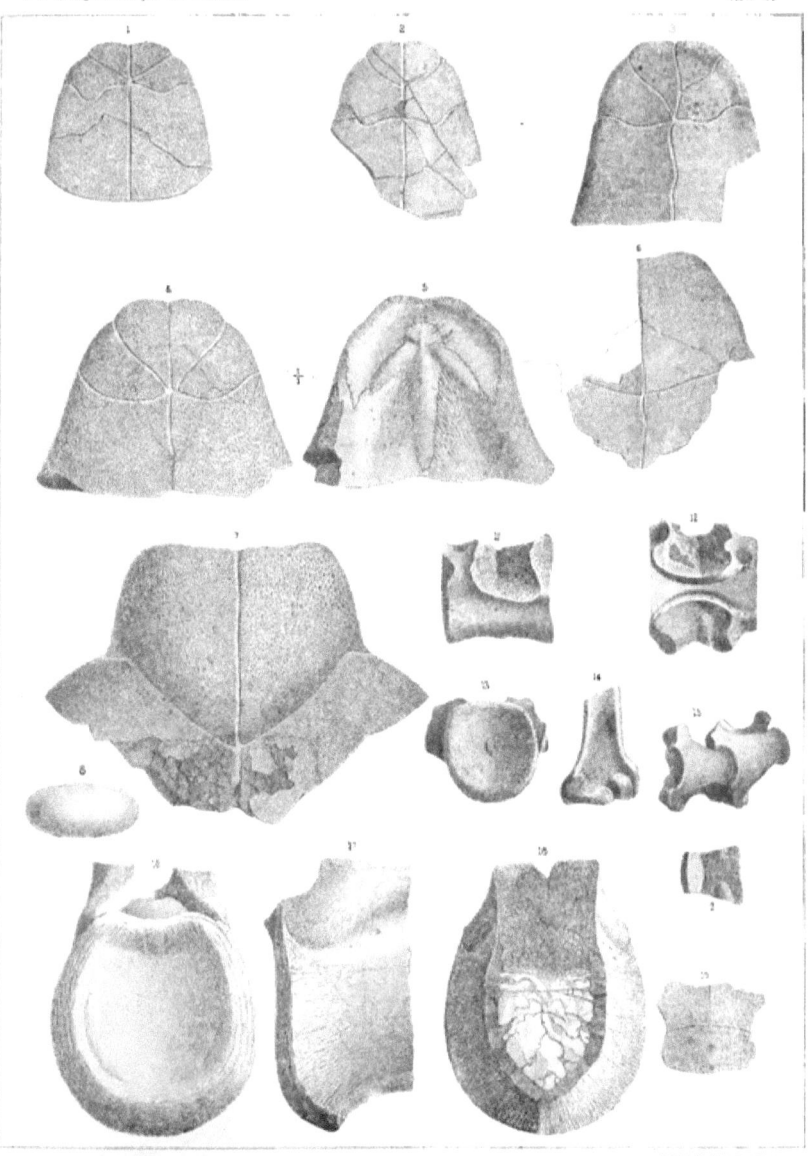

1-5 BAENA ARENOSA ½
6 BAPTEMYS ½
7 TESTUDO CORSONI ½
9 HYBEMYS.
10 STYLEMYS ½
11-12 NOTHOSAUROPS
14 15 SANIVA
16-18 ANTRODEMUS ½

EXPLANATION OF PLATE XVI.

Figs. 1–6. ANOSTEIRA ORNATA:
  Fig. 1. Portions of the carapace.
  Fig. 2. Portion of the same specimen, with portions of the plastron. Specimens collected by Dr. Carter in the vicinity of Fort Bridger.
  Fig. 3. Inner view of three costals, from a portion of the same specimen as Fig. 1, exhibiting the costal capitula.
  Fig. 4. A third marginal plate from a larger individual. Dr. Carter.
  Fig. 5. A fourth marginal plate of the left-side of another individual. From Washakin; collected by James Stevenson.
  Fig. 6. Section of a pygal plate. From a specimen found by Professor Hayden at Church Buttes.
  Fig. 7. Ilium of a turtle. Obtained at Grizzly Buttes by Dr. Carter.

Figs. 8, 9. BAËNA ARENOSA:
  Fig. 8. Ilium of the right side, outer view. Fig. 9. Sacrum, inferior view. Specimens obtained from portions of the matrix, pertaining to the specimen of the shell represented in Figs. 1, 2, Plate XIII.
  Fig. 10. Opisthocœlian caudal vertebra of a turtle. From near Lodge-pole trail. Dr. Carter.
  Fig. 11. Fragment of a costal plate of a trionyx. From near Fort Bridger. Dr. Carter.
  Fig. 12. Fragment of a costal plate of a trionyx. From Little Sandy Creek. Professor Hayden.

Figs. 13–17. GLYPTOSAURUS. All magnified two diameters.
  Figs. 13–15. Osseous dermal plates of the body. Figs. 16, 17. Plates of the head. From Grizzly Buttes. Dr. Carter.

Figs. 18, 19. OLIGOSIMUS GRANDÆVUS:
  Fig. 18. Posterior view of a caudal vertebra. Fig. 19. Lateral view. Specimen obtained by Professor Hayden's party on Henry's Fork of Green River.

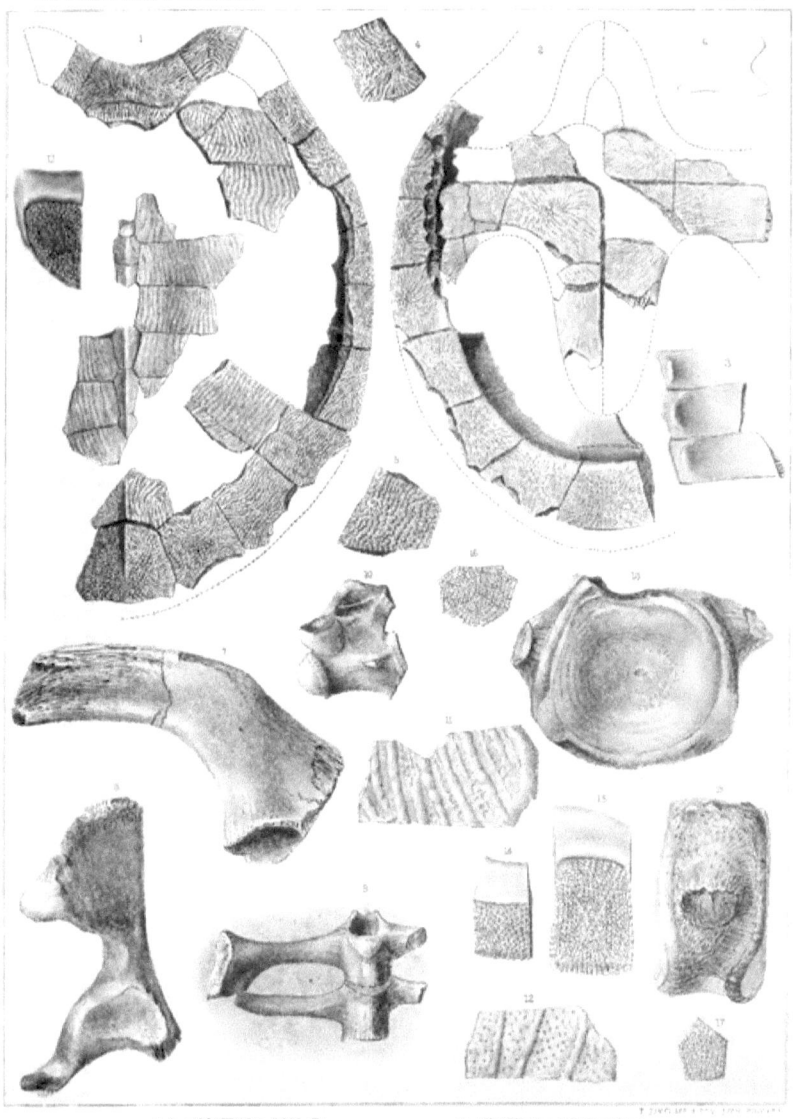

1-6 ANOSTEIRA ORNATA
7 ILIUM OF TURTLE
8, 9 BAENA ARENOSA
10 VERTEBRA OF TURTLE
11, 12 COSTALS OF TRIONYX
13-17 GLYPTOSAURUS
18, 19 OLIGOSIMUS GRANDÆVUS

EXPLANATION OF PLATE XVII.

All the figures of the natural size, except Figs. 9, 10.

Fig. 1. CLUPEA HUMILIS. From the original specimen obtained by Dr. John E. Evans, on Green River, in 1856.

Fig. 2. CLUPEA ALTA. From the "Petrified Fish Cut," on the Union Pacific Railroad, near Green River.

Fig. 3. PETALODUS ALLEGHANIENSIS. Tooth, front view, from a specimen obtained by Messrs. Meek and Hayden, in the upper carboniferous formation of Fort Riley, Kansas.

Figs. 4–6. CLADODUS OCCIDENTALIS. Tooth found by Messrs. Meek and Hayden, in the upper coal measures of Manhattan, Kansas.
Fig. 4. Back view. Fig. 5. Section of the crown. Fig. 6. Bottom of the root.

Figs. 7, 8. PTYCHODUS OCCIDENTALIS. Tooth discovered by Dr. John L. Le Conte, in the Cretaceous formation east of Fort Hays, Kansas.
Fig. 7. Upper view. Fig. 8. Lateral view.

Figs. 9, 10. XIPHACTINUS AUDAX. A pectoral spine, one-half the natural size.
Fig. 9. Inferior view. Fig. 10. Superior view.

Figs. 11–17. MYLOCYPRINUS ROBUSTUS. Pharyngeal bones, from Idaho, contained in the collection of Professor J. S. Newberry.
Fig. 11. Inferior view of a left pharyngeal, containing the three intermediate teeth.
Fig. 12. Inferior view of a right pharyngeal, containing the anterior three teeth.
Fig. 13. Same view of a smaller left pharyngeal, with the posterior four teeth.
Fig. 14. Similar view of another specimen, with the anterior three teeth and the bases of the posterior two teeth.
Fig. 15. Posterior view of a right pharyngeal of an old animal, with the second and fourth teeth.
Fig. 16. Inner view of a right pharyngeal, with the posterior four teeth.
Fig. 17. Posterior view of the same specimen.

Figs. 18, 19. ONCOBATIS PENTAGONUS. Dermal plate, from the Pliocene of Sinker Creek, Idaho.
Fig. 18. Upper view. Fig. 19. Lateral view.

Fig. 20. ENCHODUS SHUMARDI. Dentary bone, natural size, but reversed in position. From the Cretaceous of Dakota.

Figs. 21, 22. CLADOCYCLUS OCCIDENTALIS. Two scales, natural size. Found with the preceding.

Figs. 23, 24. PHASGANODUS DIRUS. From Cannonball River, Dakota.
Fig. 23. A tooth of the natural size.
Fig. 24. Dentary bone, reduced one-third.

Fig. 25. XYSTRACANTHUS ARCUATUS. A dorsal spine, from Leavenworth, Kansas.

Fig. 26. HADRONYUS SUPREMUS:
The mutilated crown of an upper premolar tooth, natural size, seen on the triturating surface. From the Miocene Tertiary of Oregon.

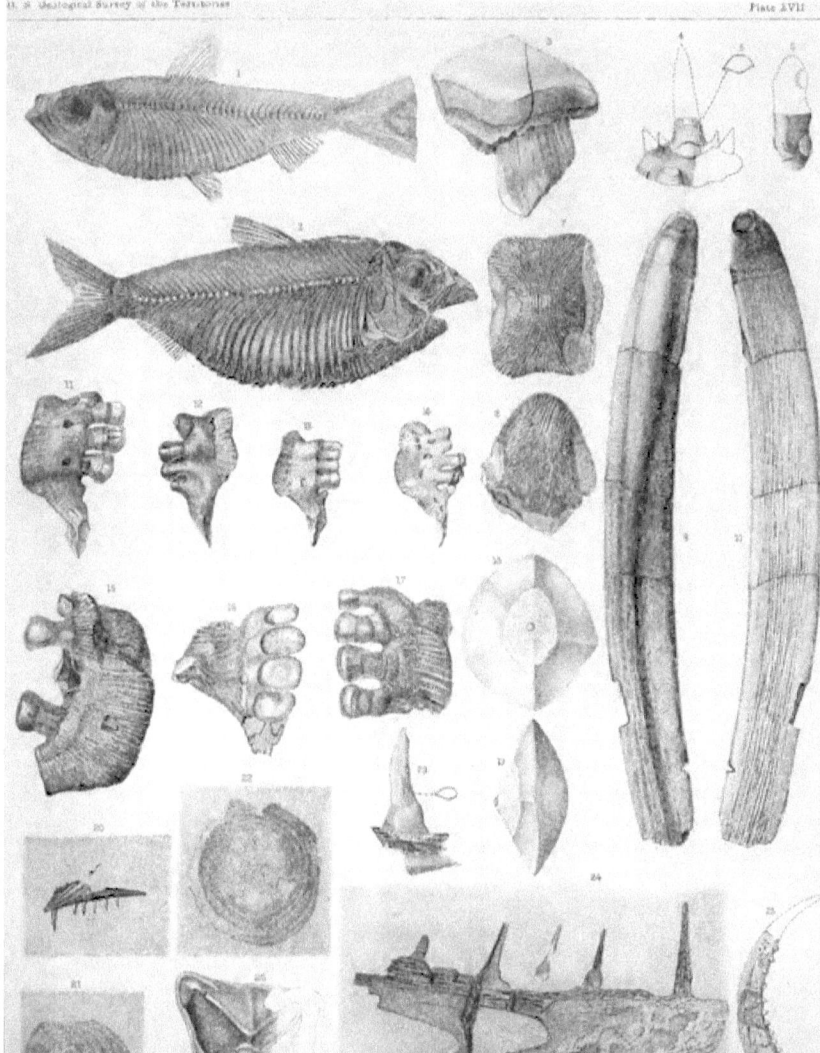

Plate XVII

EXPLANATION OF PLATE XVIII.

All the figures are of the natural size except Figs. 51, 52.

Figs. 1–14. PTYCHODUS MORTONI:
    Figs. 1, 2. Upper and posterior views of a large tooth from Kansas, obtained by Dr. George M. Sternberg.
    Figs. 3, 4. Upper and posterior views of another tooth, apparently from the same individual.
    Figs. 5, 6. Upper and posterior views of another tooth from the same locality.
    Figs. 7, 8. Upper and posterior views of another tooth from the same locality.
    Figs. 9, 10. Upper and posterior views of another tooth from the same locality.
    Figs. 11, 12. Upper and anterior views of a large tooth from near Columbus, Mississippi, found by Dr. William Spillman.
    Figs. 13, 14. Upper and posterior views of a tooth from Green County, Alabama, obtained by Professor Joseph Jones.

Figs. 15–18. PTYCHODUS OCCIDENTALIS. Specimen obtained near Fort Hays, Kansas, by Dr. John L. Le Conte.
    Figs. 15, 16. Upper and posterior views of a worn tooth.
    Figs. 17, 18. Upper views of two small teeth.

Figs. 19, 20. PTYCHODUS WHIPPLEYI. The specimen obtained in the Cretaceous formation of Texas, by Dr. Benjamin F. Shumard.
    Fig. 19. Upper view of a tooth.
    Fig. 20. Posterior view of the same tooth.

Figs. 21–25. OXYRHINA EXTENTA:
    Figs. 21–23. Views, external or anterior, of three teeth from the Cretaceous formation of Kansas, obtained by Dr. George M. Sternberg.
    Figs. 24, 25. External views of two teeth, from the Cretaceous formation near Columbus, Mississippi, obtained by Dr. William Spillman.

Figs. 26–28. OTODUS DIVARICATUS. The specimen from Texas, probably from a Cretaceous formation. Loaned for examination by Dr. William Spillman.
    Fig. 26. External or anterior view of the tooth.
    Fig. 27. Lateral view reversed.
    Fig. 28. Internal or posterior view.

Figs. 29–40. GALEOCERDO FALCATUS. External views of teeth.
    Figs. 29–31. Specimens from the Cretaceous of Kansas, collected by Dr. George M. Sternberg.
    Figs. 32–36. Specimens from the Cretaceous, near Columbus, Mississippi, collected by Dr. William Spillman.
    Figs. 37–40. Specimens from the Cretaceous, near Fort Hays, Kansas, collected by Dr. John L. Le Conte.
    Figs. 41, 42. Specimens from the Cretaceous of Texas, collected by Dr. Benjamin F. Shumard.
    Fig. 43. Specimen from the chalk of Sussex, England.

Figs. 44, 45. LAMNA:
    Fig. 44. External view of a tooth, from the Cretaceous, near Fort Hays, Kansas, found by Dr. John L. Le Conte.
    Fig. 45. External view of a similar but smaller tooth, from the chalk of Sussex, England.

Figs. 46–49. LAMNA:
    Figs. 46, 47. Specimens from the Cretaceous of New Jersey. Fig. 46. Lateral view of a tooth. Fig. 47. External view of another specimen.
    Figs. 48, 49. Specimens from the Cretaceous of Mississippi, collected by Dr. William Spillman. Fig. 48. Lateral view of a tooth. Fig. 49. External view of another tooth.
    Fig. 50. Outer view of a tooth. Specimen from the Cretaceous of Kansas, collected by Professor Hayden.

Figs. 51, 52. PALÆOSYOPS PALUDOSUS. One-half the natural size.
    Fig. 51. Side view of the face; from the same specimen as the teeth of Fig. 3, Plate IV.
    Fig. 52. Lower jaw; repeated from the same specimen as Fig. 11, Plate V.

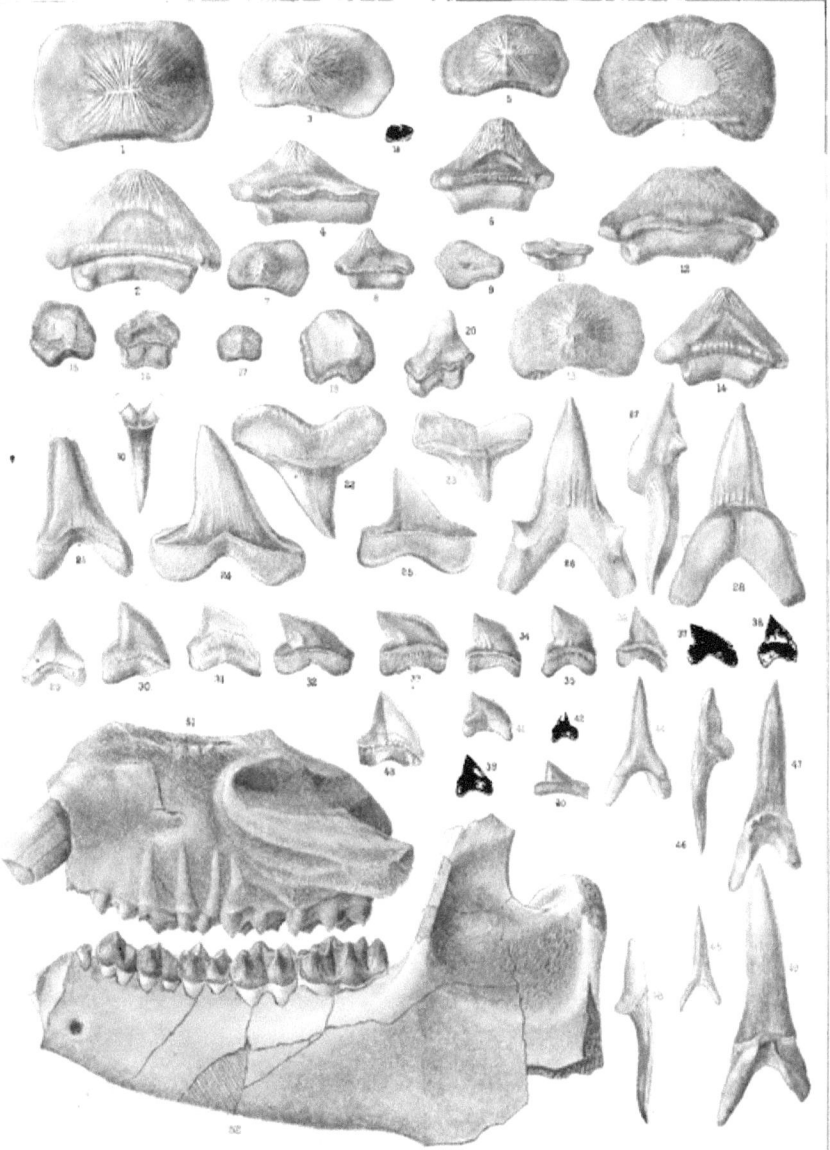

EXPLANATION OF PLATE XIX.

Figs. 1-4. PALÆOSYOPS PALUDOSUS. All half size except Fig. 4.
   Fig. 1. Front view of the left femur.
   Fig. 2. Lower extremity of the right femur.
   Fig. 3. Distal extremity of the right humerus.
   Fig. 4. The right patella, inner view, natural size. Lodge-pole trail. Dr. Carter.

Fig. 5. HYRACHYUS. An astragalus. Natural size.

Fig. 6. Distal extremity of left femur of *Testudo niobrarensis*, one-half the natural size.

Fig. 7. Distal extremity of right humerus of *Testudo nebrascensis*, from a young animal, half the natural size.

Fig. 8. Distal extremity of the right humerus of *Testudo niobrarensis*, half the natural size.

Fig. 9. Portion of a carapace of *Testudo nebrascensis*, internal surface exhibiting the ridge of attachment of the neural spines and the narrow costal capitula, natural size.

Fig. 10. Portion of right scapula of *Testudo nebrascensis*, back view, one-half the natural size.

Fig. 11. Sacral vertebræ of *Chisternon undatum*, inferior view, natural size.

Fig. 12. Lateral view of the same.

Fig. 13. Ungual phalanx of an undetermined reptile, one-half the natural size. See page 285.

Fig. 14. Dermal plate of *Tylosteus ornatus*, one-half the natural size.

Figs. 15, 16. PYCNODUS FABA. Natural size.
   Fig. 15. Portion of a left ramus of the lower jaw, with teeth. The specimen from the Cretaceous formation of Mississippi.
   Fig. 16. Fragment of the left ramus of the lower jaw, with three teeth, from the greensand of Crosswicks, Burlington County, New Jersey.

Figs. 17-20. HADRODUS PRISCUS, natural size. Specimen belonging to Dr. William Spillman, of Columbus, Mississippi, and found by him in the cretaceous formation of that State.
   Fig. 17. Front view of a supposed premaxillary bone, with two teeth.
   Fig. 18. Posterior view of the same, exhibiting at the sides the two reserve cavities for successional teeth.
   Fig. 19. Lateral view.
   Fig. 20. Inferior view.

Figs. 21, 22. EUMYLODUS LAQUEATUS. Mandible two-thirds natural size. From the Cretaceous formation of Mississippi, discovered by Dr. William Spillman.
   Fig. 21. Inner view; specimen reversed.
   Fig. 22. View of the upper or triturating surface, with the inner surface in perspective.

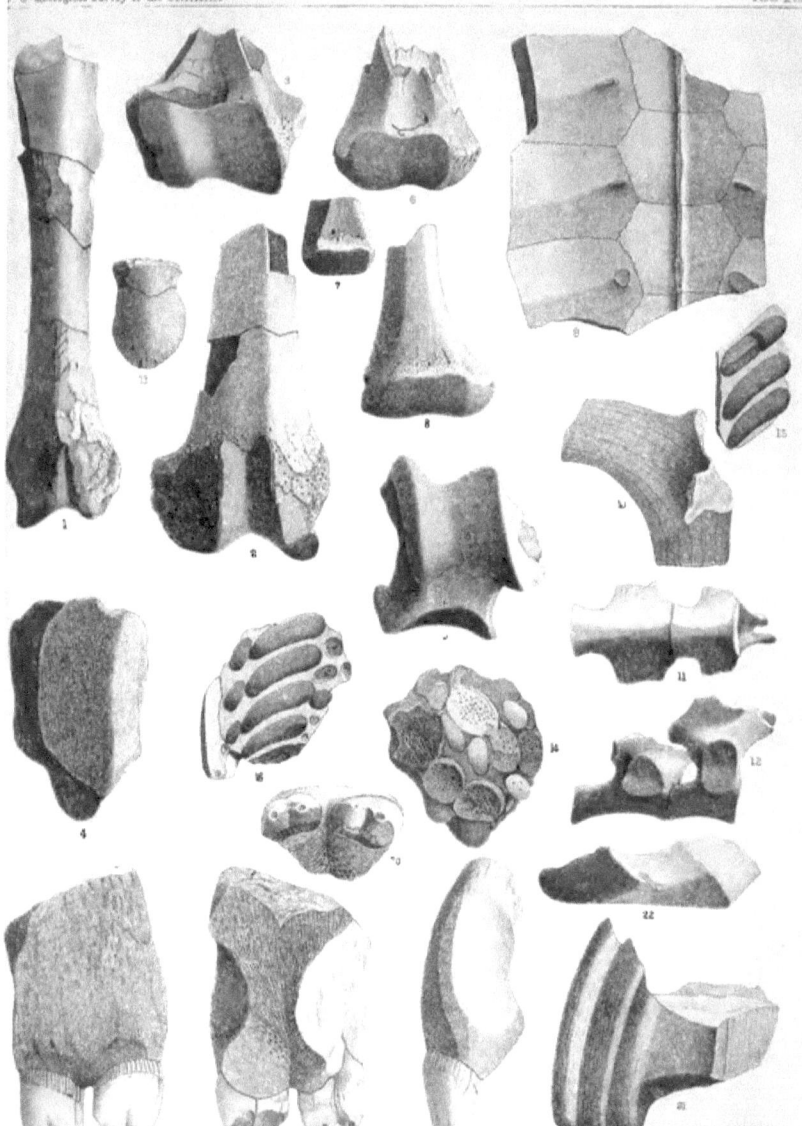

EXPLANATION OF PLATE XX.

Fig. 1-7. PALÆOSYOPS PALUDOSUS. Figures one-half size.
    Fig. 1. Tibia of the right side, front view. From Grizzly Buttes. Hayden's collection of 1870.
    Fig. 2. Calcaneum, upper view. Found by Dr. Corson on Smith's Fork of Green River.
    Fig. 3. Astragalus, upper view. Found by Dr. Carter near Millersville.
    Fig. 4. Cuboid, scaphoid, and external cuneiform. From Church Buttes. Hayden's collection.
    Fig. 5. Metatarsal. Found by Dr. Corson near Fort Bridger.
    Fig. 6. First phalanx. Found by Dr. Carter on Henry's Fork of Green River.
    Fig. 7. Second phalanx. Found by Dr. Carter near Fort Bridger.

Fig. 8. PALÆOSYOPS MAJOR :
    Portion of the right ramus of a lower jaw, one-half size. The specimen is somewhat swollen and altered in character from disease, and is one of those upon which the species was first indicated. Discovered by Dr. Carter at Grizzly Buttes.

Figs. 9-11. MERYCOCHŒRUS RUSTICUS. Natural size. From Hayden's collection of the Sweetwater River.
    Fig. 9. Lower extremity of the right tibia, front view.
    Fig. 10. Astragalus of the right side, upper view.
    Fig. 11. Calcaneum of the right side, upper view.

Fig. 12. MERYCOCHŒRUS (?) Natural size.
    Lower end of the right tibia of a smaller species than the preceding, with the specimens of which it was found.

Fig. 13. HIPPARION (?) Natural size.
    Right cuneiform bone, upper view, of a small equine animal. Specimen found with the remains of Merycochœrus just indicated.

Figs. 14-22. Remains from Texas, submitted to examination by Professor S. B. Buckley. All of the natural size.

Fig. 14. HIPPARION SPECIOSUM (?)
    Last upper molar of the right side; view of the triturating surface. From Washington County.

Fig. 15. HIPPARION ———— (?)
    A third or fourth upper molar of the left side. Found with the preceding specimen.

Fig. 16. PROTOHIPPUS PERDITUS (?)
    A second or third upper molar of the right side. From Independence, Washington County.

Figs. 17, 18. PROTOHIPPUS PLACIDUS (?)
    Fig. 17. A third or fourth upper molar of the right side. Found in association with the specimens of Figs. 14, 15.
    Fig. 18. A first upper molar of the right side, probably of the same species as the former. From Bastrop County.

Fig. 19. ANCHITHERIUM (?) AUSTRALE :
    First upper molar of the right side. Found in association with the specimen of Fig. 16.

Fig. 20. PROTOHIPPUS :
    A lower molar of the right side. From Navarro County.

PLATE XX.

Fig. 21. PROCAMELUS ———(?)
A first or second upper molar of the left side, view of the triturating surface. Specimen found in association with those of Fig. 14, 15, and 17, in Washington County.

Fig. 22. Astragalus of the left side, upper view, probably of the same species as the last, and found with it.

Fig. 23. HIPPARION(?) PARVULUS
A coronary bone, or second phalanx, of the natural size. Found at Antelope, Nebraska.

Fig. 24. FELIS AUGUSTUS?
Distal extremity of the right humerus, front view, one-half size. Specimen found on the Niobrara River, by Professor Hayden.

Figs. 25, 26. HYRACHYUS AGRARIUS. From a specimen obtained by Dr. Carter at Grizzly Buttes. Natural size.
Fig. 25. Left ramus of the lower jaw, containing the back two premolars and the two succeeding molars.
Fig. 26. View of the triturating surface of the same teeth, with the addition of part of the second premolar.

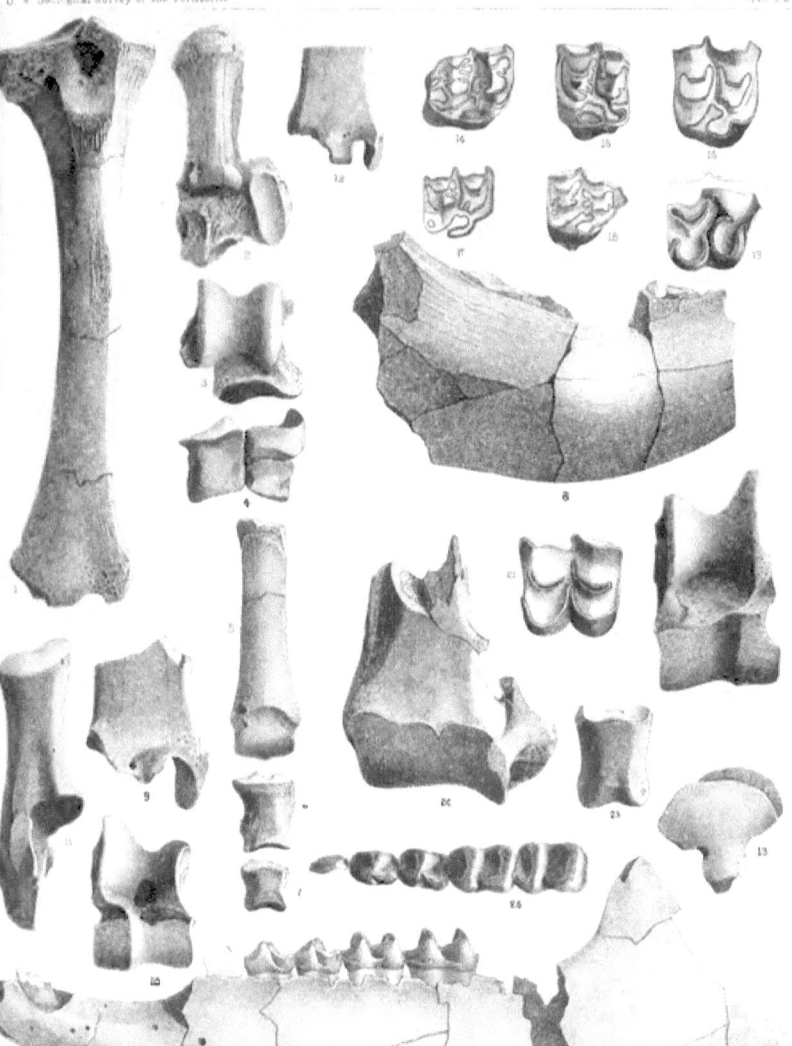

EXPLANATION OF PLATE XXI.

Figs. 1-4. MASTODON OBSCURUS:

Last lower molar of the left side, natural size. Specimen discovered by Dr. Lorenzo G. Yates, in Contra Costa County, California; and now in the museum of Amherst College.

Fig. 1. View of the triturating surface.

Fig. 2. Outer view of the same specimen.

Fig. 3. Fragment of a tusk, two-thirds the natural size, exhibiting the broad band of enamel indicated by the darker shade. Specimen found by Dr. Yates in Stanislaus County, California, and belonging to Amherst College museum.

Fig. 4. Outline of the transverse section from the smaller end of the same specimen, of the natural size.

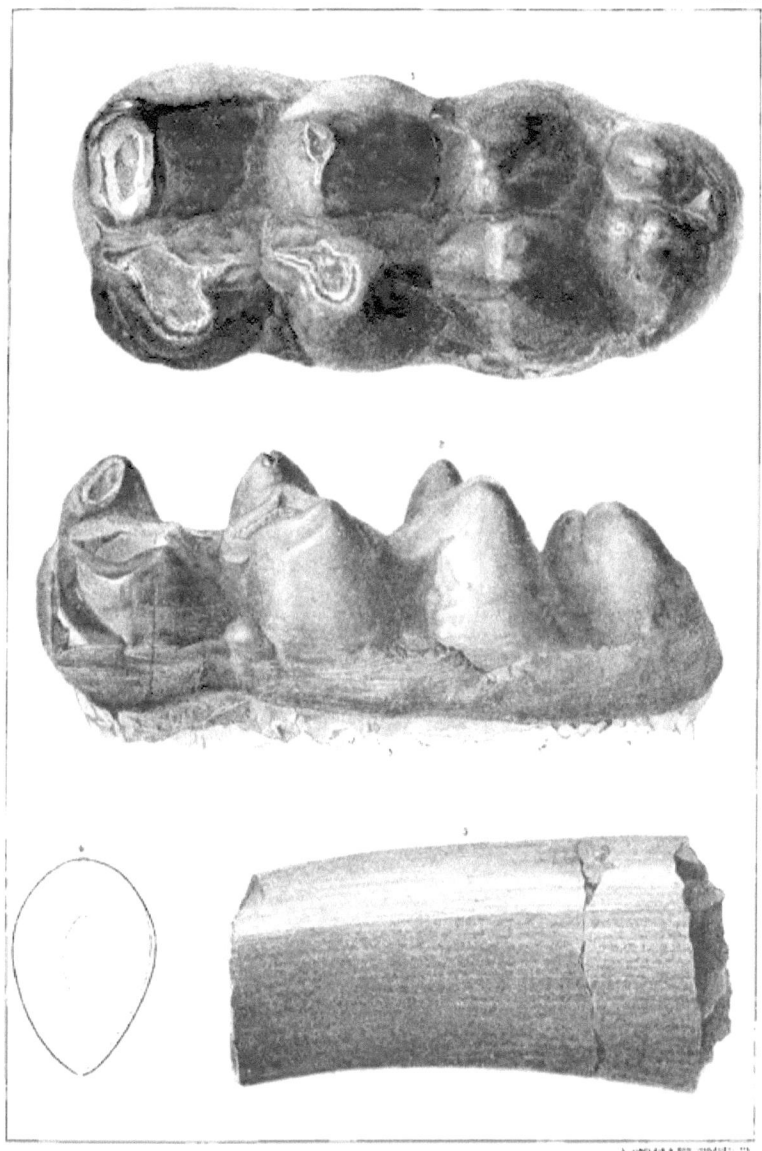

1, 2. MASTODON, CONTRA COSTA COUNTY, CAL.　3, 4. MASTODON, STANISLAUS COUNTY, CAL.

EXPLANATION OF PLATE XXII.

Fig. 1-4. MASTODON OBSCURUS:

    Fragments of a lower jaw, from near Santa Fé, New Mexico, presented to the Smithsonian Institution by W. F. M. Arny.

    Fig. 1. Portion of the jaw containing the greater part of the last molar tooth. Fig. 2. Portion of the symphysis. The two fragments placed in their relative position, and reduced to one-sixth the natural size.

    Fig. 3. Inferior view of the symphysial fragment, exhibiting exposed portions of the incisors. One-fourth the natural size.

    Fig 4. The last inferior molar, natural size, seen on the triturating surface. The back portion, consisting of another division and the heel, are broken away.

Figs. 5, 6. MASTODON AMERICANUS. An anomalous molar tooth, natural size.

    Fig. 5. View of the triturating surface.
    Fig. 6. Side view.

Fig. 7. GRAPHIODON VINEARIUS. A tooth of the natural size. Specimen from the Miocene of Martha's Vineyard, belonging to the museum of the Smithsonian Institution.

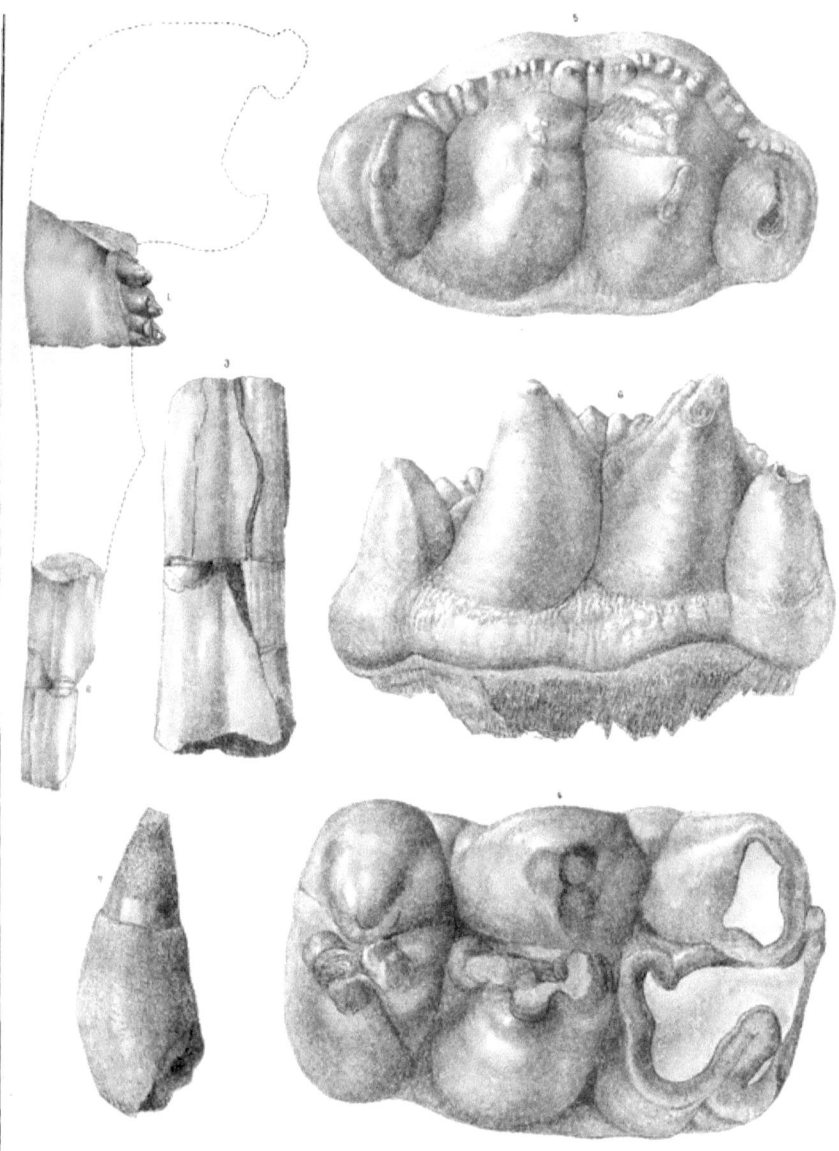

EXPLANATION OF PLATE XXIII.

All the figures of the natural size except Fig. 16, which is one-half size.

Figs. 1, 2. PALÆOSYOPS MAJOR:

    Fig. 1. The complete series of molar teeth of the left side of the lower jaw, except the first premolar. The second and third premolars are reversed from those of the opposite side. Specimen discovered by Dr. Carter in Dry Creek Cañon, forty miles from Fort Bridger.

    Fig. 2. A series consisting of the molars and last two premolars contained in detached fragments of a lower jaw. Specimens obtained by Dr. Carter on Dry Creek. The molars are larger and more worn than in the preceding specimen.

Figs. 3–6. PALÆOSYOPS PALUDOSUS. Specimens upon which the species was originally established. Hayden's collection of 1870.

    Fig. 3. A third lower premolar of the left side.
    Fig. 4. A last lower premolar of the right side.
    Fig. 5. A first lower molar of the left side.
    Fig. 6. Anterior part of a second upper molar of the left side.

Figs. 7–11. PALÆOSYOPS MAJOR. Specimens found by Dr. Corson in Dry Creek Cañon.

    Fig. 7. The left upper canine tooth.
    Fig. 8. The second upper premolar of the left side.
    Fig. 9. The last upper premolar of the same side.
    Fig. 10. The second upper molar of the same side.
    Fig. 11. The last upper molar of the same side.

Fig. 12. PALÆOSYOPS MAJOR:

    Series of premolars from the second to the last, inclusive, of the right side. From Dry Creek. Dr. Carter.

Fig. 13. PALÆOSYOPS (LIMNOHYUS) LATICEPS (?)

    A second upper molar of the right side. A comparatively smooth tooth. Specimen discovered by Dr. Corson, in association with the large canine tooth of Figs. 1–3, Plate XXV.

Figs. 14–16. PALÆOSYOPS MAJOR:

    Fig. 14. A last lower molar of the right side. Contained in a jaw fragment obtained by Dr. Carter at Dry Creek Cañon.

    Fig. 15. An inferior incisor, lateral view, belonging to the same individual as the specimens of Figs. 7–11.

    Fig. 16. Upper view of a cranium, one-half the natural size. The specimen discovered by Dr. Carter at Dry Creek Cañon.

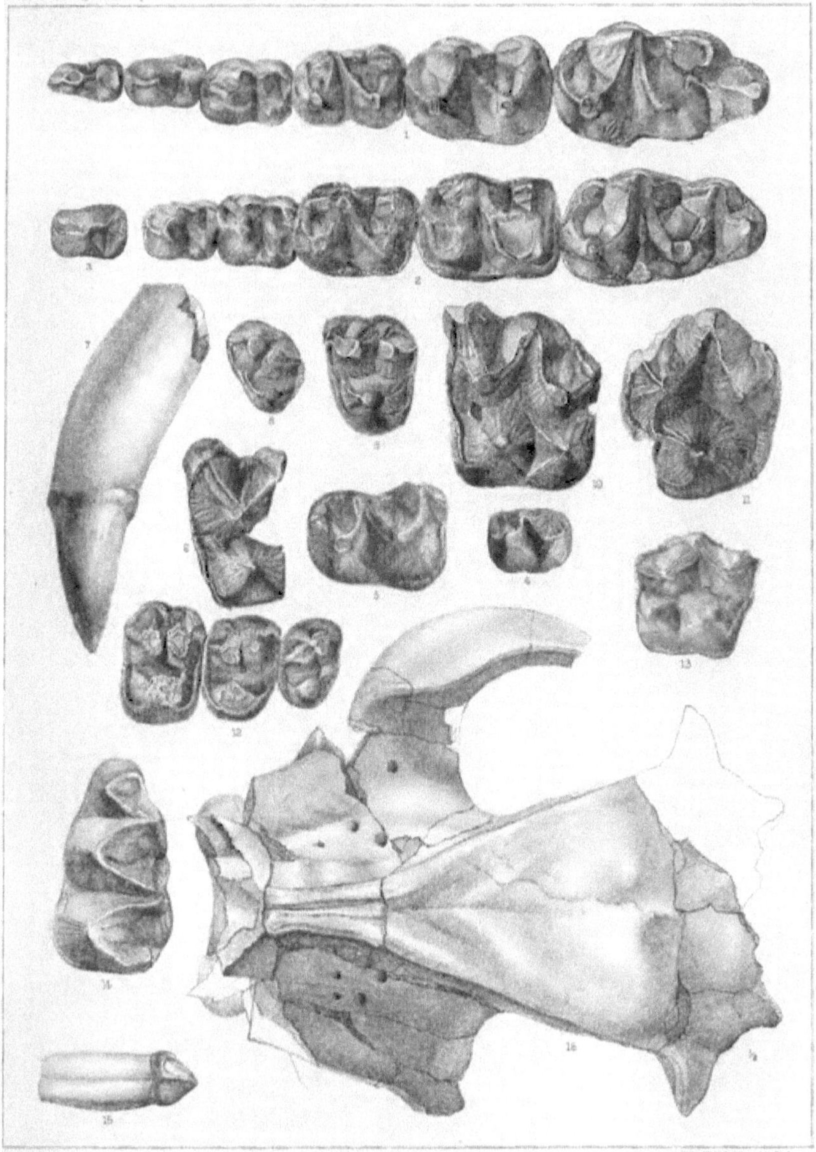

PALAEOSYOPS.

EXPLANATION OF PLATE XXIV.

Figs. 1–5. PALÆOSYOPS MAJOR:

    Fig. 1. View of a left side of a cranium, one-half the natural size. Specimen discovered by Dr. Carter on the buttes of Dry Creek Cañon.

    Fig. 2. View of the left side of a crushed facial specimen, one-half the natural size. Specimen found by a Shoshone Indian, and brought to Dr. Carter.

    Fig. 3. View of the triturating surface of a penultimate upper molar of the right side, natural size. From the same skull as Fig. 1.

    Fig. 4. Portion of the right ramus of the lower jaw of the same animal, one-half the natural size.

    Fig. 5. An upper lateral incisor, natural size. Specimen found by Dr. Corson in the buttes of Dry Creek Cañon.

Figs. 6, 7. PALÆOSYOPS PALUDOSUS (?) Natural size.

    Fig. 6. Fore part of the upper jaw, containing the first three premolars and part of the fang of the canine.

    Fig. 7. Triturating surfaces of the premolars.

Fig. 8. PALÆOSYOPS HUMILIS:

    A last upper molar of the left side, natural size. Found by Dr. Corson on the buttes of Dry Creek Cañon.

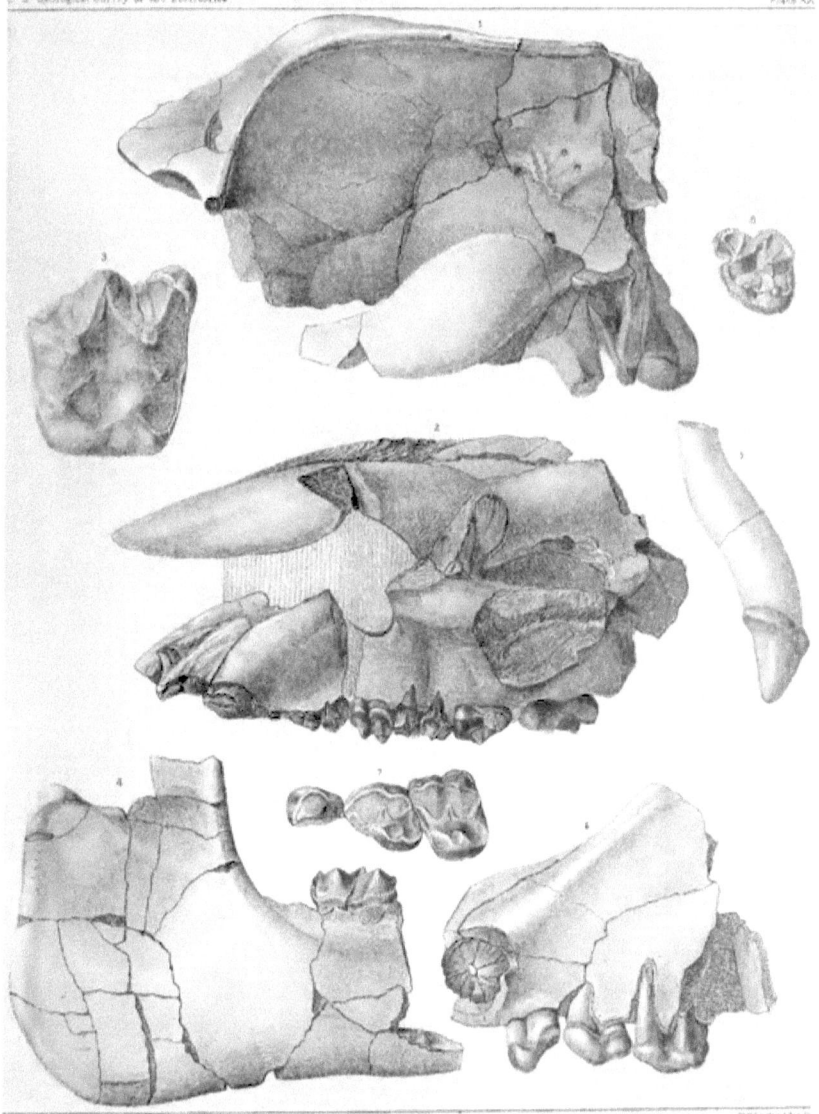

PALAEOSYOPS.

EXPLANATION OF PLATE XXV.

UINTATHERIUM ROBUSTUM. All the specimens discovered by Drs. Corson and Carter at Dry Creek Cañon. Natural size, except Figs. 8 and 11, which are one-half size.

Figs. 1-5. A supposed upper canine tooth. Discovered by Dr. Corson in company with a fragment of the same tooth of the other side, the specimen represented in Figs. 13, 14, and the molar of Palæosyops represented in Fig. 13, Plate XXIII. Originally referred to a supposed carnivore, with the name of *Uintamastix atrox*.
    Fig. 1. Outer view of the right canine. The restored outline of the lance-head-like point is, perhaps, a little exaggerated.
    Fig. 2. Inner view of the point of the same specimen.
    Fig. 3. Front view.
    Fig. 4. Outline of a transverse section of the lance-head-like point.
    Fig. 5. Outline of a section near the base of the specimen.

Figs. 6-12. Specimens found together, with portions of the skull and other bones of the skeleton, ten miles distant from the former. Discovered by Drs. Carter and Corson.
    Fig. 6. Inner view of the last upper molar of the right side.
    Fig. 7. View of the triturating surface of the same tooth.
    Fig. 8. Outer view of the same tooth inserted in a jaw-fragment, half the natural size.
    Fig. 9. Inner view of the last lower molar of the right side.
    Fig. 10. View of the triturating surface of the same tooth.
    Fig. 11. Outer view of the lower-jaw fragment, containing the same tooth, one-half the natural size.
    Fig. 12. Triturating surface, much worn, of the first upper molar, of the right side.

Figs. 13, 14. A supposed upper premolar of the same animal. Discovered by Dr. Corson in company with the large canine tooth of Figs. 1-5.
    Fig. 13. Inner view of the tooth.
    Fig. 14. Triturating surface.

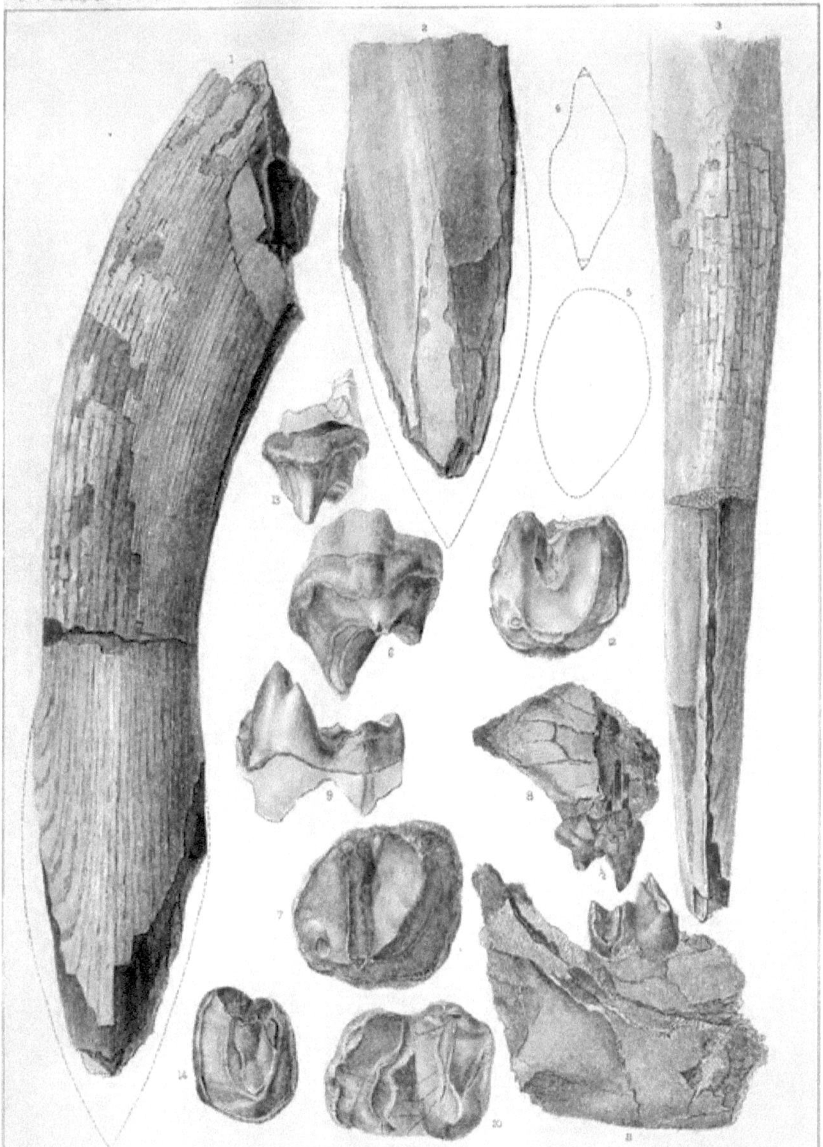

UINTATHERIUM

EXPLANATION OF PLATE XXVI.

Figs. 1–8. UINTATHERIUM ROBUSTUM:

    Fig. 1. View of the right side of a mutilated cranium, one-half the diameter of nature. Specimen upon which the genus was characterized. Discovered by Dr. Carter about fifty miles from Fort Bridger.
    Fig. 2. An atlas, of the same species. Inferior view, one-fourth the diameter.
    Fig. 3. A right humerus. Found by Dr. Carter in the same locality as the specimen of Fig. 1. Anterior view, one-fourth the diameter.
    Fig. 4. Proximal extremity of a femur, probably pertaining to a larger species of the same genus, or perhaps to a larger variety. One-fourth the diameter.
    Fig. 5. Distal extremity of another femur, probably of *U. robustum*. One-fourth the diameter.
    Fig. 6. Calcaneum of the left side. Upper view, one-half size.
    Figs. 7, 8. Astragalus, one-half size.
        Fig. 7. Upper view. Fig. 8. Inferior view.

Figs. 9, 10. HYRACHYUS EXIMIUS:

    Left lower penultimate molar tooth, natural size.
    Fig. 9. Outer view. Fig. 10. Upper view.

Fig. 11. HYRACHYUS NANUS:

    Right ramus of the lower jaw, retaining the back four molar teeth. Natural size.

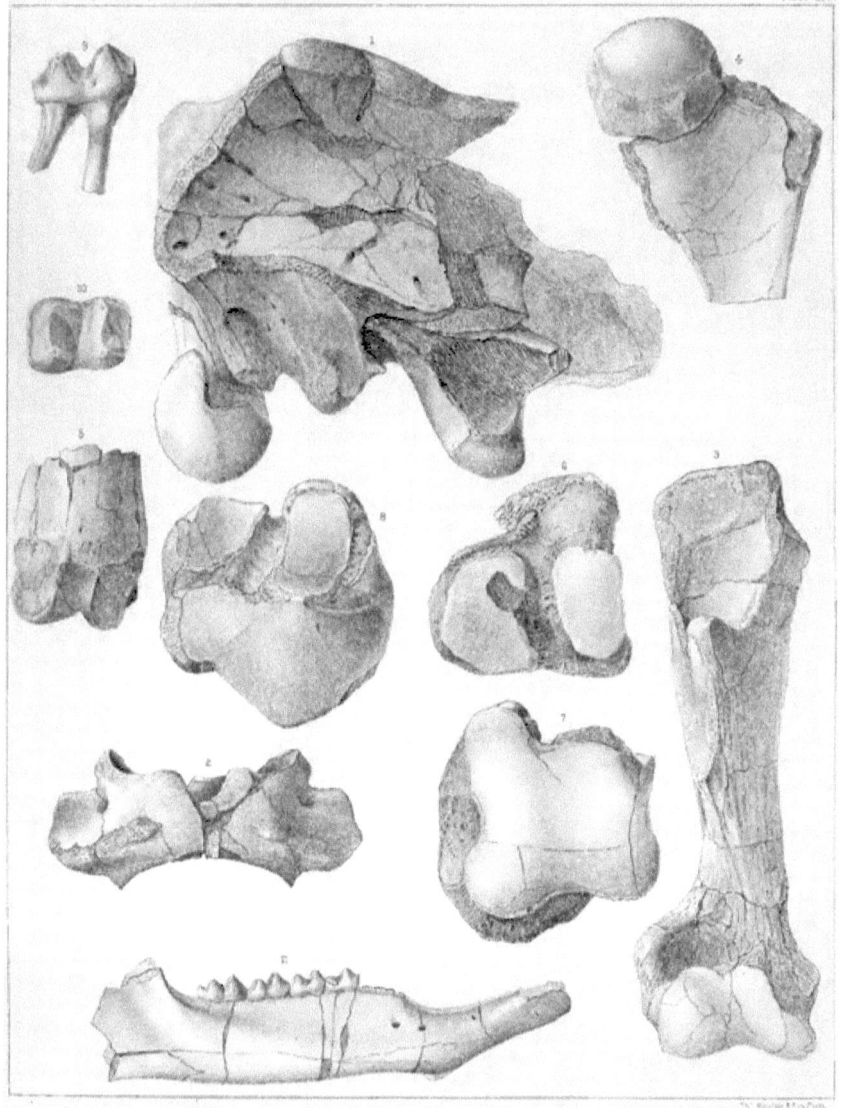

1-8 UINTATHERIUM. 9-11 HYRACHYUS.

EXPLANATION OF PLATE XXVII.

Figs. 1, 2. HIPPOSYUS FORMOSUS:
>An upper molar tooth, probably the second true molar of the left side, magnified three diameters.
>Fig. 1. Outer view of the crown.
>Fig. 2. View of the triturating surface.

Figs. 3, 4. WASHAKIUS INSIGNIS:
>Fig. 3. Portion of the right ramus of the lower jaw, containing the last two molars, magnified three diameters.
>Fig. 4. View of the triturating surfaces of the teeth, magnified eight diameters.

Fig. 5. HYOPSODUS MINUSCULUS:
>View of the triturating surfaces of the last premolar and the molars of the left side, magnified four diameters.

Figs. 6–10. UINTACYON EDAX:
>Fig. 6. Right side of the lower jaw, containing the intermediate three premolars, part of the first molar, and the second molar, natural size.

Figs. 7–10. The teeth, magnified three diameters.
>Fig. 7. Triturating surface of the second molar.
>Fig. 8. Outer view of the same tooth.
>Fig. 9. Upper view of the premolars.
>Fig. 10. Outer view of the same.

Figs. 11–13. UINTACYON VORAX:
>Fig. 11. Fragment of the left side of the lower jaw, containing part of the first molar and the second molar, natural size.
>Fig. 12. Upper view of the second molar.
>Fig. 13. Outer view of the second molar.

Figs. 14, 15. MYSOPS FRATERNUS:
>Fig. 14. Right side of lower jaw, with the last three molars, magnified two diameters.
>Fig. 15. View of the triturating surfaces of the molars, magnified eight diameters.

Figs. 16–18. PARAMYS DELICATIOR:
>Fig. 16. Lower molar of the right side, the second or third of the series, seen on the triturating surface, magnified three diameters.
>Fig. 17. Upper molars of the same animal, apparently the intermediate pair. Outer view, magnified three diameters.
>Fig. 18. View of the triturating surfaces of the same teeth, magnified three diameters.

Figs. 19, 20. MICROSYOPS (?)
>An upper molar tooth, magnified four times.
>Fig. 19. Outer view.
>Fig. 20. View of the triturating surface.

Figs. 21, 22. HYRACHYUS NANUS (?)
>A last upper premolar, magnified two diameters.
>Fig. 21. Outer view.
>Fig. 22. View of the triturating surface.

PLATE XXVII.

Fig. 23. Fragment of the left side of the lower jaw, containing two premolars, apparently the third and fourth, of an undetermined carnivore, natural size. From the Bridger Eocene of Wyoming.

Figs. 24, 25. MEGALOMERYX NIOBRARENSIS (?) A lower molar tooth, natural size. From the Tertiary of L'Eau qui Court County, Nebraska. Specimen in the museum of Swarthmore College.
    Fig. 24. Triturating surface.
    Fig. 25. Outer view.

Figs. 26-29. PROCAMELUS VIRGINIENSIS. Natural size. Specimens from the Miocene of Virginia, and belonging to Mr. C. M. Smith, of Richmond, Virginia.
    Fig. 26. Outer view of the last lower molar of the right side.
    Fig. 27. Triturating surface of the same.
    Fig. 28. The last premolar and first molar of the right side, outer view.
    Fig. 29. Triturating surfaces of the same.

Figs. 30-34. UINTATHERIUM ROBUSTUM:
    Fig. 30. Last upper molar of the right side, outer view, natural size.
    Fig. 31. Last lower molar of the right side, outer view, natural size.
    Fig. 32. Portion of the left ramus of the lower jaw, one-half the natural size. Fig. 33. Mutilated coronoid and condyle of the same specimen as the former.
    Fig. 34. Upper view of the atlas, from the same specimen as Fig. 2, Plate XXVI, one-fourth the diameter of nature.

Fig. 35. SANIWA ENSIDENS. Tooth magnified eight diameters.

Figs. 36, 37. SANIWA MAJOR:
    Two dorsal vertebræ, natural size.
    Fig. 36. Inferior view.
    Fig. 37. View of right side.

Figs. 38, 39. CHAMELEO PRISTINUS. Fragment of the lower jaw, magnified three diameters.
    Fig. 38. Outer view.
    Fig. 39. Inner view.
    Fig. 40. Undetermined tooth of a reptile, magnified two diameters. From the Bridger Eocene formation of Wyoming. It may be the tooth of a crocodile or a lacertian. It is an isolated specimen, partially imbedded in a greenish sandstone, with fresh-water shells. The crown is compressed mammillary, and strongly striate, from an acute-bordered summit.

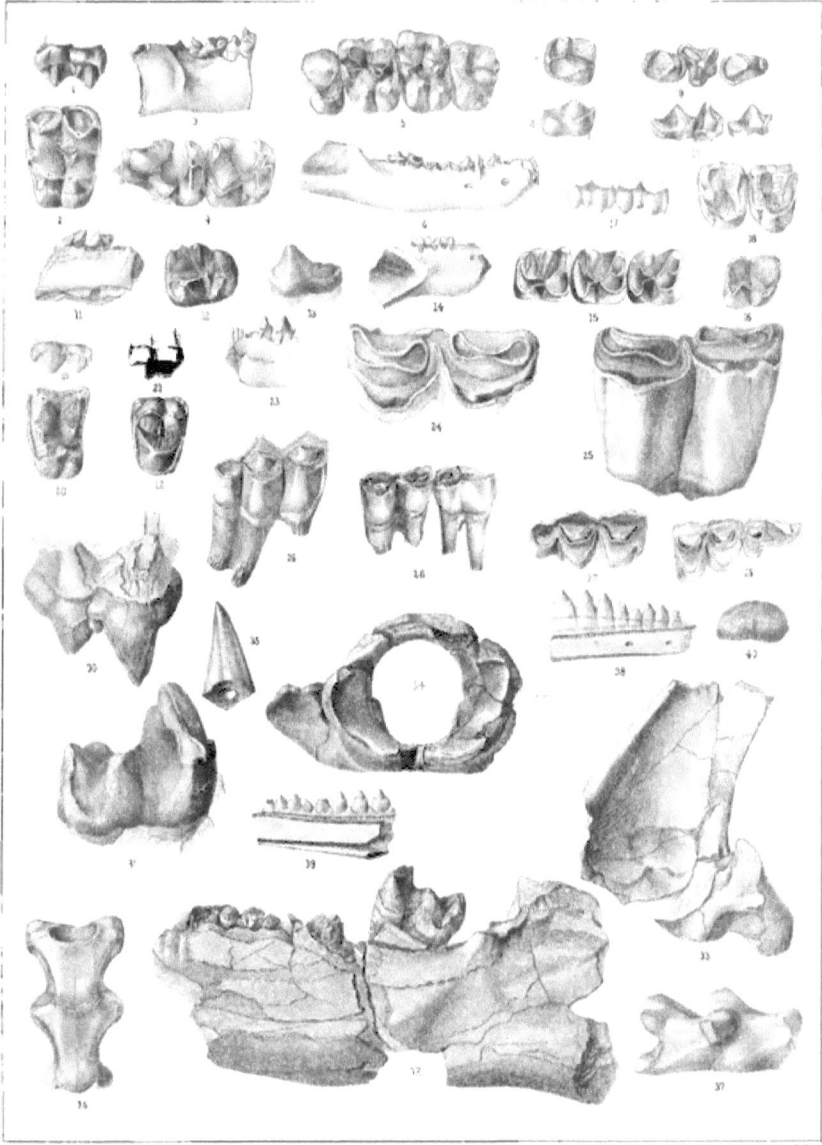

EXPLANATION OF PLATE XXVIII.

Figs. 1, 2. UINTATHERIUM ROBUSTUM:

    Fig. 1. Outline taken from Professor Marsh's Fig. 1, Plate II, of *Dinoceras mirabilis*, in the Am. Jour. Science, 1873, enlarged so as to accord with one-sixth of the size of the fragments introduced in the figure, which correspond with those of Figs. 1 and 8, Plate XXV, and Fig. 1, Plate XXVI.

    Fig. 2. View of the base of the cranial specimen also represented in Fig. 1, Plate XXVI. One-sixth the diameter of nature.

    Fig. 3. Large osseous protuberance, one-half the size of nature, resembling the similar osseous protuberances of the specimen of *Megacerops*, represented in Figs. 2, 3, Plate I. The specimen is from the Mauvaises Terres of White River, Dakota, and was originally suspected to belong to *Titanotherium*.

Figs. 4-8. BISON LATIFRONS:

    Figs. 4, 5. Cranium from Pilarcitos Valley, California, discovered by Messrs. Calvin and Wilfred Brown, and presented to the Academy of Natural Sciences of Philadelphia. One-fifth the natural size.

    Fig. 4. Upper view. Fig. 5. Posterior view.

    Figs. 6, 7. The second and last upper molars seen on their triturating surfaces. Natural size. Specimens from California, belonging to Wabash College, Indiana.

    Fig. 8. An upper second molar of the left side, considerably worn, and seen on its triturating surface. Natural size. From Luzerne County, Pennsylvania.

Fig. 9. MASTODON AMERICANUS:

    A first lower premolar of the right side, natural size. Found with the preceding.

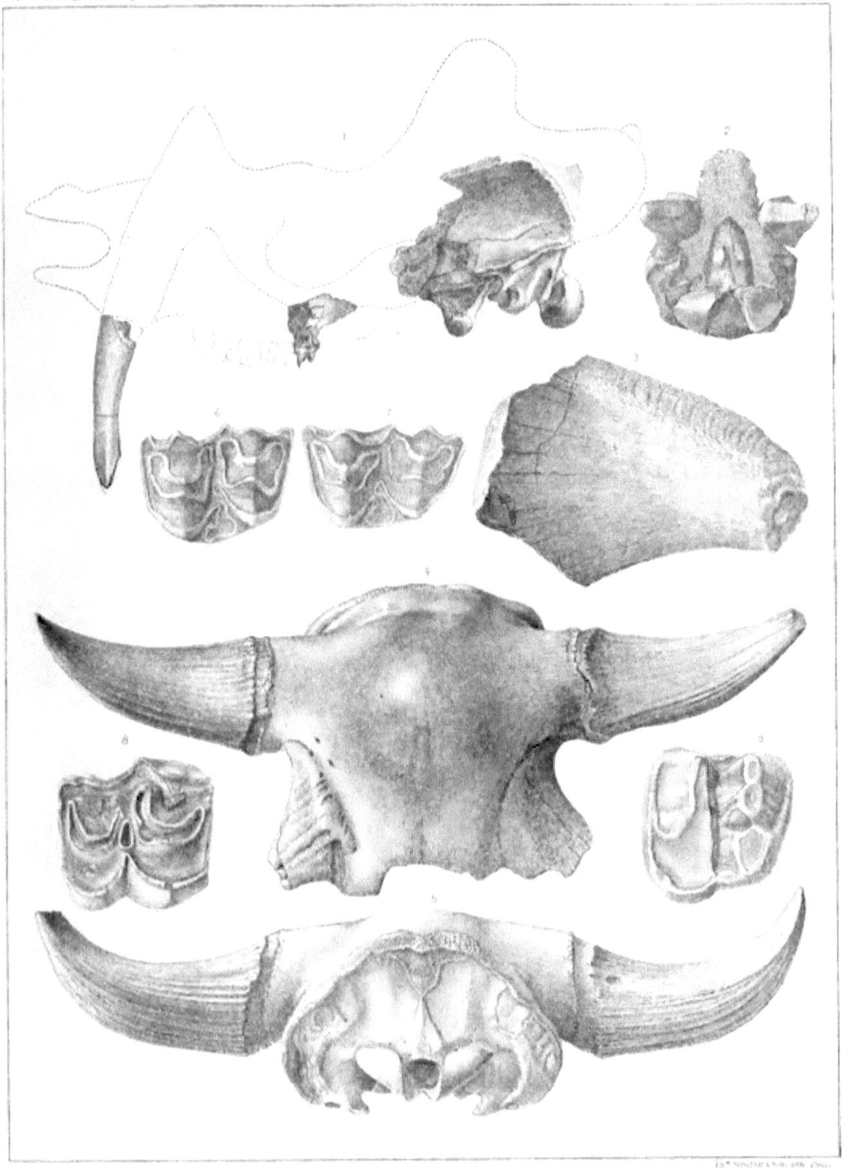

EXPLANATION OF PLATE XXIX.

All the figures one-half the natural size.

Fig. 1. TRIONYX UINTAENSIS:

The nearly entire carapace or upper shield, partially represented. Specimen discovered by Major Robert S. La Motte, in the buttes of Dry Creek, ten miles from Fort Bridger, and presented by him to the Academy of Natural Sciences of Philadelphia.

Figs. 2–4. TESTUDO CORSONI:

Specimens discovered by Dr. Joseph K. Corson, in association with portions of the plastron, and the specimen of the carapace represented in Fig. 1, Plate XXX.
Fig. 2. Anterior view of the proximal extremity of the right humerus.
Fig. 3. Outer view of the same specimen.
Fig. 4. Distal extremity of the right femur, front view.

Fig. 5. PALÆOSYOPS PALUDOSUS:

Femur of the left side, anterior view. Specimen obtained by Dr. Carter on Grizzly Buttes.

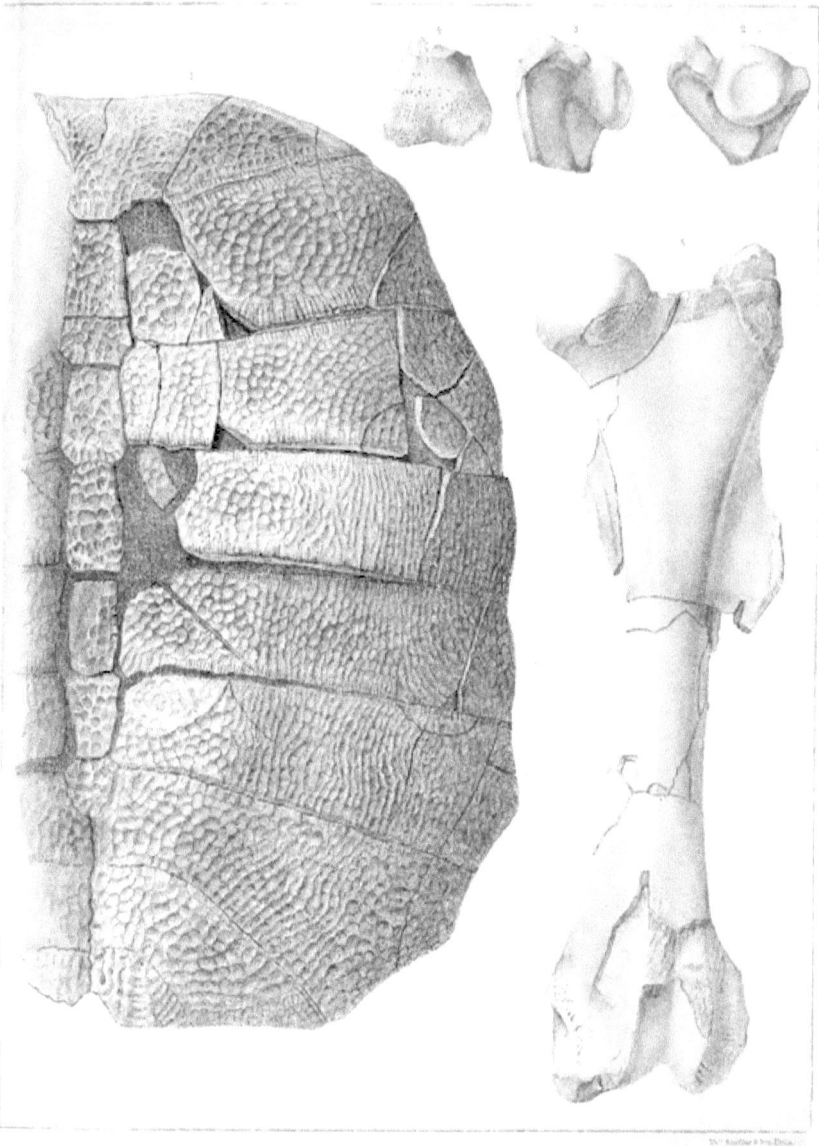

1 TRIONYX UINTAENSIS 2 4 TESTUDO CORSONI 5 PALAEOSYOPS PALUDOSUS

EXPLANATION OF PLATE XXX.

Figs. 1–4. TESTUDO CORSONI:

    Fig. 1. Intermediate portion of the carapace, one-half the natural size, exhibiting the series of vertebral plates, from the first to the eighth and part of the ninth, and contiguous portions of the costal plates. Specimen discovered by Dr. Joseph K. Corson on the buttes of Dry Creek, and presented by him to the Academy of Philadelphia.

    Fig. 2. Plastron, or lower shield, one-third the natural size. Specimen discovered by Dr. Corson on Grizzly Buttes, and presented to the Academy.

    Fig. 3. Anterior process of another plastron, one-half the natural size. From a specimen discovered by Dr. Corson in the same locality as the last.

    Fig. 4. Anterior process of a nearly complete plastron, one-half the natural size. From a specimen discovered by Mrs. Dr. Carter on the buttes of Dry Creek, and presented by her to the Academy of Philadelphia.

Fig. 5. CLADOCYCLUS OCCIDENTALIS:

    Large scale, imbedded in a lead-colored calcareous shale, natural size. Specimen obtained by Professor Hayden from the Cretaceous formation of Sage Creek, Dakota.

1-4. TESTUDO CORSONI.   5. CLADOCYCLUS

EXPLANATION OF PLATE XXXI.

Fig. 1. Restored skull of *Palæosyops*. The cranium and face are introduced from the specimens of Figs. 1, 2, Plate XXIV, and Fig. 51, Plate XVIII; and the lower jaw from the specimen of Fig. 52 of the latter plate, and Fig. 4, Plate XXIV. About half the natural size of the skull of *P. paludosus*.

Fig. 2. CANIS INDIANENSIS:

Right ramus of the lower jaw, one-half the natural size. Specimen from San Leandro, California. Discovered by Dr. Lorenzo G. Yates, and now in the museum of Wabash College, Crawfordsville, Indiana.

Fig. 3. FELIS IMPERIALIS:

Fore part of the upper jaw, with the second premolar, one-half the natural size. Accompanying the preceding specimen.

Fig. 4. LUTRA PISCINARIA:

Tibia of the right side, two-thirds the natural size. From Sinker Creek, Idaho, and belonging to the Smithsonian Institution.

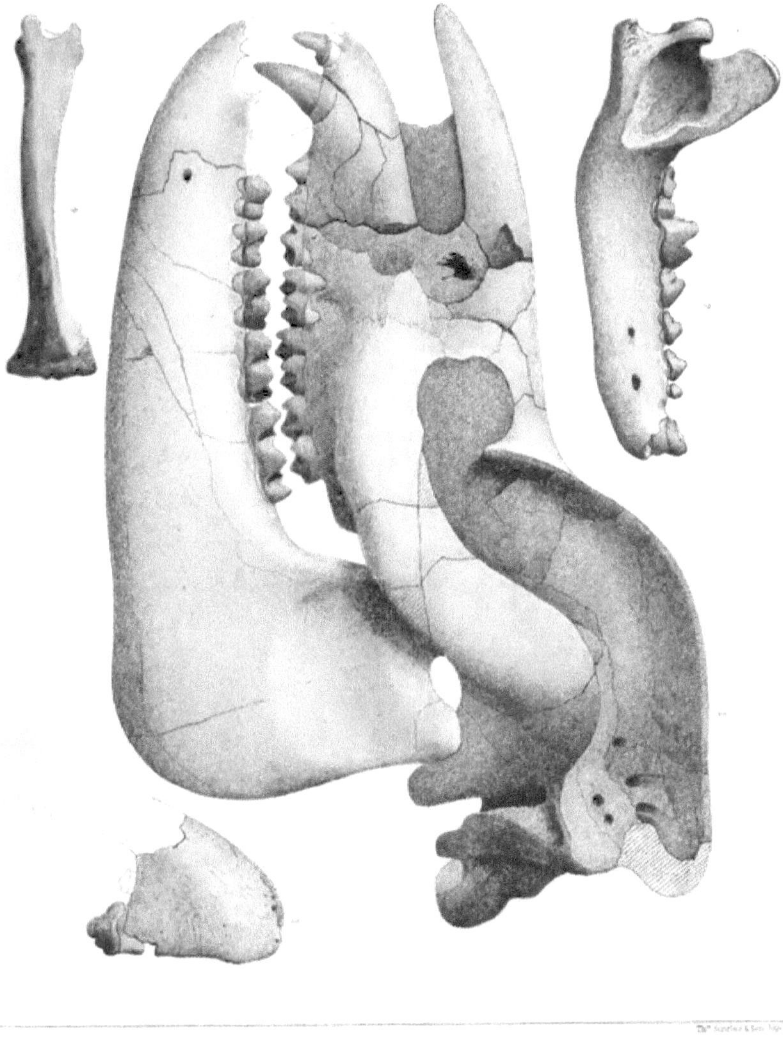

EXPLANATION OF PLATE XXXII.

All the figures of the natural size.

Figs. 1–6. AMIA (PROTAMIA) UINTAENSIS:
    Fig. 1. Centrum of a dorsal vertebra, anterior view. Fig. 2. View of the same beneath. From Dry Creek Cañon.
    Fig. 3. Centrum of an atlas, anterior view. Fig. 4. Inferior view of the same. From Dry Creek Cañon.
    Fig. 5. A series of three posterior dorsal centra, inferior view. From Dry Creek Cañon.
    Fig. 6. Basi occipital, posterior view. Fig. 6a. Inferior view of the same. From Dry Creek.

Figs. 7–11. AMIA (PROTAMIA) MEDIA:
    Fig. 7. Centrum of a dorsal vertebra, upper view. Fig. 8. Posterior view. Fig. 9. Inferior view. Junction of Sandy and Green Rivers.
    Fig. 10. Centrum of a posterior dorsal vertebra, back view. Fig. 11. Inferior view. Dry Creek.

Figs. 12, 13. LEPIDOSTEUS NOTABILIS:
    Fig. 12. Centrum of a dorsal vertebra, inferior view. Fig. 13. Posterior view. From near Washakie, Wyoming.

Figs. 14, 15. LEPIDOSTEUS ATROX:
    Fig. 14. Centrum of an anterior dorsal vertebra, inferior view. Fig. 15. Posterior view. From the junction of Big Sandy and Green Rivers.

Figs. 16, 17. LEPIDOSTEUS ——— (?) See page 190.
    Fig. 16. Centrum of a posterior dorsal vertebra, seen beneath.
    Fig. 17. Posterior view of the same.

Fig. 18. LEPIDOSTEUS SIMPLEX:
    The basi occipital and three vertebral centra, seen beneath. From near Washakie Station, Wyoming.

Figs. 19–22. HYPAMIA ELEGANS. A vertebral centrum. From Dry Creek.
    Fig. 19. Upper view. Fig. 20. Lateral view. Fig. 21. Posterior view. Fig. 22. Inferior view.

Figs. 23, 24. AMIA (PROTAMIA) GRACILIS. A centrum from near the middle of the dorsal series. Henry's Fork of Green River, Wyoming.
    Fig. 23. Posterior view. Fig. 24. Inferior view.

Fig. 25. LEPIDOSTEUS ——— (?)
    Fragment of the right dentary bone. See page 190.

Fig. 26. LEPIDOSTEUS SIMPLEX. A tooth. See page 191.

Figs. 27–30. LEPIDOSTEUS ——— (?) Scales from Big Sandy and Green River.

Figs. 31–34. LEPIDOSTEUS SIMPLEX. Scales. From near Washakie Station. See page 191.

Figs. 35–38. LEPIDOSTEUS. Scales. Little Sandy Creek. See page 192.

Figs. 39–42. LEPIDOSTEUS. Scales. Near Fort Bridger.

Fig. 43. LEPIDOSTEUS. Scale. See page 192.

PLATE XXXII.

Figs. 44–46. PIMELODUS ANTIQUUS:
    Figs. 44, 45. Fragments of pectoral spines.
    Fig. 46. Portion of a dentary bone, seen from beneath.

Figs. 47–51. PHAREODUS ACUTUS. Jaw fragments, from the junction of Big Sandy and Green Rivers, Wyoming.
    Fig. 47. Portion of the right premaxillary.
    Fig. 48. Portion of left premaxillary.
    Fig. 49. Portion of right dentary.
    Fig. 50. Portion of left dentary.
    Fig. 51. Portion of a maxillary.

Figs. 52, 53. TRYGON (?). Caudal spine of a Ray, From the Miocene of Virginia. Fig. 52. Anterior view of basal portion of the spine. Fig. 53. Section of the same. Belonging to Mr. C. M. Smith.

Figs. 54, 55. MYLIOBATES (?). Caudal spine. Found with the preceding.
    Fig. 54. Anterior face of basal portion.
    Fig. 55. Section of the same.

Figs. 56, 57. PROTAUTOGA CONIDENS:
Portions of premaxillaries, with teeth, from the Miocene of Virginia. Belonging to C. M. Smith.
    Fig. 56. Fragment of the left premaxillary, containing the first tooth.
    Fig. 57. Right premaxillary, inner view, exhibiting, besides the outer row of large teeth, an inner row of small ones.

Fig. 58. ACIPENSER ORNATUS:
A dermal plate. From the Miocene of Virginia. Belonging to Mr. C. M. Smith.

Fig. 59. ASTERACANTHUS SIDERIUS:
Basal portion of an ichthyodorulite, lateral view.

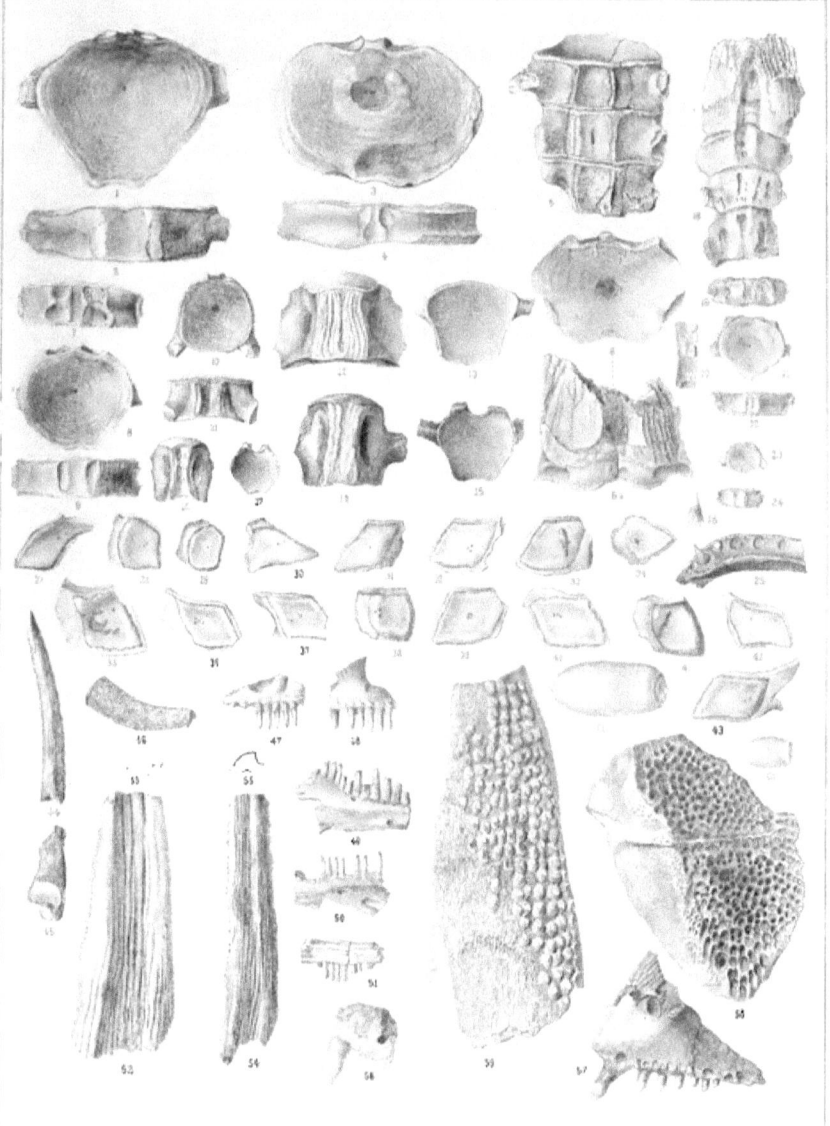

EXPLANATION OF PLATE XXXIII.

All the figures of the natural size.

Figs. 1, 2. EQUUS OCCIDENTALIS:
    Fig. 1. The anterior four upper molars of the left side, seen on their triturating surfaces. The teeth are contained in a jaw-fragment, obtained by Dr. George H. Horn from an asphaltum deposit near Buena Vista Lake, California. Specimen in the museum of the Academy of Philadelphia.
    Fig. 2. A second upper left molar, seen on the triturating surface. From Tuolumne County California.

Figs. 3–18. EQUUS MAJOR:
    Figs. 3, 4. A first upper molar tooth of the right side. Fig. 3. Outer view. Fig. 4. Triturating surface. Specimen from Hardin County, Texas.
    Figs. 5, 6. A first upper molar of the right side. Fig. 5. Outer view. Fig. 6. Triturating surface. From Illinois Bluffs, Missouri.
    Figs. 7, 8. A last upper molar of the right side. Fig. 7. Outer view. Fig. 8. Triturating surface. From Hardin County, Texas.
    Fig. 9. A last lower molar of the left side, view of the triturating surface. Found with the last.
    Fig. 10. A fifth lower molar of the left side, triturating surface. Found with the last.
    Fig. 11. A second or third upper molar of the right side, triturating surface. From Galveston Bay, Texas. Presented to the Academy by Dr. Thomas H. Streets.
    Fig. 12. A first lower temporary molar, triturating surface. From Hardin County, Texas.
    Fig. 13. An upper last temporary molar of the left side. Found with the last.
    Fig. 14. An upper second or third molar of the left side. From the "phosphate beds" of Ashley River, South Carolina.
    Fig. 15. A second or third lower molar of the right side. From the same locality as the last.
    Fig. 16. An upper second or third molar of the right side. From Luzerne County, Pennsylvania.
    Fig. 17. A second lower molar of the left side. Found with the last.
    Fig. 18. An upper fourth or fifth molar of the left side. From Texas.

Fig. 19. EQUUS. Portion of an upper molar of the left side of an undetermined species. From the lignite beds of Shoalwater Bay, Washington Territory.

Figs. 20, 21. Two phalanges of undetermined animals, both found, in association with the equine and other remains, in an asphaltum deposit in Hardin County, Texas. They are both saturated with bitumen. Fig. 20. Lateral view of the specimen. Fig. 21. Inferior view of the second specimen.

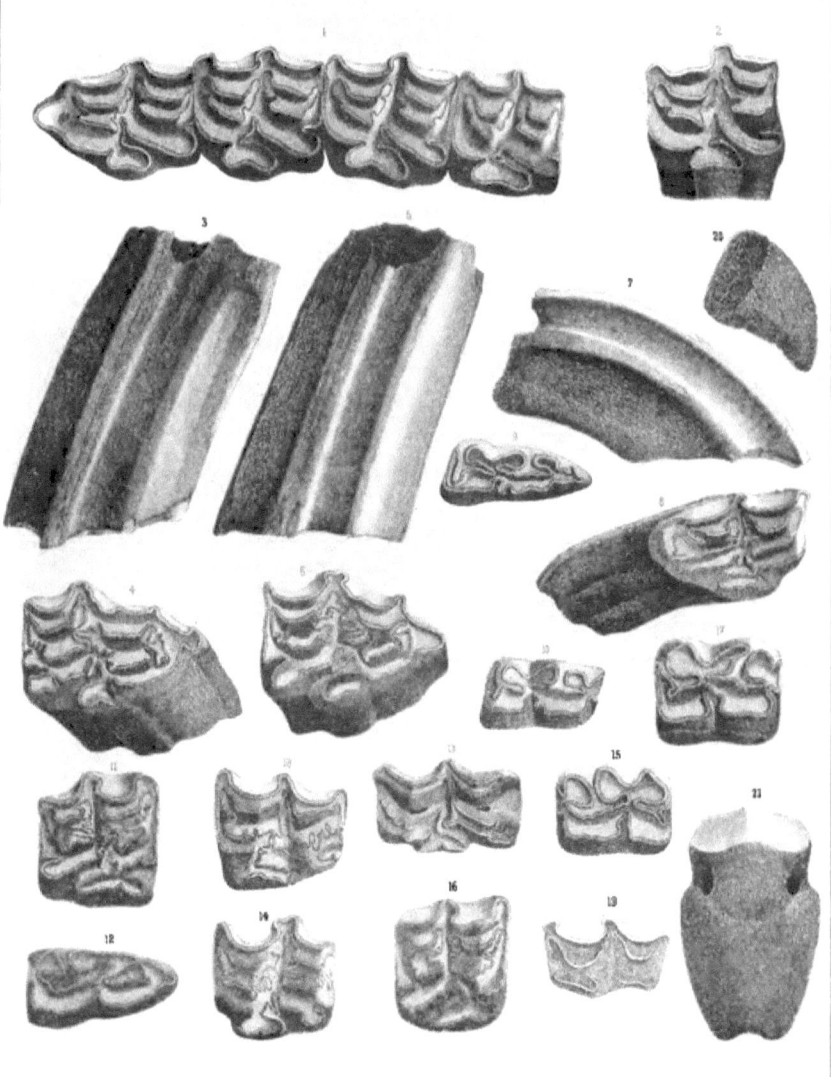

EXPLANATION OF PLATE XXXIV.

All the figures two-thirds the size of nature except Figs. 12–22, which are of the natural size.

Figs. 1 to 5 and 10. CLIDASTES INTERMEDIUS. From the Cretaceous of Alabama. Museum of the Academy of Natural Sciences.
    Fig. 1. Outer view of the fore part of the left mandible.
    Fig. 2. Back part of the right mandible, outer view reversed.
    Fig. 4. Reserve tooth, concealed in the excavated base of the last of the series in the specimen of Fig. 1, seen from within.
    Fig. 5. Reserve tooth, concealed in the excavated base of the second of the series of the same specimen, seen from within.

Figs. 6 to 9 and 11. CLIDASTES AFFINIS. From the Smoky Hill River, of Kansas. Belonging to the museum of the Smithsonian Institution.
    Fig. 6. Outer view of the left mandible.
    Fig. 7. Inner view of back part of the right mandible, exhibiting the glenoid articulation.
    Fig. 8. Upper view of two fragments of the cranium.
    Fig. 9. The basi-sphenoid bone.

Fig. 10. C. INTERMEDIUS. The axis seen below and with the fore part downward.

Fig. 11. C. AFFINIS. The left humerus, posterior view.

Fig. 12. LESTOSAURUS CORYPHÆUS:
    Greater portion of a palate-bone, with teeth, natural size. From the Smoky Hill River, Kansas.

Fig. 13. CLIDASTES AFFINIS. Tooth contained within a jaw-fragment. From Smoky Hill River, Kansas.

Fig. 14. Crown of a similar tooth. From L'Eau qui Court County, Nebraska. It is compressed, conical, curved, with acute borders and smooth surfaces. Fig. 15. Section of the same tooth.

Figs. 16–22. Teeth of mosasauroids, natural size, together with the preceding specimen from L'Eau qui Court County, Nebraska. Presented to Swarthmore College by George S. Truman.
    Fig. 16. Crown of a shed tooth, with striated enamel. Fig. 17. Transverse section of the same, at the base.
    Fig. 18. Shed crown of a large tooth, with striated enamel, anterior view.
    Fig. 19. Shed crown of a tooth, with distinct subdivisional planes. Fig. 20. Outlines of sections of the same at the base and above the base.
    Fig. 21. Crown of another shed tooth, intermediate in character with the two preceding.
    Fig. 22. Outline of a section of the same at the base.

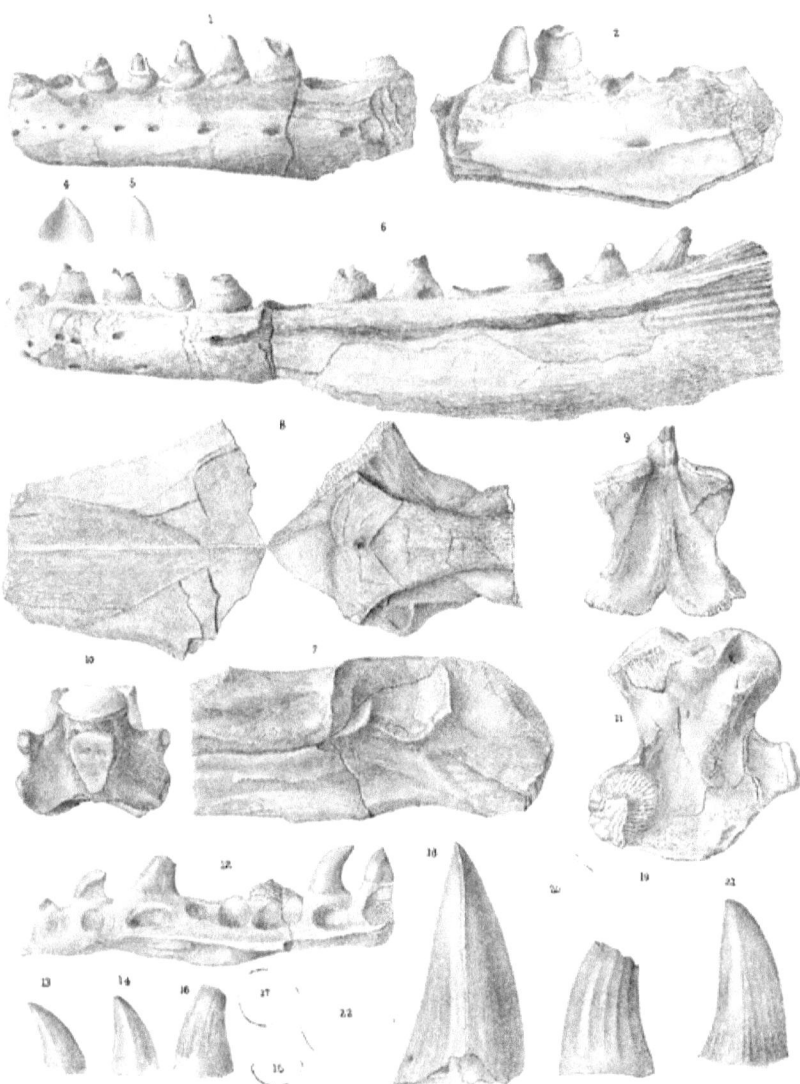

EXPLANATION OF PLATE XXXV.

All the figures one-half the natural size except Fig. 14, which is of the natural size.

Figs. 1–11. TYLOSAURUS DYSPELOR. Specimens from the Cretaceous of New Mexico, and belonging to the Smithsonian Institution.

    Fig. 1. Articular ball of a posterior dorsal centrum.
    Fig. 2, 3. The same of two other specimens, exhibiting a successive increase of compression from above downward.
    Fig. 4. Articular ball of a caudal centrum.
    Fig. 5. Left lateral view of the same.
    Fig. 6. Articular ball of a more posterior caudal centrum.
    Fig. 7. Left lateral view of the same specimen, exhibiting the reduction in the size of the diapophysis.
    Fig. 8. Left lateral view of a more posterior caudal vertebra, devoid of diapophyses.
    Fig. 9. Supposed femur, posterior view.
    Fig. 10. Supposed fibula.
    Fig. 11. Supposed tibia.

Figs. 12, 13. TYLOSAURUS PRORIGER. Specimens from the Cretaceous of Kansas, belonging to the Smithsonian Institution.

    Fig. 12. Extremity of the snout, or of the premaxillary.
    Fig. 13. Posterior articular surface of the left splenial bone of the lower jaw.
    Fig. 14. Tooth of a mosasauroid, natural size, from the Cretaceous of L'Eau qui Court County, Nebraska. The crown is compressed, conical, with acute borders and smooth surfaces. The base is compressed oval, and it exhibits on its inner side a small concavity for the accommodation of a successor.

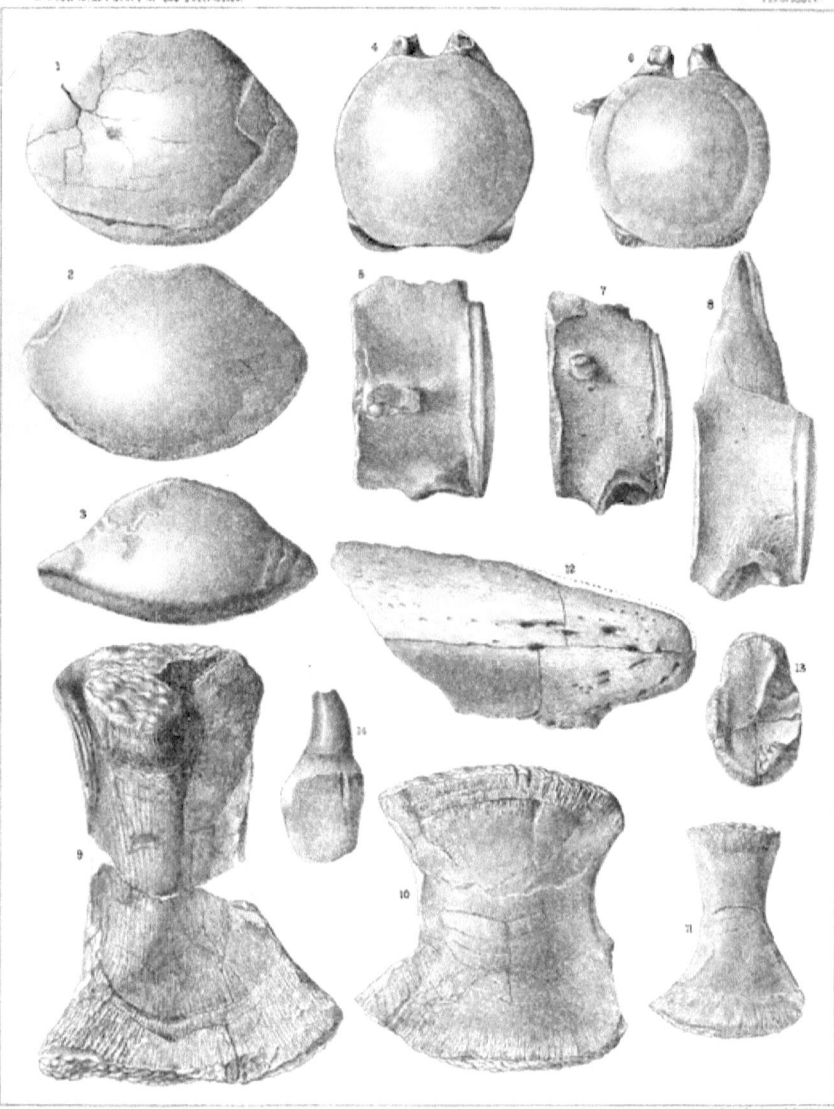

EXPLANATION OF PLATE XXXVI.

Figs. 1-3. TYLOSAURUS PRORIGER:

    Figs. 1, 2. A caudal vertebra, one-half the natural size. From the Cretaceous of Kansas. Smithsonian Institution.
    Fig. 1. Left lateral view. Fig. 2. Posterior view.
    Fig. 3. A tooth which accompanied the former specimen, lateral view, natural size.

Figs. 4-14. LESTOSAURUS CORYPHÆUS. All the figures one-half the natural size. From the Cretaceous of Kansas. Museum of the Smithsonian Institution.

    Fig. 4. Inferior view of a dorsal vertebra. Within the position of the right zygapophysis a rudimental zygosphene is observed.
    Fig. 5. Inferior view of a second specimen.
    Fig. 6. Inferior view of the body of a cervical vertebra.
    Fig. 7. Right lateral view of another cervical vertebra.
    Fig. 8. Left lateral view of an anterior caudal vertebra.
    Fig. 9. Same view of a more posterior specimen.
    Fig. 10. Posterior view of the same.
    Fig. 11. Left lateral view of the bodies of two posterior vertebræ.
    Fig. 12. Posterior view of the second of the latter.
    Fig. 13. Limb-bone, probably an ulna or a fibula.
    Fig. 14. Probably a radius or a tibia.

Fig. 15. MOSASAURUS:

    A caudal vertebra, from L'Eau qui Court County, Nebraska. Museum of Swarthmore College. Presented by George S. Truman. Inferior view one-half the natural size.

Fig. 16. TYLOSAURUS DYSPELOR. Inferior view of the same caudal centrum as that of Fig. 4, of the preceding plate. Half the natural size.

    Figs. 17-21. Limb-bones of a turtle, from the Cretaceous of Smoky Hill River, Kansas. Smithsonian Institution. Three-fourths the natural size.
    Fig. 17. Upper extremity of the right humerus, anterior view.
    Fig. 18. The right femur, anterior view.
    Fig. 19. Portion of a left scapula, inverted in position. The broken process to the left is the precoracoid. Posterior view.
    Fig. 20. Portion of the coracoid. The articular surface at the upper end is for the scapula.
    Fig. 21. Portion of an undetermined limb-bone.

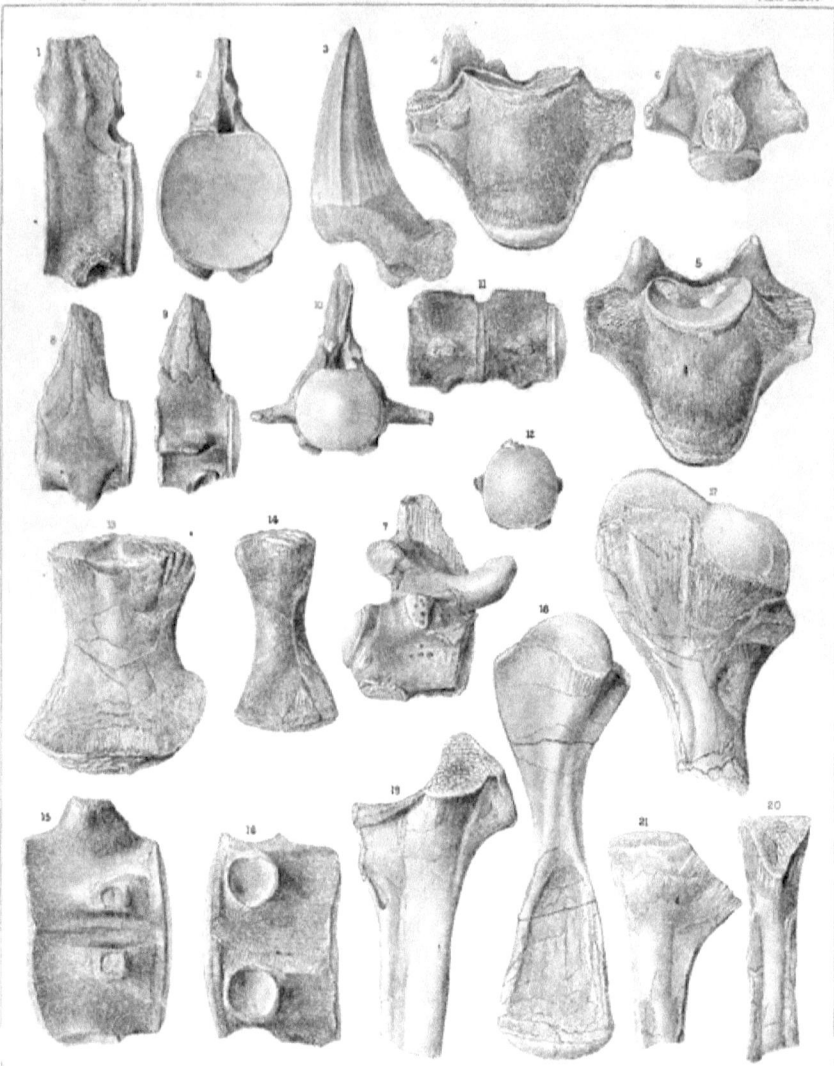

EXPLANATION OF PLATE XXXVII.

Figs. 1–3. AUCHENIA HESTERNA:

    Specimens from the Quaternary of California, and belonging to the cabinet of Wabash College, Crawfordsville, Indiana.
    Fig. 1. Outer view of the series of lower molar teeth of the left side, one-half the natural size.
    Fig. 2. Triturating surfaces of the same series, natural size.
    Fig. 3. A second upper molar of the left side, view of the triturating surface, natural size.

Fig. 4. BISON:

    Last lower molar of the left side, triturating surface, natural size. Specimen found with remains of *Megalonyx Jeffersoni*, in Illinois.

Fig. 5. AUHODUS HUMILIS. Magnified one and a half times. View of the triturating surface of a tooth. From the Cretaceous of New Jersey.

Figs. 6–12. EDAPHODON MIRIFICUS. One-half the natural size. Specimens from the Cretaceous of New Jersey, and belonging to the cabinet of Rutgers College, New Brunswick, New Jersey.
    Fig. 6. The mandibles seen on their oral surface.
    Fig. 7. Outer view of the left mandible.
    Fig. 8. Inner view of the left mandible.
    Fig. 9. Posterior outline of the same, with outlines of the dental columns.
    Fig. 10. The maxillæ seen on their palatine or oral surface.
    Fig. 11. Outer view of the left maxilla.
    Fig. 12. Posterior outline of the same, with outlines of the dental columns.

Figs. 13, 14. EUMYLODUS LAQUEATUS:

    Left lower maxilla, one-half the natural size.
    Fig. 13. Oral surface, exhibiting the dental tubercle.
    Fig. 14. Outer view. Specimen from the Cretaceous of Mississippi, and discovered by Dr. William Spillman.

Fig. 15. PONTOBASILEUS TUBERCULATUS. Fragment of a tooth, with restored outline, natural size.

Figs. 16, 17. MANATUS INORNATUS. A lower right molar, natural size. From the phosphate beds of Ashley River, South Carolina.
    Fig. 16. Upper view.
    Fig. 17. Outer view.

Figs. 18, 19. PYCNODUS ROBUSTUS. Tooth of the natural size.
    Fig. 18. Triturating surface.
    Fig. 19. Posterior view. Specimen from the Cretaceous of New Jersey.

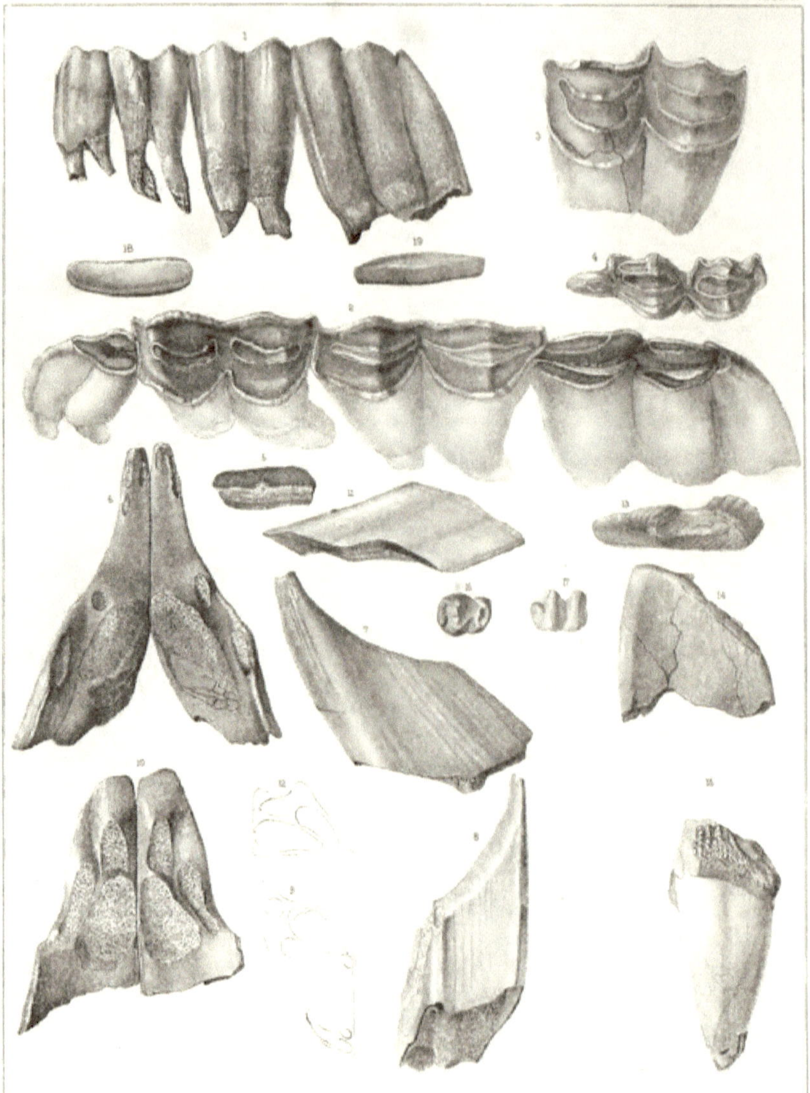

www.ingramcontent.com/pod-product-compliance
Lightning Source LLC
Chambersburg PA
CBHW020831020526
44114CB00040B/533